Friction and Wear

Friction and Wear

A. D. Sarkar

Consultant Tribologist
and Lecturer

1980

ACADEMIC PRESS

A Subsidiary of Harcourt Brace Jovanovich, Publishers
London · New York · Toronto · Sydney · San Francisco

ACADEMIC PRESS INC. (LONDON) LTD
24/28 Oval Road
London NW1

United States Edition published by
ACADEMIC PRESS INC.
111 Fifth Avenue
New York, New York 10003

British Library Cataloguing in Publication Data

Sarkar, Ananda Dulal
  Friction and wear.
  1. Friction   2. Mechanical wear
  I. Title
  621.8′9   TA418.72   80–40526

  ISBN 0–12–619260–X

Printed in Great Britain
by J. W. Arrowsmith Ltd., Bristol

Preface

Wear has been defined as the progressive loss of material when two surfaces undergo relative movement under load. Material–material interaction such as this also gives rise to the phenomenon of friction, which is discussed in Chapter 2. Wear has been classified into various types; namely, adhesive and abrasive wear, fretting, and loss of material from a body due to impingement attack by solid or liquid particles. These aspects have been dealt with in Chapters 3 to 6. Chapter 7 summarizes the aspects leading to the loss of material due to rolling motion. There are separate chapters on the wear of specific metals and non-metals, the latter including fabric, wood etc. as well as plastics and rubber. The work attempts to discuss principles of wear and is aimed at undergraduate and postgraduate students and at investigators or practising tribologists in industry. There are sufficient examples of application in industrial situations, as Chapters 13 and 14 show, which discuss friction materials and plain bearings. The latter should demonstrate that the quasi-empirical laws of wear are being used with success, but efforts must continue to provide information so that we can calculate wear rates of couples and design for a minimum of wear. The fundamentals of wear processes are incompletely understood, but it is clear that a whole range of parameters must be taken into account apart from load and speed. Some of these are material combinations, the manufacturing technique, surface finish and the prevailing environment in service. Apart from engineering and domestic situations, there is a need to understand the tribology of teeth and of the articulating surfaces of human joints. The available literature regarding these has been reviewed in Chapters 15 and 16.

I have not tried to convert all the units into SI mainly because SI units have not been accepted universally as yet. I have for the most part left the units used by the original authors intact but have included a conversion table.

My thanks are to various publishers and authors for permission to reproduce diagrams, all of which have been acknowledged individually. I am particularly grateful to Ampep Ltd UK, Engineering Science Data Unit Ltd UK and Climax Molybdenum Co USA for supplying me with information from their own experience. I thank Mr Douglas Scott for reading my manuscript. I am indebted to Mr Arthur Bourne of Academic Press and members of staff for their assistance whenever I needed it. It is a pleasure to acknowledge the help of Miss Anne Henkel of Academic Press, who processed the manuscript with such meticulous care.

February 1980   *A. D. Sarkar*
                                                          Liverpool

Contents

To my mother
Govinda Bhabini
with deep regret for my neglect

Nomenclature

A	Area	d	Depth of cut; diameter
B	Constant	f	Degree of fragmentation; deceleration
C	Constant specific heat		
D	Diameter	g	Acceleration due to gravity
E	Energy; Young's modulus	h	Height
F	Tangential force; frictional force	k	Thermal conductivity
		m	Mass
G	Energy	p	Pressure
H	Heat flow; heat of sublimation; hardness	q	Wear rate
		r	Radius
I	Moment of inertia	t	Time
J	Mechanical equivalent of heat	v	Velocity
		z	Thickness
K	Constant; specific wear rate		
L	Length	α	Angle; constant; probability
M	Mass	β	Probability
N	Number of cycles; r.p.m.; co-ordination number	ε	Strain
		θ	Temperature; angle
Q	Activation energy; heat	μ	Coefficient of friction
R	Relative wear resistance	ν	Poisson's ratio
S	Sliding distance	ξ	Coefficient of thermal expansion
T	Torque		
U	Velocity	ρ	Density
V	Volume	σ	Normal stress
W	Weight; work of adhesion	τ	Shear stress
		ψ	Erosion factor
a	Radius	Ω	Thermal shock resistance
b	Width of cut	ω	Angular velocity

Conversion factors

The following table gives the SI equivalents of some units frequently found in the literature.

1 in.	=	25·4 mm
1 lb	=	0·454 kg
1 lbf	=	4·45 N
1 kgf	=	9·81 N
1 lbf in.$^{-2}$	=	6·89 kN m^{-2}

1   Introduction

Wear is inevitable when two surfaces undergo slip or sliding under load. Apart from engineering mechanisms, there are many other areas where loss of material by this mode is commonplace; for example in teeth, garments, shoes etc. There is now a growing realization that the days of the throw-away society are numbered and we must learn to conserve. Wear is antagonistic to conservation and, although it will be impossible to prevent the process, we must try to minimize it. It is necessary, therefore, to make full use of all the empirical knowledge which is available and carry out planned investigations to improve our basic understanding of the phenomenon of wear.

1.1   Types of wear

All surfaces possess hills and valleys so that a load at the interface is only supported at suitably oriented high spots, resulting in micro-areas, which when added together form the true area of contact (Fig. 2.5). It follows that the true area of contact is much smaller than the apparent area so that the interface will be subjected to high contact stresses. We expect these areas to yield plastically and be removed by shearing due to sliding. This has been termed adhesive wear and is the most destructive. We shall see later that a thin surface layer soon work hardens during sliding so that contact may become largely elastic.

The effect of an elastic encounter is to provide a situation where the micro-contacts may fail by what has been termed fatigue wear. Fatigue due to sliding manifests itself in the form of cracks transverse to the direction of motion and layers of material spall off from the surfaces. A sliding interface can trap wear debris or grit and dust from the

1

surrounding environment, giving rise to abrasive wear. Wear also occurs when solid or liquid particles impinge upon a surface. This is known as erosive wear. Erosive wear is an environmental effect, as is corrosive wear, the latter probably playing a major role in the loss of tooth material in human beings.

Sliding can be unidirectional or reciprocating. When the amplitude of the latter is very low such that the interface undergoes small amplitude oscillatory motion, loss of material occurs by what is known as fretting. One way of ensuring long life is to separate the interfaces with a lubricant so that the components are unable to interact directly. It is necessary, however, for the interface to be in a state of fluid film lubrication, a situation when the components of a friction couple are fully separated. Unfortunately, at low sliding speeds or at heavy loads, lubrication is only partially effective. This is known as the boundary lubrication mode and destructive interaction between the opposing surfaces occurs at points of lubricant starvation. Another form of lubrication is elastohydrodynamic in nature and can occur, for example, between the contact surfaces of two mating gears, separated by a thin film of fluid. Although the surfaces are separated by a film of lubricant, high pressure is transmitted through this to the underlying surfaces and wear by the fatigue mode is a strong possibility.

1.2   Laboratory study

By far the most popular laboratory equipment is a pin–disc machine where a cylinder, referred to as the pin, is loaded vertically against a disc which rotates. Provision is always made to vary the speed of the disc or the load on the pin. Friction at the pin–disc interface can be measured and the amount of wear can be assessed by one of the following means:

(1) Weight variation with time.
(2) Height variation with time.
(3) Measurement of the change in the dimensions of the wear scar at suitable intervals during sliding.

To measure the wear scar as a means of expressing loss of material due to sliding under load, the pin is manufactured with conical or hemispherical ends. In another variation, one cylinder rotates and a second static cylinder is held across the former under load. This is known as the crossed cylinders configuration. Some typical wear machines have been described elsewhere [1] and the formulae for

Table 1.1 Wear volume V calculated from the dimensions of wear scars

Configuration	Specimen	Volume, V	Remarks
Cone on disc	Cone	$(\pi/24)d^3 \cot \theta$	θ = semi-apex angle of the cone; d = diameter of the wear scar
Hemisphere on disc	Hemisphere	$(\pi/64R)d^4[1+(d^2/12R^2)]$	R = radius of the hemisphere; d = diameter of the wear scar
Crossed cylinders	Small cylinder	$\left(\dfrac{\pi}{64R}\right)\left(\dfrac{a}{R}\right)^{1/2} d^4\left[1+\dfrac{d^2}{16R^2}\left(7-\dfrac{R}{a}\right)\right]$	a = radius of the small cylinder; R = radius of the large cylinder; d = major diameter of the elliptical wear scar on the small cylinder

calculating volumes from wear scars for various geometry of pins are shown in Table 1.1.

Apart from pin–disc and similar machines, simulated studies are quite common, for example, the use of disc test or four ball machines among others. The reader may wish to familiarize himself with these from the reference given at the end of this chapter before proceeding further with this work.

Reference

1. Sarkar, A. D. (1976). "Wear of Metals", Pergamon, Oxford.

2 Friction

Frictional resistance is redundant energy and engineers continuously strive to reduce this. Running a metal shaft on a metal bearing is not realistic simply because of the surface damage and eventual seizure of the components. A lubricant is used to separate the components and its beneficial role is to provide low friction which is essential for efficiency and economy. As the ceiling for speed and load is raised, the choice of materials to provide the friction couple and the use of a suitable lubricant becomes the province of specialists. We know that h.c.p. metals slip readily along the basal planes and are sought after because of their low frictional resistance. Graphite bearings are well known components where external lubrication is unnecessary. On the other hand f.c.c. metals slip but soon work harden, so that a journal and a bearing made from such metals will without exception lock at asperity contacts and the interface will seize. Even lubricated mechanisms must avoid such combinations because of their probable contact during boundary regimes. The sheer wastage of energy due to friction is obvious but we tend to ignore the beneficial role that friction plays in our day to day activities. If our feet or the soles of our shoes did not have a high friction as they grip the road surface, it would be impossible for us to stand firm. Typically on clay-rich soil in the rainy season in India or on icy roads in Europe, this difficulty is ever present. If we had not been able to grip dry roads with our feet our evolutionary behaviour would probably have been different as far as mobility is concerned. Whereas the ideal mode for a dynamic system such as an automobile is that it should have sliding or rolling interfaces completely free of friction, in practice interfacial friction must be present to bring it to rest. A brake shoe relies entirely on the presence of friction for its efficiency. Examples are many but it is clear that from the tribological point of view an understanding of friction is very important, because of both its positive and damaging roles in the relative motion of components. As yet there is no general quantitative relationship between friction and wear for metals, but as both these phenomena start from asperity interactions it should be possible to find one.

2.1   Historical background

The role of friction must have been known to primitive man because it would appear that he was aware of the fact that a spherical object will move more readily than a flat one. One does not know if the act of rubbing pieces of wood together to produce fire was discovered by accident or from some vague notion that rubbing generates heat since, in cold climates, human beings rub their hands habitually to produce warmth. *Homo sapiens* has much to learn about his society from observations on the pattern of behaviour of his ancestors and there is an active interest [1, 2] in the history of tribology. It is now known that palaeolithic men used bearings of bone for implements which produced fire. A similar assembly is still found in India, which is a drill operated by a bow giving a reciprocating rotary action. It is not known why bone was chosen for the bearing surface in the drills but as materials they are known to be durable and to provide low friction. A very interesting account, among others, has been given by Davison [3] based on his own careful observation of an ancient Egyptian painting, *c*.1880 BC. The painting shows a statue weighing about 60 tons (W) being dragged by 172 men, the pride of place being given to the one pouring a lubricant, possibly oil. The author [3] shows that the statue is on a sledge which is pulled over lubricated wooden boards. Each board measured about 192 in. × 18 in. and calculation shows that the boards were subjected to a bearing pressure of about 20 lb in.$^{-2}$ which, incidentally, is the pressure on modern tracked vehicles. Assuming that the average tangential pull by a man is $F = 120$ lb,

$$\mu = \frac{F}{W} = \frac{120 \times 172}{60 \times 2240} = 0 \cdot 15 \qquad (2.1)$$

The coefficient of friction, μ, for hard wood sliding on itself, but well lubricated, is similar to that given by Equation 2.1. If we take a possible value of wood on wood without a lubricant as 0·45, the number of men necessary to pull the statue will be 516, an increase by a factor of 3. It has been documented that in the state of Egypt about 100 000 men were employed each year to carry out the task of transporting heavy objects. Without a lubricant, 300 000 men for a state in *c.* 2000 BC can be a large expense even if the monarchs fed them only one meal a day and paid them no wages!

The technology of low friction elements has been known for a long time. For example, rolling bearings were used by the Greeks around 330 BC. The early Egyptians used grease-packed bearings on their chariots and these plain bearings were protected from grit by covers made from leather. There is evidence for the use of wood bearings in Italy around 25 BC. There are many examples such as these of the historical development of tribological components, but it must remain in the realm of conjecture as to how the role of lubrication was discovered or how human beings changed from slides to roller bearings.

Elucidation of the science of friction is attributed much later to Leonardo da Vinci in the fifteenth century. Leonardo da Vinci's perspicacity is well known in many fields but, regarding friction, he observed that:

(1) the frictional resistance F is proportional to the weight W of the object which is being moved;

(2) the frictional force is independent of the apparent area of contact.

The above two observations are known as Amontons' laws of friction, stated fully in 1699. It appears that Amontons observed these two phenomena independently and was not aware of any documentation relating to Leonardo da Vinci, *vis-à-vis* friction. There is a third law which states that the interfacial resistance between two surfaces is independent of the velocity of sliding. The law is attributed to Coulomb who enunciated it in 1780 and holds only over a limited range of speed. Amontons' two laws are obeyed quite well and the exceptions will be mentioned in the next section and elsewhere.

2.2 Laws of friction

The proportionality of F on W giving a characteristic value of the static coefficient of friction, μ, and the kinetic component, μ_k, is probably true in a very broad sense in ambient atmosphere. We must immediately be cautious, however, and state that μ is not an intrinsic property of a material but depends primarily on the solids constituting the friction couple. Thus, as will be discussed later, a material combination which produces intermetallic compounds at the interface may give rise to low friction.

On the other hand, similar metals sliding together may produce a high value of μ. We know that sorbed gases may lower the frictional resistance, but the same couple, fully outgassed and then tested, may increase the value of μ by one or two orders of magnitude. Other things being equal, the frictional resistance is independent of the apparent area of contact because it is the true area established at the interface which determines the friction bonds. Basic understanding of the kinetic component of friction is for the time being inadequate, but experience shows that the third law of friction can be obeyed within a limited speed range only. The range varies between metals but for titanium sliding on itself it is 10^{-8} to 10^2 cm s^{-1}. If the relative movement at the interface is inordinately high, the coefficient of friction is known to fall considerably. The reason given is that incipient fusion of the contact points occurs, thus providing a layer of viscous lubricant and hence a low friction.

2.3   Molecular theory

Although subject to serious qualifications, the three laws of friction provide a welcome framework within which the theories of dry friction can be developed. We know that two surfaces under load are held together and a finite amount of force is necessary to separate them. It is convenient now to examine the information which enables us to attempt to explain why frictional bonds are formed.

We should begin by recognizing that not all the energy given to a solid is available for work. That is, the interface opposes the intended function and a certain amount of the total energy is not available as useful work. It has been established that this non-useful energy at a tribological interface is dissipated into one or more of the following phenomena:

(1) elastic hysteresis;
(2) plastic deformation;
(3) fracture and creation of new surfaces.

Both (1) and (2) will be considered in detail later, and fracture is implicit in the deformation of asperities. Prior to this, the molecular theory due to Tomlinson will be summarized, because it established a similar dependence of friction, some half a century ago, upon certain parameters which are incorporated into current theories.

The molecular theory starts by recognizing that a solid is in a state of equilibrium when the attractive and the repulsive forces between the atoms are equal and opposite. Consider now one solid approaching another. If the distance of approach is close enough, the repulsive forces between the atoms will cause the surfaces to separate, which will manifest itself as frictional resistance acting in a negative manner to the external energy given to the system. It is recognized that surfaces in reality are rough so that only part of the surface of one will approach closely that of the other. This can be likened to those asperities which are tall and are penetrated by a second smooth surface which approaches them under load. However, in the particular case of the atomic model, only the atoms from the peaks of the asperities are repelled by the smooth surface, and there is no penetration. The repulsion is a measure of friction and the implication is that there is no surface damage. It will be shown in Chapter 3 that wear, according to the molecular theory, occurs by plucking of atoms from the surface so that a change in its topography occurs. However, regarding friction, Tomlinson's model proposes that the resistance to motion can be appreciable simply because a set of atoms reaches close enough into the field of repulsion of another set.

Let the interface of a couple be under a normal load W. Let the sum of the forces of attraction be ΣF_A and that due to repulsion between the approaching atoms be ΣF_R. Then at equilibrium,

$$W + \Sigma F_A = \Sigma F_R \tag{2.2}$$

For a real friction couple, the force of attraction between the atoms of the solids is negligible compared to the load exerted at the interface. Equation 2.2 can, therefore, be modified to,

$$W = \Sigma F_R \tag{2.3}$$

The model thus is that a pair of atoms comprising one from each surface exerts a repulsive force as the two solids approach and Equation 2.3 describes the static condition and is also applicable to a sliding situation, although the atom pairs keep changing because of a continual shift in the relative position of the two surfaces. The number of repelling pairs must remain the same for ΣF_R not to change from moment to moment.

Suppose that, because of the repulsion, the amount of energy dissipated by an atomic pair per encounter is E, which is the frictional resistance offered by a pair of approaching atoms. Suppose that the

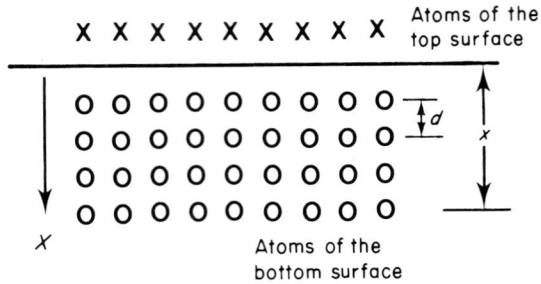

X X X X X X X X X   Atoms of the top surface

Fig. 2.1   The atomic model of friction.

distance between the centres of these atoms is d and that the top surface has penetrated the bottom surface by an amount x (Fig. 2.1). If the mean repulsive force is p and the number of atomic contacts at a load W is n,

$$W = np \tag{2.4}$$

If the top surface moves a distance x, the total number of times the atomic pairs will separate is N, given by

$$N = n(x/d) \tag{2.5}$$

The schematic diagram in Fig. 2.1 is idealized and to account for a real situation, there is only a probability that certain atom pairs will separate. This modifies Equation 2.5 as

$$N = \alpha \frac{nx}{d} \tag{2.6}$$

Multiplying both sides by E,

$$NE = \alpha \frac{nx}{d} E \tag{2.7}$$

Now the work done in penetrating a distance x is

$$\mu Wx = NE, \tag{2.8}$$

where μ is the coefficient of friction. Substituting for NE from Equation 2.7 in Equation 2.8,

$$\mu = \alpha \frac{nE}{dW} \tag{2.9}$$

Substituting for W from Equation 2.4,

$$\mu = \alpha\,(E/dp) \tag{2.10}$$

Suppose an atom is displaced from its prior equilibrium position as it enters the repulsive field of another. It will try to return to its position of original equilibrium once the opposing atom has moved on. Let the distance traversed in this return to its original position be l. If the mean cohesive force between a pair of atoms is E_c,

$$E_c l = E$$

Substituting this value of E in Equation 2.10,

$$\mu = \alpha\,\frac{E_c}{p} \cdot \frac{l}{d} \tag{2.11}$$

The interatomic distance d at equilibrium for a pair of particular atoms is fixed. The distance l involved is unknown but the term l/d is a dimensionless quantity. The term α is in fact the probability of an atom encountering another from the counterface and is in conformity with the statistical nature of peak height distribution and hence of atoms in these summits. The terms E_c and p are associated with the elastic constants of the material under study.

It should be noted that Equation 2.11 does not incorporate the normal load. That is, the coefficient of friction is independent of the applied normal load, which agrees with Amontons' first law which may be stated to be that the coefficient of friction is independent of the normal load. Since Tomlinson's model is concerned with atomic repulsion, one could say that contact is elastic. The implication in the analysis is that the frictional resistance F is proportional to the load W. We will see later that, for elastic contact, $F \propto W^n$ where n is between 0·8 and 0·96. Just the same, our understanding of friction and wear phenomena is incomplete and one could say that Tomlinson's atomic model opened the door to current theories.

2.4 Plastic interface

A serious limitation of the atomic model is that if contact is elastic and on an atomic scale, there will be no surface damage once a solid is placed upon another and then lifted for examination. It is common experience that some form of mechanical change in the interface occurs due to a tribological interaction, for example, metal transfer, work hardening, etc. Penetration of solids should therefore be on a large enough scale for the unaided eye to be able to detect an alteration of the prior state of surfaces.

The theory of Bowden and Tabor [5] is neat, and although there are occasional attempts to modify it, the basic concept remains. There are two core components to this theory, viz. adhesion and ploughing. To understand this, consider a rider being placed upon a flat surface (Fig. 2.2(a)) under a negligible load. Suppose a load W is now applied which is heavy enough for the softer flat component to yield (Fig. 2.2(b)). Yielding of the soft component will result in the formation of junctions and the rider will stick to the plate. If now a tangential force F is applied, the rider will move (Fig. 2.2(c)) provided F exceeds the force of adhesion between the rider and the plate. As the rider moves horizontally, the indented volume, assuming that there is no lateral spread, will form a wall of material which must be pushed by the rider to sustain motion. Assuming an ideal situation, frictional resistance develops because of the additive contribution of two terms. That is,

$$F = S + P, \tag{2.12}$$

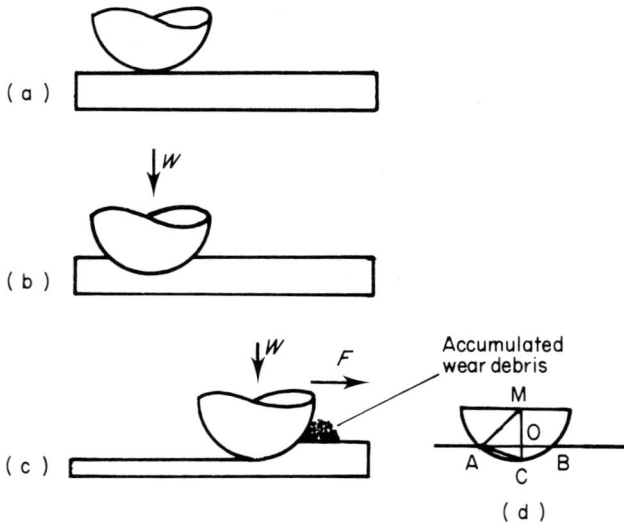

Fig. 2.2   A hemispherical rider under a load W being dragged along a flat surface.

where S = shearing resistance at the interface and P = the ploughing effort to push the wall of material.

The terms S and P now need examining in detail. To begin with, suppose for the time being that the geometric cross-section of the indented hemispherical cup provides the true area of contact, which we shall denote by A. The interfacial junctions created by plastic flow will have an intrinsic shear stress τ, so that the shearing force which will overcome the adhesion between the rider and the plate is

$$S = A\tau \tag{2.13}$$

The next question to answer is which member of the friction couple does provide the strength of the junction. It is suggested [5] that the softer of the two yields and adheres to the hard counterface so that it is the former which decides on the strength of the interface which is generated by a normal load. This is very neat and useful, but the interface is more complex than this and such phenomena as material transfer and work hardening must enter into the shear strength of the junctions.

The ploughing term P is simply the pushing of a wall of material by a hemispherical segment as shown in Fig 2.2(c). Considering a metal couple, it is important therefore to know the yield stress σ_Y of the metal which is being pushed. To work out the ploughing force we need to obtain the area of the segment of the cross-section of the sphere. Figure 2.2(d) shows the end view of Fig. 2.2(c) and the segment involved is AOBC. If the height of the segment is OC = h and the chord AB = d, to a first approximation [6], the area AOBC = $2hd/3$. Considering the right-angled triangle MAO in Fig. 2.2(d), neglecting higher powers of h, $h = d^2/8r$, where r is the radius of the whole circle. From these relationships, the area of the segment is $d^3/12r$.

Therefore, the resistance to motion due to the ploughing action is

$$P = (d^3/12r)\sigma_Y \tag{2.14}$$

Note that the ploughing component P provides only the force necessary to induce yielding of a segmental area, but there will be a resistance term due to the accumulated debris unless this is dispelled, for example by brushing as the rider moves on. Adding Equations 2.13 and 2.14, the net frictional force is

$$F = A\tau + (d^3/12r)\sigma_Y \tag{2.15}$$

If we consider a conical asperity, with a semi-apex angle θ, so that d is the diameter of the indented area, the ploughing term is $(d^2/4) \cot \theta \sigma_Y$. It is noted that the predominant geometric feature in both cases is d, which increases with the depth of indentation. Thus a material with a low yield stress under a heavy normal load will give a large value of d for both the spherical and conical riders. The ploughing term will then increase with soft metals, but σ_Y is also low in these cases. It is quite likely that there is an interaction between S and P; that is a factor which affects S may also influence P. Therefore, it may not be correct to separate the shearing and ploughing terms, assuming them to be independent parameters. A mathematical treatment of friction, taking all the factors into account, may prove intractable but Bowden and Tabor's statement is that for most situations, P can be neglected and the frictional resistance is given by

$$F = A\tau \tag{2.16}$$

2.4.1 Coefficient of friction
By definition, $\mu = F/W$. F is given by Equation 2.16, neglecting the ploughing term. $W = A\sigma$ where σ is the compressive stress on the area A. Therefore

$$\mu = A\tau/A\sigma$$

or

$$\mu = \tau/\sigma \tag{2.17}$$

Equation 2.17 is in accordance with one of the laws of Amontons, which states that the coefficient of friction is independent of the applied normal load. The shear strength of the interface is an imponderable item but it is assumed that $\tau = \tau_c$, the critical shear stress of the material which has yielded. Equation 2.17 can therefore be modified to,

$$\mu = \tau_c/\sigma \tag{2.18}$$

The stress below a rider is quite complex, but a simplified statement [5] is that of the total stress at the interface, about two-thirds is in the form of hydrostatic pressure which does not contribute to the yielding of the appropriate metal. The remainder is available for plastic flow so that

$$\sigma_Y \simeq \sigma/3 \tag{2.19}$$

According to the theory of plasticity, $\tau_c = \sigma_Y/2$. Thus Equation 2.18 can be rewritten as $\mu = (\sigma_Y/2)/(3\sigma_Y)$; that is, $\mu = 0.17$.

For most metal couples, in ambient atmosphere, the coefficient of friction under dry sliding condition is stated [5] to be of the order of unity. The above argument giving $\mu \approx 0.2$, therefore, needs modification. To do this it is necessary to consider the phenomenon of junction growth.

2.4.2 *Junction growth*

The single rider forms a junction with the flat plate (Fig. 2.2(b)) producing an interfacial area, A. The junction may be elastic in nature when it will completely disappear once the load is removed. If it is plastic, the hemispherical rider will adhere more permanently to the flat surface.

Experiments [7–9] show that the combined effect of a normal stress and an applied tangential pull is to increase the area of contact, generally in the direction of motion. At this stage, only micro-displacement of the upper surface relative to the lower body in the junction region occurs and as yet there is no macroscopic sliding. Using spherical riders on flat specimens and keeping the normal load constant, the micro-movement of the interface of the junction has been observed with the aid of multiple beam interferometry [9]. It is observed that as the applied tangential force is increased steadily, there is a progressive increase of the area of the junction. With an increase in the magnitude of the micro-displacement, there is a corresponding increase in the instantaneous coefficient of friction, μ_i. At the onset of macroscopic sliding the junction area has reached a maximum steady value and the coefficient of friction is μ, the static component. The yield criterion for junction growth is

$$\sigma^2 + \alpha\tau^2 = \sigma_Y^2, \tag{2.20}$$

where σ is the applied normal stress and τ is the applied tangential stress. α is a constant and σ_Y is the yield stress of the material forming the junction. Examination of Equation 2.20 step by step as the junction area increases in a monotonic manner is now necessary. Initially $\tau = 0$ and we do not begin to apply the yield criterion. At most values of σ, yielding of some of the asperities will occur and junctions will form having a potential value of coefficient of friction $\mu_i < \mu$. As τ assumes a value, the junction increases in size and according to Equation 2.20, σ

decreases. To maintain plasticity, therefore, τ must be increased with a concomitant increase in area and decrease in σ. This continues until τ is large enough to overcome the static coefficient of friction μ and large-scale sliding begins. Thus, the postulate is that as an asperity is squashed to form a junction, there will be a small but finite amount of friction. This will continue to increase as the area increases because the instantaneous frictional resistance F_i is related to the instantaneous area A_i as $F_i = \tau_c A_i$, τ_c being the critical shear stress of the material forming the junction. $\mu_i \to \mu$ at gross sliding and obviously μ is greater than the calculated value of $0 \cdot 17$, which does not give cognizance to the true area of contact which is getting larger as τ continues to increase up to its value at gross sliding. In more specific terms than given by the yield criterion of Equation 2.20, the equation [5] for junction growth is, since $\sigma_Y = \phi(\tau_c)$,

$$\sigma^2 + \alpha_0 \tau_i^2 = \alpha_0 \tau_c^2 \tag{2.21}$$

α_0 has a value of about 9 and τ_i is the applied tangential stress which varies with time and lies between zero and τ. τ is the stress applied finally when sliding on a macro-scale commences. When that happens, $\sigma \to 0$ and $\tau \to \tau_c$.

2.4.3   *Work hardening*
So far we have discussed the nature of the interface using a single spherical rider. All the postulates, it must be emphasized, have centred around the indispensable premise that the area produced is completely plastic. Plastic yielding results because of a compressive stress, but nevertheless, the use of σ as in Equation 2.17 to express μ can be misleading. This is because not all the compressive stress is available for plastic yielding and we write that the true area of contact A_t due to plastic interaction is

$$A_t = W/3\sigma_Y \tag{2.22}$$

Since the hardness of the material whose frictional resistance is being evaluated is $H \approx 3\sigma_Y$

$$A_t = W/H \tag{2.23}$$

and

$$\mu = \tau_c/3\sigma_Y \quad \text{or} \quad \mu = \tau_c/H \tag{2.24}$$

One could argue that work hardening will result in an increase in the value of τ_c of the junctions and hence a rise in the magnitude of μ. However, the hardness or the yield strength of the material increases as well so that μ will remain relatively unchanged as expressed by Equation 2.24.

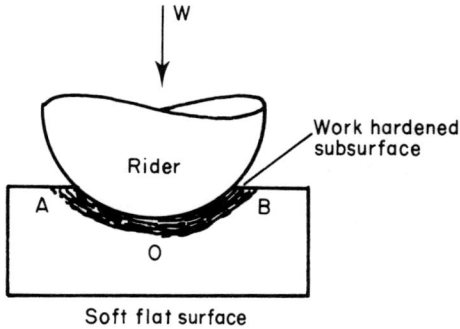

Fig. 2.3   Work hardened sublayer due to a normal load W.

When a hard rider indents a soft metal, a sublayer of the latter may work harden. There is a strong possibility that the effect of a lateral pull will be to cause shear at the interface AOB (Fig. 2.3) which is weaker than the hard material above it. This, probably, is the reason why under certain conditions of material combination and load, the wear debris assumes a lump form. If the flat specimen did not work harden, it would lose material by adhesion to the hard rider if the interaction was plastic.

2.5 Contaminant

We have seen that there are a number of reasons why μ does not always assume a constant value of about 0·2. One reason is that the asperities which approach do not make fully metallic contact since all surfaces in ambient atmosphere are contaminated. Contaminants can take a number of forms. For example, surfaces exposed to an industrial environment may be contaminated with variable layers of oil or grease. The surfaces may simply be covered with a monomolecular layer of gas and most metals will oxidize while left exposed to room environment. All this means that when two components come in contact under load, metal to metal interaction will be impeded. One expects the frictional resistance in this case to be different from that described by Equation 2.24.

Consider a steel hemispherical rider resting on a similar material in the form of a flat which is covered with a thin layer of a different material of low intrinsic shear strength. If the coating remains mechanically locked to the flat surface, the coefficient of friction will be influenced by the following parameters [10]:

(1) τ_c, the critical shear stress of the steel;

(2) τ, the applied tangential stress;

(3) τ_f, the critical shear stress of the coating.

Assume that $\tau_f < \tau_c$, which is usually the case. If the coating is thick, the rider will slide as soon as τ exceeds τ_f. That is, the interfacial adhesion is decided by the critical shear stress of the coating.

If the area of contact at the interface is A, the frictional resistance F is,

$$F = A\tau_f \qquad (2.25)$$

We can find a dependence of τ_f on τ_c such that

$$\tau_f = \alpha\tau_c, \qquad (2.26)$$

where $\alpha < 1$.

The yield criterion is still applicable since junction growth will occur as τ increases until it equals or exceeds τ_f when macroscopic sliding takes place. Therefore, substituting the value of τ_c from Equation 2.26 in Equation 2.21,

$$\sigma^2 + \alpha_0\tau_f^2 = \alpha_0(\tau_f/\alpha)^2 \quad \text{or} \quad \sigma^2 = \alpha_0\tau_f^2(1-\alpha^2)/\alpha^2 \qquad (2.27)$$

Now $\mu = F/W$ and $W = A\sigma$

$$\therefore \quad \mu = A\tau_f/A\sigma = \tau_f/\sigma$$

Substituting for σ from Equation 2.27 in the above expression,

$$\mu = \alpha\,[\alpha_0(1-\alpha^2)]^{-1/2} \qquad (2.28)$$

α_0 is usually taken as 9, so that

$$\mu = \frac{\alpha}{3(1-\alpha^2)^{1/2}} \qquad (2.29)$$

For surfaces free from any protective coating, $\alpha = 1$ and μ is infinitely large. The physical meaning of $\alpha = 0$ is that the coating is fully effective, that is, steel on steel contact does not occur. μ then is zero, but this is never true because the contaminant itself has finite shear strength. Taking various values of α from zero to unity, a plot of μ against the

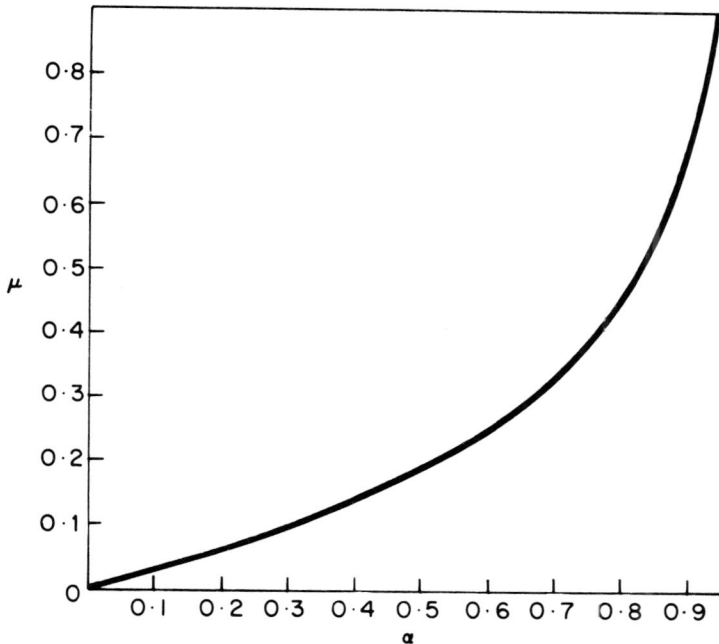

Fig. 2.4   Variation of the coefficient of friction, μ, with the degree of surface contamination, which decreases with increasing values of α (see Equation 2.29).

degree of contamination (Fig. 2.4) shows a typical pattern of μ against α. As expected, μ is very large when the coating is not allowed to protect the metallic surfaces. Note that for a not uncommon value of $\mu = 1$, for metal couples, only about 5% of the surface is contaminated at $\alpha = 0.95$. This may suggest that monomolecular layers of atmospheric gases or oxide films as found on metals left in room atmosphere, giving $\mu \simeq 1$, are not as effective as oils, greases or graphite applied externally from the viewpoint of lubrication. One can suggest that the coherency of both the gas envelope and of an oxide film is destroyed under the combined action of normal and tangential stresses.

2.6   Two rough surfaces

So far we have considered a single hemisphere interacting with a perfectly flat surface. All real surfaces of course are covered with undulations so that if one solid approaches another, contact will occur

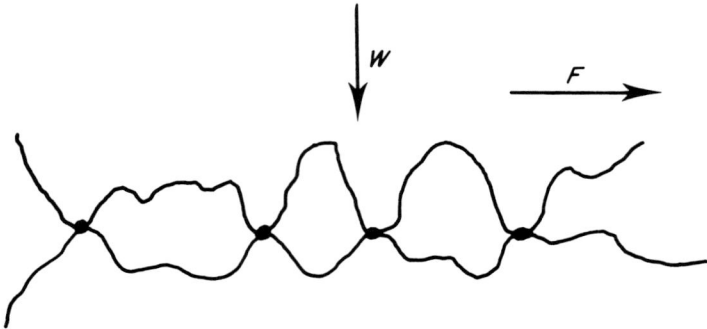

Fig. 2.5   Two rough surfaces making contact at four points (W = normal
load; F = load applied tangentially).

between the appropriate summits of the opposing components. This is
shown in Fig. 2.5 where contact has just occurred. With most metals,
the first areas thus produced would continue to yield laterally and
increase in size. If the first summits subside due to lateral spreading,
contact will occur between fresh asperities as soon as they are favourably disposed to meet the opposing surface. Once the sum of the
discrete areas thus produced is large enough to support the load W,
plastic flow will cease. This increase in area is entirely due to the static
load and must not be confused with junction growth due to a tangential
stress.

Suppose a number of junctions formed due to static loading results in
areas A_1, A_2, A_3 ... A_n which are not as a rule equal. The previous
concepts for a single rider on a flat surface are again applicable, but for
clarity it is essential to emphasize the true area of contact A_t as opposed
to the apparent area A. Thus $A_t = A_1 + A_2 + A_3 + ... + A_n$ and $A_t =
W/3\sigma_Y$. Alternatively, $A_t = W/H$.

Similarly, all previous effects such as junction growth, effect of
contaminants, etc. apply but the additive process of all the interacting
asperities must be considered. The true areas can be evaluated experimentally and it has been shown [11] that $A_t \ll A$. As an example, for
steel on steel, $A \simeq A_t \times 10^4$ for a particular situation.

2.7   Elastic recovery

The nature of the interaction between elastic bodies will be discussed in
appropriate chapters, but elastic recovery is of importance during
contact between metallic components [12]. Since the areas produced by

interpenetration are small, we can assume plastic flow of some of the summits at least. That is, at realistic loads, a number of contact areas formed by plastic interaction will be dispersed in a matrix which is elastic. We can estimate the plastic fraction, assuming that all the true areas are due to plastic yielding of asperities. Then the ratio of the sum of the true areas A_t to the apparent area A will increase with increasing load. Suppose a brass block 5 mm × 5 mm square is held against a very hard steel block of the same dimensions. Taking the hardness of the brass as 90 kg mm^{-2}, the variation of A_t/A with the applied static load is shown in Fig. 2.6. Note that under static loading a load of 2250 kg is required for the true area of the brass to equal its apparent area!

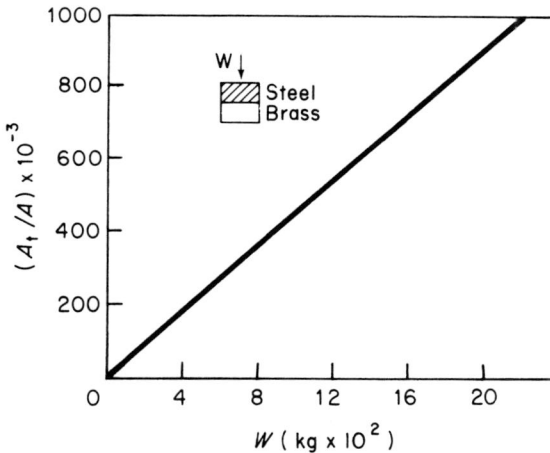

Fig. 2.6   Variation of the ratio A_t/A with the normal load, W. A_t = true area of contact; A = apparent area of contact. The upper block is steel which indents the softer brass block. The interfacial area is 25 mm^2, and the hardness of the brass is 90 kg mm^{-2}. Only the lower brass block flows plastically.

We can only speculate at the moment regarding the true magnitude of A_t/A, but it is quite correct to say that at very low loads a few plastic junctions will remain dispersed in an elastic matrix. In that case as the load is removed the surrounding elastic areas will recover their original shape and may disrupt the plastic junctions in the process. It follows that a couple with high elasticity will produce predominantly elastic interactions and weak bonds between the two surfaces because of the disruption during elastic recovery. The true area of contact due to plastic flow will also be low if the hardness is high. All this will result in a low value of the coefficient of friction. The effect of work hardening adds to the complexity of the phenomenon and it is not known precisely if elastic recovery lowers the amount of friction to a marked degree in practice.

2.8 Stick–slip

Tribological deformation embodies interaction at surface layers only, but spreads rapidly into the body of the component resulting in a subcutaneous characteristic regime. The latter in all probability plays the dominant role in friction and wear, but at the moment the information regarding this layer is inadequate. The existence of fractional intimacy only between two surfaces under compressive loading, however, has been proved [13], but the idea of adhesion was conceived long before that. It is probable that in this century the formation of true areas of contact by adhesion of asperities, and hence sticking of two surfaces, was first reported by Hardy and Hardy [14] in 1919 and again a decade later by Tomlinson [4] in 1929. Using the information that parts of an interface will adhere, the phenomenon of sticking together of two surfaces and their separation has been followed experimentally [15, 16]. Most machines in laboratories are designed to follow the frictional resistance with time as separate from wear. The trace of friction, unlike that of wear, as plotted by the weight loss technique, for example, does not produce a steady smooth line but fluctuates instead about a mean value as sliding continues. An idealized pattern of the variation of friction takes the appearance of a saw-tooth, and an analysis of the nature of this has provided basic information on the mode of interaction between surfaces undergoing relative motion under load.

A systematic variation of the coefficient of friction can be obtained if experiments are conducted using a specially designed apparatus. If a rider is mounted non-rigidly on an elastic frame and is loaded against a moving flat surface, a characteristic trace of friction against time is obtained, as shown in Fig. 2.7. As the flat surface commences motion, the rider moves with it and the friction rises from A to B. At the interval A to B, the surfaces of the rider and the flat specimen are sticking together and, at B, the friction is a maximum. The value at B is of course the maximum value of the static friction. The static component of friction disappears at B because it is then that the two surfaces separate,

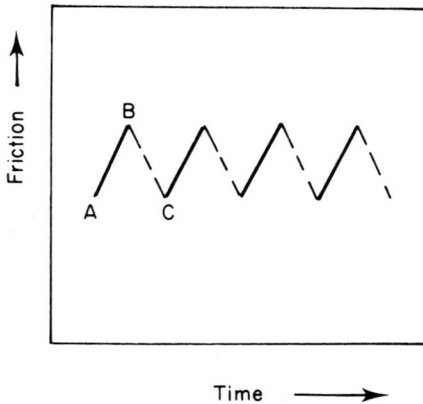

Fig. 2.7 Variation of friction with time (schematic—from the work of Bowden *et al.* [15, 16]).

that is, the interface slips so that the falling coefficient of friction tends to assume the value of the kinetic component at C. The nature of the apparatus is, however, such that the interface sticks, separates and repeats the process. In a well-conducted experiment, the friction at A equals that at C, the kinetic component.

Measurement of electrical resistance shows that the area of contact of the junctions increases from A to B. We recall that the combined effect of a normal and tangential force at an interface is to cause an increase in the area of contact, and hence in friction. The period B to C is of course that when the top surface slides relative to the bottom, so that the interfacial area slips and sticks again. The trace shown in Fig. 2.7 is typical of a soft rider sliding on a hard surface and there are significant variations in the stick–slip sequence. For example, when a hard high melting point rider is loaded against a soft surface, the sliding is jerky but sudden slip, as in Fig. 2.7, is not so obvious. For couples of similar materials the frictional resistance fluctuates in a random manner and is high, no rapid slip being detectable.

The nature of the friction machine should also modify the magnitude and the pattern of the relative slip as presented by plotting μ against sliding distance. In the experiment which produces a pattern such as is given by Fig. 2.7, the speed is low and the top rider has freedom to move bodily together with the bottom surface up to a certain distance. As will be discussed in Section 2.8.1, a long contact time of individual junctions

allows the magnitude of the frictional resistance to increase monotonically from some ambient value which may be zero. It follows that at
high sliding speeds, the friction coefficient will have a potentially low
value.

Slip means relative motion at the junctions, and we have already
differentiated between the static and kinetic coefficients of friction
designated as μ and μ_k respectively. The amount of slip ΔS should be a
function of μ and μ_k such that

$$\Delta S = \phi(\mu - \mu_k) \tag{2.30}$$

and, as will be shown in Section 2.8.1, $\mu > \mu_k$.

A detailed treatment of the interrelationship between μ and μ_k and
the speed of the machine is complex [17] but a simplified discourse is
useful for us to explain some macroscale events due to the microscopic
interplay between opposing asperities under load. Considering Fig. 2.7
again, as the speed of a machine is increased, the time interval between
A to B gets smaller and the overall magnitude of the coefficient of
friction is also diminished. Thus at very rapid relative interfacial
movement, the distinction between the process of stick and of slip
becomes less accentuated and $\mu \to \mu_k$ so that $\Delta S \to 0$. A very low value of
ΔS means that sticking and slipping of asperities are indistinguishable
and the machine runs smoothly. We must note at this stage, however,
that μ may tend not to be similar to μ_k at all speeds for all combinations
of metals or for all types of machines. It is important to accept this, as it
will be described later that in certain applications excessive speed may
cause a marked disparity between μ and μ_k, in which case the machine
will run in a jerky manner.

A side effect of friction is a rise in surface temperature. The temperature of the interface increases during the rapid slip process of the
junctions and falls as the asperities begin to stick. These thermal flashes,
as they are called, can last for only about a thousandth of a second, and
the rise and fall during slip and stick behaviour can continue
throughout the sliding period. There are two effects of this rise in the
interfacial temperature. One is that the yield stress of the asperities is
lowered so that they adhere readily. Secondly, a high temperature in
room atmosphere will facilitate the formation of oxide films. The two
effects are antagonistic because a film of oxide impedes the formation of
strong metallic junctions.

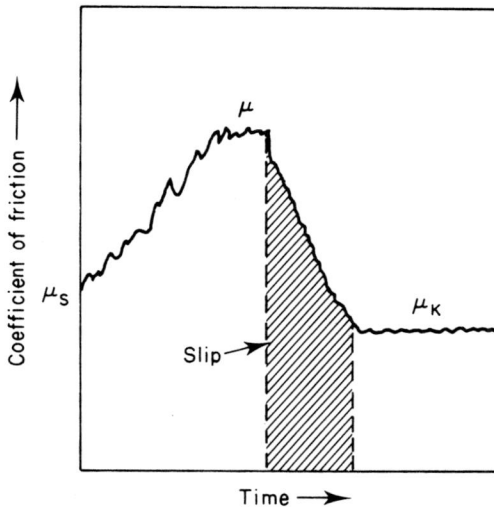

Fig. 2.8 Variation of coefficient of friction with time. μ_s is the value at rest which increases to μ due to junction growth. The junctions slip and the kinetic component μ_k persists (schematic).

From the preceding discussions we could represent the variation of μ with time as in Fig. 2.8. There is always a static value of μ given by μ_s before the machine starts. As the interface moves, μ_s rises to the maximum value of the static coefficient of friction, μ, and as the junctions slip, at a characteristic steady speed, the kinetic component μ_k, probably lower than μ_s, will persist. If the machine stops and restarts and runs under the same conditions, the pattern as shown in Fig. 2.8 should be repeated. The value of μ_k is lower, the higher the speed of the machine.

On an asperity scale, sticking is inevitable if adhesion is favourable. Adhesion is dependent on the mechanical property of the metal but can be altered if contaminants are present. Asperity interaction is a microscopic event and can be verified experimentally. For certain metal combinations a laboratory machine will develop severe vibrations because of the stick–slip process between, say, the pin and the disc.

Vibration of a system is also sensitive to its friction–speed characteristic [18, 19]. Generally, if for a couple the coefficient of friction decreases with surface speed, stick–slip on a large scale may occur. A well-known example is the cast iron brake block on railway vehicles

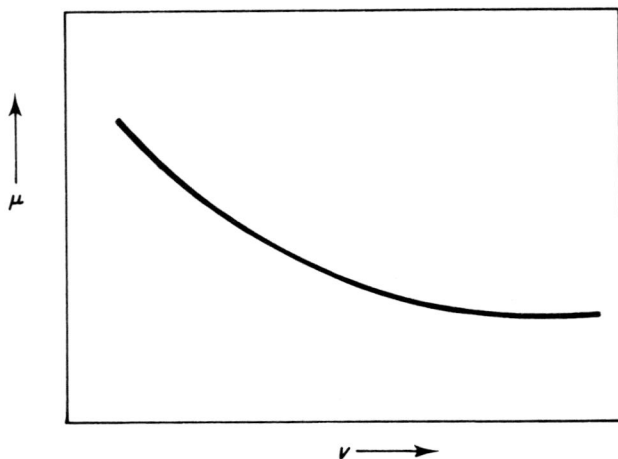

Fig. 2.9   A falling frictional coefficient, μ, with surface speed, v (schematic).

which produces a typically falling characteristic of the coefficient of friction with speed (Fig. 2.9). At very high speeds the slope is zero, so that if the brake is applied, the motion of the interface is still smooth since $\mu \rightarrow \mu_k$. The vehicle slows down but now the gradient is steep and $\mu \neq \mu_k$ so that ΔS is appreciable according to Equation 2.30, a condition conducive to chatter. Incidentally, this is the reason why brake materials are sought which will produce a shallower friction–speed gradient than that of cast iron against steel.

It has been established that the propensity to stick–slip is independent of the geometry of the interface or the applied load. One would expect some influence of the latter on the phenomenon because the true area of contact will increase with increasing load. However, couples such as brake assemblies have generally high contact pressure anyway so that experiments are conducted in the high load range. In that case the true area does not increase as steeply as when the load is increased in the low load range. The reason for a lack of dependence on the geometry of the interface is as expected, because it is the true area of contact which adheres. We should be able to explain why the frictional resistance decreases with speed in the next section. Experience shows that most metal combinations exhibit a falling friction–speed characteristic and vibration will be expected of them, albeit of varying magnitudes.

2.8.1 *Kinetic friction*

It is clear that asperity interaction can in the limit cause vibrations in a friction couple. A precondition is, of course, that the junctions will stick and remain together for a while and then slip with a degree of rapidity. It is not essential that the interaction is plastic because two opposing asperities may also interlock elastically and then slip with the same ultimate effect on the nature of sliding of two bodies.

Consider an asperity making a circular contact with a flat surface. It is known that a normal static load will continue to increase the area of contact because of creep. We can represent this by drawing a few expanding circles as in Figs. 2.10(a), (b) and (c). The area will extend further if a tangential stress is now applied but in the direction of motion (Figs. 2.10(d) and (e)). We have seen that the static coefficient of

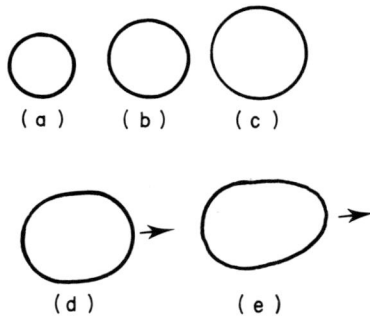

Fig. 2.10   Increase in the area of contact due to creep and to an applied tangential force in the direction of the arrows.

friction increases monotonically as the area increases. Creep increases with time. Therefore, at slow speeds, time is available for the area to increase both radially and possibly also tangentially. The opposite being the case at high sliding velocities, most metals show a falling μ–speed characteristic. Increased temperature due to high sliding speed should facilitate creep, but low friction is still the case because of one or both of the phenomena, viz. low yield stress and oxidation. A low yield stress will increase the area by creep but the applied tangential stress necessary to rupture the junctions will be low. Increased temperature also results in oxidation of the surfaces. The effect of oxide films on lowering the frictional resistance is well known.

There is some information available regarding the time necessary to attain a certain size of contact area. A stationary contact will grow under load to its near maximum size in a few seconds. Asperity interaction in tribological components must be at best of the order of milliseconds and the external tangential pull is high. The junction size, therefore, is probably that determined simply by the hardness of the metal and the applied normal load so that growth of the area by creep and lateral pull is minimal. All this supports the observed fact that the kinetic friction is always less than the static component for a particular situation and that μ_k decreases with surface speed for most metals, becoming asymptotic at a limiting value. It has been shown [21] that junctions adhere and remain so for a distance of about 10^{-4} cm as sliding begins. The friction starts decreasing thereafter up to a distance of 10^{-3} cm when the kinetic component is reached. It is imprudent to generalize but, from results in published literature, it would appear that an order of magnitude of the average size of metallic junctions is in the range 10^{-4} to 10^{-5} cm.

2.9   Energy of work

The simple theory of friction shows that μ is more or less the same regardless of the metals used. The limitations have already been summarized and the modified relationship given by Equation 2.29 goes some way to meet certain objections. Nevertheless, even allowing for contamination, there are many exceptions to the rule. Even different metal couples with comparable mechanical properties produce the frustrating situation where the coefficient of friction varies [22]. Attempts to obtain a more satisfactory theory of friction continue and one possibility is to explain the phenomenon of friction in terms of energy involved in the work done in causing interfacial adhesion followed by separation of the junctions [23, 25]. For two surfaces, identified by the subscripts 1 and 2, the work of adhesion E_{12} is given by

$$E_{12} = G_1 + G_2 - G_{12} \tag{2.31}$$

where G_1 = surface free energy of 1, G_2 = surface free energy of 2, G_{12} = free energy of the interface as a result of contact of 1 with 2.

Similar metals have a large work of adhesion and say, for 1,

$$E_1 = 2G_1 \tag{2.32}$$

Equation 2.31 assumes that the surfaces are chemically clean. This is not so in practice and strictly

$$E_{12} = G_1 + G_2 - G_0 - G_{12}$$

where G_0 is an arbitrary term used for the pressure exerted in the surfaces by a contaminant, the presence of which is to lower the free surface energy of the solids. For our discussion, however, we shall ignore G_0 to avoid undue complication.

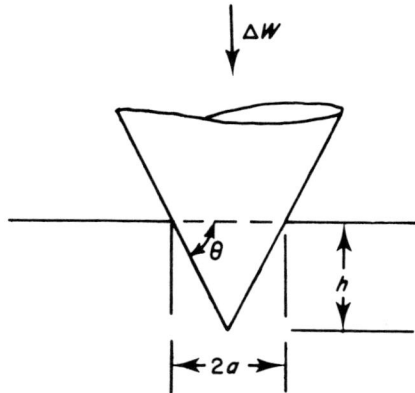

Fig. 2.11   A hard conical asperity of base angle θ penetrating a solid to a depth h.

To obtain an expression for μ in terms of surface energy parameters, it is necessary to state the energy balance equation for the interface in question. Thus, consider a hard conical asperity (Fig. 2.11) under a load ΔW indenting a soft solid to a depth h. Let the asperity have a base angle θ as shown and let the diameter of indentation be 2a. If the interaction is wholly plastic, the energy terms involved will be as follows:

    (1) energy is required in the plastic deformation of the soft metal whose hardness is H. The magnitude of this energy is $\int_0^h \pi a^2 H \, dh$;

    (2) work is done to move a distance h by a load ΔW and this is given by $\Delta W h$;

    (3) the cone indents an area of contact in the solid. In so doing, the surface energy of the whole system is reduced by an amount decided by the work of adhesion per unit area (Equation 2.31). The indented area is πa^2; therefore the amount by which the surface energy is lowered is $E_{12} \pi a^2$.

Let the total energy of the system be G. Then the energy available for the system is $G - E_{12}\pi a^2$. This must equal the energy of work done in moving a depth h, less the energy expended in causing plastic flow. Thus

$$G - E_{12}\pi a^2 = \Delta Wh - \int_0^h \pi a^2 H \, dh. \tag{2.33}$$

Putting $a = h \cot \theta$ (Fig. 2.11),

$$G = E_{12}\pi h^2 \cot^2\theta + \Delta Wh - \pi H \cot^2\theta \int_0^h h^2 \, dh$$

or

$$G = E_{12}\pi h^2 \cot^2\theta + \Delta Wh - \frac{\pi}{3} H \cot^2\theta h^3. \tag{2.34}$$

Consider now the situation when a top surface is placed upon another component. Isolating the behaviour of one single conical asperity from the top surface with dimensions as in Fig. 2.11, indentation will continue until a depth h has been penetrated. At that moment there is no further movement and the total energy of the system is in an equilibrium state, when $dG/dh = 0$. To obtain dG/dh, differentiate Equation 2.34, giving

$$\frac{dG}{dh} = 2\pi E_{12}h \cot^2\theta + \Delta W - \pi Hh^2 \cot^2\theta$$

At equilibrium dG/dh is zero. Therefore,

$$\Delta W = \pi Hh^2 \cot^2\theta - 2\pi E_{12}h \cot^2\theta$$

Dividing both sides by $\pi h^2 \cot^2\theta$,

$$\frac{\Delta W}{\pi h^2 \cot^2\theta} = H - \frac{2}{h}E_{12}$$

Resubstituting $h \cot \theta = a$ in the above relationship,

$$\frac{\Delta W}{\pi a^2} = H - \frac{2E_{12} \cot \theta}{a} \tag{2.35}$$

To obtain a relationship incorporating μ we can use the simple relationship that the frictional resistance at the interface is $\tau \pi a^2$ where τ is the shear strength of the soft metal. Since, in practice, h is very small, the ploughing term can be neglected so that

$$\mu = \tau \pi a^2 / \Delta W = \tau / (\Delta W / \pi a^2)$$

Substituting for $(\Delta W / \pi a^2)$ from the above expression in Equation 2.35,

$$\mu = \frac{\tau}{H - (2E_{12} \cot \theta)/a} \qquad (2.36)$$

Equation 2.36 expresses the coefficient of friction in terms of the hardness of the metal and the degree of adhesion between the two solids.

It can be seen that μ is infinitely large when

$$H = \frac{2 \cot \theta}{a} E_{12} \qquad (2.37)$$

Taking a typical value of θ as $6°$ and $a = 2 \times 10^{-2}$ cm, the term $(2 \cot \theta)/a = 10^3$. For $\mu \to \infty$ we can assume that the measurements on the friction couples are carried out in high vacuum. We can denote E_{12} as E_∞ and for the coefficient of friction in air E_{12} can be designated as E_{air}. From Equation 2.37, $E_\infty = H \times 10^{-3}$. Putting $\tau = \tau_c = \sigma_Y/2 = H/6$ in Equation 2.36, and taking μ in air as 1, $E_{air} = 8·34H \times 10^{-2}$. The above calculations show very little difference between E_∞ and E_{air}.

It is difficult to comment on values obtained by this kind of calculation for metals where the creation of new surfaces, say, by shearing of junctions, is preceded by plastic deformation. A great deal of energy is expended in this plastic work and any increase in surface energy should be small. The energy model will probably be difficult to apply to metals.

On the other hand with brittle materials such as ceramics, strength to fracture can be related to surface energy. Measurements on mica show that it has a surface energy of 300 erg cm^{-2} in air. This can be increased by a factor of nearly 16 when the experimental pressure is reduced to 10^{-6} Torr, and can be much higher if the vacuum is increased further. The coefficient of friction is high in vacuum and the surface energy terms also increase.

References

1. Dowson, D. (1973). *Mech. Eng.* **95**, 12.
2. Davison, C. stC. (1957). *Engineering* **183**, 2.
3. Davison, C. stC. (1961). *Technology and Culture* **2**, 11.

4. Tomlinson, G. A. (1929). *Phil. Mag.* **7**, 905.
5. Bowden, F. P. and Tabor, D. (1964). "The Friction and Lubrication of Solids", Part 2, Oxford University Press.
6. Sarkar, A. D. (1976). "Wear of Metals", Pergamon Press, Oxford.
7. McFarlane, J. S. and Tabor, D. (1950). *Proc. Roy. Soc. A* **202**, 244.
8. Parker, R. C. and Hatch, D. (1950). *Proc. Phys. Soc. B* **63**, 185.
9. Courtney-Pratt, J. S. and Eisner, E. (1957). *Proc. Roy. Soc. A* **238**, 529.
10. Tabor, D. (1959). *Proc. Roy. Soc. A* **251**, 378.
11. Bowden, F. P. and Tabor, D. (1939). *Proc. Roy. Soc. A* **169**, 391.
12. McFarlane, J. S. and Tabor, D. (1950). *Proc. Roy. Soc. A* **202**, 224.
13. Holm, R. (1946). "Electrical Contacts", H. Gerbers, Stockholm.
14. Hardy, W. B. and Hardy, J. K. (1919). *Phil. Mag.* **38**, 32.
15. Bowden, F. P. and Leben, L. (1939). *Proc. Roy. Soc. A* **169**, 371.
16. Bowden, F. P., Leben, L. and Tabor, D. (1939). *The Engineer* **168**, 214.
17. Morgan, F., Muskat, M. and Reed, D. W. (1941). *J. Appl. Phys.* **12**, 743.
18. Blok, H. (1940). *J. Soc. Aut. Eng.* **46**, 54.
19. Bristow, J. R. (1942). *Nature* **149**, 169.
20. Barwell, F. T. (1956). "Lubrication of Bearings", Butterworths, London.
21. Rabinowicz, E. (1951). *J. Appl. Phys.* **22**, 1373.
22. Tamai, Y. (1961). *J. Appl. Phys.* **32**, 1437.
23. Machlin, E. S. and Yankee, W. R. (1954). *J. Appl. Phys.* **25**, 576.
24. Rabinowicz, E. (1961). *J. Appl. Phys.* **32**, 1440.
25. Rabinowicz, E. (1958). *Trans. ASLE* **1**, 96.

3   Adhesive Wear

If a flat surface of a hard steel is pressed against a block of brass and examined at high magnification after separation, discrete areas of embedded brass particles are at once evident. Obviously, these patches are those true contact spots or junctions which formed as a result of plastic flow and broke when the pressure was removed. Junctions can break if the applied load is not high enough to produce strong cold welds at the points of contact. It is also known that elastic recovery of the matrix subsequent to the removal of the normal load can disrupt the integrity of the junctions. The net result is that the brass loses a certain amount of material which adheres to the steel surface. Since there has been a loss of material, by definition, the brass has undergone wear. To be more specific, wear has occurred by adhesion of brass on to the counterface. That is, the brass component undergoes adhesive wear and continues to do so if it is repeatedly loaded against the steel and then separated from it. In real tribological situations, the interface moves and a simple wear run with a brass pin on a steel disc will show that the counterface is smeared with a layer of brass. Obviously, junctions form by plastic flow and the contact area extends in the overall direction of motion before rupturing. Fresh junctions form and break up and this happens over a large number of contact points. As the contact areas merge due to prolonged sliding, the transferred brass appears continuous although the discrete nature of the deposit may be evident even under moderate magnification. Adhesive wear is associated with the most severe form of surface damage apart from loss of material and most metal combinations will adhere together, the strength of the interfacial bond depending on the nature of the friction couple. The only possible way of avoiding adhesion is to separate the interface by a solid or a liquid lubricant. Lubrication may not be always effective and metal to metal contact can then occur resulting in gross surface damage and wear. We should mention at this stage that most systems in engineering, human articulating joints or floor coverings, among others, are probably subjected to mixed modes of interaction. For example, a bearing may be under the influence of adhesion, abrasion and a hostile chemical environment. The suggestion is that real situations are complex, but it is essential that we attempt to isolate each type of wear for a full understanding of it. If this is achieved, the next stage will be to examine a situation where more than one type of wear is operative.

3.1   Friction couple

It is a basic characteristic of a friction couple that junctions form and separate so that relative movement at the interface is maintained. Whether the same pair of asperities will meet more than once will be decided by the laws of chance, but junctions will continue to form throughout the life time of a pair of friction elements unless effectively separated. In certain situations, the function of the couple can only be fulfilled by the formation of effective junctions such as a brake shoe on a disc. If the encounter between two asperities is elastic, the bonds will simply separate under an applied tangential stress although fracture at an asperity level may occur due to repeated contacts by a fatigue process. Adhesive wear will only take place when the applied load results in a plastic interaction and a very important aspect of experimental studies is to observe the nature of the surface or subsurface. If a soft metal such as tin slides on a hard steel, the former will be expected to flow and adhere onto the latter, creating an interface of low critical shear stress, τ_c. Deposition of tin will continue with sliding and wear debris may form as a result of ploughing of the soft surface by the asperities of the hard steel or by flaking off of patches from the deposit. If a pin of aluminium is slid on a bush of the same metal, both the components flow plastically followed by work hardening. The result is that failure on a microscopic scale may occur below this work hardened layer where the shear strength is lower. The rate of metal loss is high and the severe nature of surface damage is shown by the large size of the wear product.

There are many intermediate situations between those given above, some of which are, at a first glance, anomalous. For example, a soft metal such as copper can pluck material from a harder steel counterface. Much of the explanation regarding the phenomena at the surface and subsurface of a tribological interface involves a close metallurgical investigation.

3.2   Atomic model

Because of the complex nature of the subsurface in metals and possibly many non-metals due to sliding, a single formula describing wear has not always been successful. Attempts are frequently made nevertheless to relate the rates of wear of metals and non-metals with such engineering variables as load and speed together with a particular mechanical property such as the yield stress or hardness. Hardness of materials is an easily measurable parameter and correlation of wear rates with this will be of particular value during the design stage to aid selection of processes and materials. Unfortunately, the idea that the harder a material is the less it will wear does not prove to be correct in all situations. This is because of the complexity of the subcutaneous layer of a component which develops due to sliding on another solid. The complex nature of this layer will probably only be unravelled by extensive microscopic examination coupled with X-ray and electron diffraction studies. We shall examine available evidence in due course, but firstly, it is most informative to study the theories of wear which have appeared in published literature during the last five decades.

We recall Tomlinson's model in Chapter 2 which proposed that forces of repulsion between two approaching surfaces constituted the friction component. It was implied that a repelled atom would be dislodged and would attempt to return to its original position. This need not be the case each time, and it is equally probable that as an atom is dislodged from its equilibrium position, it may adhere to another atom in the opposite surface instead of returning to its original position. This has been termed wear by atom plucking.

The theory of metal removal by plucking of atoms from a surface during sliding is developed by considering the energy dissipated by an atom pair as it forms by the plucking process. This energy is $E_c l$, as was shown in Chapter 2, where E_c is the cohesive force between two atoms and l is the distance travelled by an atom while it is free. If ρ is the density of the metal which is wearing, the mass m of an atom can be expressed as $m = \rho(\pi/6)d^3$, where d is the diameter of an atom. If in a sliding process n atom pairs interact to dissipate an amount of energy whose mean value is E,

$$n = E/E_c l$$

If the total mass of all the participating atoms in this work is M,

$$M = nm$$

Note that the mass is not $2nm$, since only one atom is plucked per pair. Substituting for n,

$$M = \frac{Em}{E_c l} = \frac{E\rho\pi d^3}{6E_c l} \tag{3.1}$$

Substituting for $E_c l$ from Equation 2.11,

$$M = \alpha\frac{\pi}{6}\frac{E\rho d^2}{p\mu} \tag{3.2}$$

The model assumes that the flow stress σ_Y of a metal can be expressed as $\sigma_Y = p_{max}/(d^2/4)$, where p_{max} is the maximum repelling force.
   Since $p = \frac{1}{2}p_{max}$,

$$\sigma_Y = 8p/d^2$$

Substituting for p/d^2 in Equation 3.2,

$$M = \alpha\frac{4\pi}{3}\frac{E\rho}{\mu}\frac{1}{\sigma_Y} = \alpha\frac{4\pi}{H}\frac{E\rho}{\mu}, \tag{3.3}$$

where H is the hardness of the metal.
   The most interesting aspect of the atomic model is that the mass of metal removed during the surface interaction is inversely proportional to the hardness of the metal. That is, Equation 3.3 states categorically that the deciding parameter for the amount of wear is the hardness of the metal. Hardness of metals is easily measurable and the parameter will appear in most laws of wear, as will be seen later. We note also that Equation 3.3 suggests, *inter alia*, that the higher the frictional resistance, the lower the amount of wear of the metal. Intuitively, this would appear to be paradoxical.

3.3   Asperity interaction
Although interesting, the idea of material removal from solids by plucking of atoms has not been generally accepted. The reason is that

wear debris is generated more or less simultaneously with the onset of motion and in all cases the debris consists of aggregates of metallic particles. The not so unreasonable hypothesis is thus that wear particles are produced by a process which involves much more than simple atom to atom encounters.

3.3.1   A wear law

The law propounded by Archard [1] expresses the volume loss, V, of metals in terms of the sliding distance, S, and the yield stress, σ_Y of a metal. We know that surfaces have protuberances so that even at infinitesimally light loads, contact will occur on at least three points at the interface. As the load is increased, these three microscopic areas of contact will grow radially and the compliance between the bodies will decrease, resulting in further interaction at the neighbouring asperities. In other words, the effect of increasing the normal load is to expand the true area of contact, A_t, by two processes:

(1) the existing areas expand;
(2) fresh areas form elsewhere.

To follow the sequence of events during interaction, consider a hemispherical asperity pressed against a similar one under a load, W (Fig. 3.1). Suppose also that the top hemisphere is very hard and let wear be confined to the non-work hardening soft hemisphere at the bottom.

The first effect of applying the load, W, is for the top rider to sink into the soft support which has flowed plastically. Let the area be circular with a diameter $2a$. Suppose now that a tangential pull separates the adherent soft metal neatly, leaving a clean circular surface at the bottom asperity. There is no need to wonder if a wear particle is detached or the

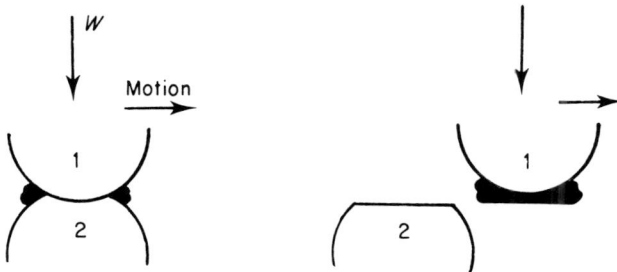

Fig. 3.1   Asperity 1 sinks and causes plastic flow in 2 which is soft. 1 moves on carrying with it adherent wear debris.

soft material remains attached to the top surface as shown in Fig. 3.1. What happens is that the bottom surface has lost an amount of material whose volume, V, can be derived as follows.

There will be more than one contact similar to the one shown in Fig. 3.1. for a friction couple. The true area of contact for n junctions is $n\pi a^2$. We should appreciate that the radius of contact is not the same for all the junctions and it will be more accurate to regard a as the mean value of all the radii of contact. To be pedantic, the area of contact may not be truly circular but it is so assumed to make the mathematics manageable. For plastic contact,

$$A_t = W/3\sigma_Y,$$

where W is the applied normal load and σ_Y is the flow stress of the material which is wearing. Therefore, the number of junctions, n, is

$$n = \frac{W}{3\sigma_Y \pi a^2} \tag{3.4}$$

Considering Fig. 3.1, a distance $2a$ has been traversed to produce a certain amount of wear for the soft metal. This distance will be traversed by the whole of the top surface relative to the bottom component. The number of asperities involved in the process is n. Therefore, the number, n_0, for unit sliding distance is $n_0 = n/2a$. Substituting for n in Equation 3.4,

$$n_0 = \frac{W}{6\pi\sigma_Y a^3} \tag{3.5}$$

As sliding proceeds junctions will continue to form and separate and be distributed randomly over the apparent area. It does not follow that the lower asperities will always lose material. To understand this, consider the bottom asperity as shown in Fig. 3.1. It may work harden due to repeated encounters and future interaction may then be elastic in nature. The probability of a junction remaining intact for a long time due to an elastic interaction is quite high. It is also known that the material picked up by the top rider may be deposited subsequently elsewhere giving rise to what is known as back transfer. This is a gain by the bottom surface rather than wear, and because there are so many possibilities, it is impossible to evaluate how many junctions will produce wear. For the time being, therefore, assume that there is a probability β that certain contacts will wear from n_0 junctions per unit sliding distance.

For a tribological mechanism, taking S as the total sliding distance, the total volume of wear, V, from the soft surface is,

$$V = \beta n_0 \times (\text{volume loss from a junction}) \times S \qquad (3.6)$$

Since the asperity is hemispherical, its volume is $(2/3)\pi a^3$. Substituting this value in Equation 3.6,

$$V = \beta n_0 (2/3)\pi a^3 S$$

Now, substituting for n_0 from Equation 3.5,

$$\frac{V}{S} = \beta \frac{W}{3(3\sigma_Y)}$$

Since $3\sigma_Y \simeq H$, the hardness of the soft metal,

$$\frac{V}{S} = \beta \frac{W}{3H} \qquad (3.7)$$

This is the adhesive law of wear.

Suppose now that a conical asperity interacts with a hard flat surface and loses an amount of metal, forming a circular area of diameter $2a$. Using a similar argument as for a hemispherical rider, the volume of material lost is $\frac{1}{3}\pi a^3 \tan\theta$ where θ is the base angle of the cone. The rate of wear now is,

$$\frac{V}{S} = \beta \frac{W}{3H} \frac{\tan\theta}{3} \qquad (3.8)$$

The rate of wear is inversely proportional to the hardness of the metal and directly proportional to the applied normal load.

A relationship as shown in Equation 3.7 is very attractive because it relates an engineering variable and a mechanical property with the rate of wear. There is, however, a shape factor, which is $\tan\theta/3$ for a conical asperity. The law using the hemispherical rider (Equation 3.7) shows that the rate of wear is independent of the geometry of the component, which seems to be substantiated in laboratory studies.

3.3.2 *Alternative approach*
An alternative approach [2] using both conical and hemispherical asperities arrives at similar types of relationship and is worth describing.

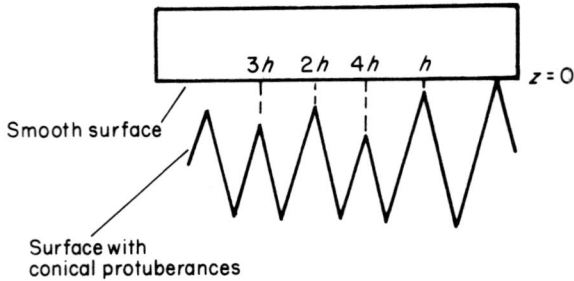

Fig. 3.2 A flat smooth surface contacting conical asperities (after Yoshimoto and Tsukizoe [2]).

Considering conical asperities, the model assumes that a perfectly smooth flat surface is pressed against a solid which has protuberances in the form of cones (Fig. 3.2). All the cones have the same base angle θ but they are at different levels relative to a horizontal datum line. Suppose in the first instance the top surface rests only on one cone and the datum line then is $z = 0$. Figure 3.2 shows that, in that configuration, the apices of the asperities are $0, h, 4h, 2h, 3h$ etc. from the datum line. Let n be the total number of asperities and suppose that the smooth component does not wear at all. The cones are soft with a flow stress σ_Y and they wear under load and relative movement. The effect of a normal load W will be to flatten the asperity in contact and we assume that the interaction produces a circular area of diameter $2a$. Let the smooth surface move laterally covering a sliding distance $2a$. If the height of the part of the cone which is removed due to sliding is h_0, the volume ΔV_0 lost during this encounter is

$$\Delta V_0 = \tfrac{1}{3}\pi a^2 h_0 = \tfrac{1}{3}\pi a^3 \tan \theta,$$

since $h_0 = a \tan \theta$. Since the distance slid is $2a$, the volume removed per unit sliding distance is

$$\Delta V = \Delta V_0/2a \quad \text{or} \quad \Delta V = (\pi a^2 \tan \theta)/6 \tag{3.9}$$

Since there are n asperities which are involved, the respective distances of their apices from the hard smooth surface are $(0, 1, 2, 3, 4, \ldots n-1)h$ respectively. Considering the contribution made by all the asperities under load, for a larger distance S, the volume of material lost is

$$V = \frac{\tan \theta}{6}\left[\sum_0^{n-1} \pi a^2\right] S$$

Since $\sum_0^{n-1} \pi a^2 = A_t = W/3\sigma_Y$,

$$V = \frac{\tan\theta}{6}\frac{W}{3\sigma_Y}S = \frac{\tan\theta}{6}\frac{W}{H}S$$

or

$$\frac{V}{S} = \frac{1}{2}\tan\theta\frac{W}{3H} \tag{3.10}$$

Comparing Equations 3.10 and 3.8, the volume rate of wear is a function of $(W/3H)$ but the constant terms are different. One weakness of the model of Fig. 3.2 is that the diameter of contact is assumed to be $2a$ for all asperities. The same assumption has been used by Archard so that this is a common factor of the two models. The implication that all the participating cones will produce wear products is unacceptable, and in the absence of theory a probability term must be included.

3.3.2.1 Hemisphere

The probability term is again omitted for hemispherical asperities [2] using an exactly similar model and suppositions as in Fig. 3.2. The volume of a hemispherical cap removed is $(2/3)\pi a^3$ so that the amount of volumetric wear per unit sliding distance is

$$(2/3)\pi a^3/2a = \tfrac{1}{3}\pi a^2.$$

The total volume V removed for a sliding distance S, is

$$V = \tfrac{1}{3}[\sum_0^{n-1} \pi a^2]S$$

and as before,

$$\frac{V}{S} = \frac{W}{3H} \tag{3.11}$$

3.4 Fractional film defect

The premise of the postulates regarding the laws of wear as derived in the preceding sections assumes clean intermetallic contact. This is never true in practice nor in planned sliding experiments in laboratory conditions. This suggests that the wear laws must be modified.

The effect of a contaminant will be to reduce the magnitude of metallic contact and we should expect a lower wear rate than that predicted by Equation 3.7. As a first step, we could incorporate a factor α in the wear equation such that,

$$\frac{V}{S} = \alpha\left(\beta\frac{W}{3H}\right) \tag{3.12}$$

α can have a value between zero and unity and has been termed [3] the fractional film defect. A modified wear law is then deduced as follows. The yield criterion, assuming plane strain conditions, is given by

$$\sigma^2 + 3\tau^2 = \sigma_Y^2,$$

using the notation of Chapter 2.

We know that for a true area of contact A_t, the frictional resistance $F = \tau A_t$. Since $\mu = \tau A_t / W$, $\tau = \mu(W/A_t)$ and $A_t = W/3\sigma_Y$, so that $\tau = \mu 3\sigma_Y$. We also know that $\tau = \mu\sigma$. Using this value of τ, the yield criterion can be written as

$$\sigma^2(1 + 3\mu^2) = \sigma_Y^2 \quad \text{or} \quad \sigma = \sigma_Y/(1 + 3\mu^2)^{1/2} \tag{3.13}$$

Adhesive wear occurs as a result of plastic yielding and we assume that there is no work hardening. In that case $\sigma = W/A_t$ or $A_t = W/\sigma$, a misleading expression.

Substituting for σ from Equation 3.13 in the above relationship,

$$A_t = \frac{W}{\sigma_Y}(1 + 3\mu^2)^{1/2} \tag{3.14}$$

Writing Equation 3.12 as

$$\frac{V}{S} = \alpha\left(\frac{\beta}{3}\frac{W}{H}\right) = \alpha\left(\frac{\beta}{3}A_t\right)$$

and substituting in the above for A_t from Equation 3.14,

$$\frac{V}{S} = \alpha\beta\left(\frac{W}{3\sigma_Y}\right)(1 + 3\mu^2)^{1/2}$$

or

$$\frac{V}{S} = \alpha\beta\left(\frac{W}{H}\right)(1 + 3\mu^2)^{1/2} \tag{3.15}$$

At a first glance, Equation 3.15 is attractive because it shows that the higher the coefficient of friction, the greater the amount of wear. If we plot V/S against μ using Equation 3.15 and assuming α, β, W and H to be constants (Fig. 3.3), there is always some wear even in the absence of

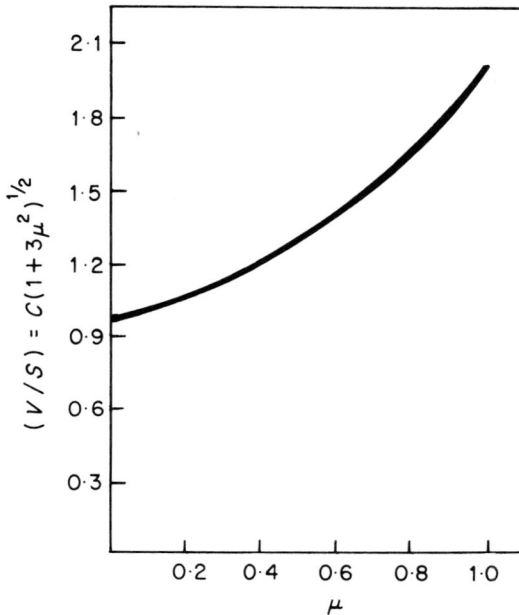

Fig. 3.3 A plot of wear rate against the coefficient of friction μ. $C = \alpha\beta(W/H) = $ constant (see Equation 3.15).

friction. This is a contradiction because zero friction means that there is no physical contact and the question of wear does not arise in that case. At finite values of μ, the volumetric loss of metal increases. This is feasible, but the evidence to date suggests a complete lack of correlation between the rate of wear of metals and the coefficient of friction. The term α equals A_t/A by definition. For a constant load and speed, that is, for a constant interfacial temperature, $A_t \rightarrow 0$ when the interface is fully contaminated and $V/S = 0$, which is as it should be. A_t and hence α increase as the surfaces become cleaner and the amount of wear increases as postulated. It is difficult to measure α and we must emphasize that the term β is still pertinent. The reason is that the wear law given by Equation 3.7 applies to perfectly clean surfaces and β is the probability of forming a wear product in the absence of contamination. In practice Equation 3.15 is seldom used. In wear runs the probability β is calculated from experimental results and is quoted for the operating environmental condition. For example, 60/40 brass sliding on steel in room atmosphere gives a value of β of the order of 10^{-4}.

3.5 Oxidation

Most metals oxidize readily, especially if the ambient temperature is high. The vast majority of metals are left therefore, with surface oxide layers which play an important part in the protection against surface damage of friction couples. When a layer of oxide is less than 3000 Å thick, it is referred to as a surface film, and when it exceeds this thickness it is called a scale [4]. The films may be coherent and protective, as in the case of aluminium. Protection is only possible if the film is undamaged so that the metal substrate is prevented from coming in contact with the oxygen of the surrounding atmosphere. Scales can prevent further oxidation for the same reason, provided they are strongly adherent. A fair guide to the efficacy of an oxide film is the Pilling–Bedworth rule which is based on the volume of the oxide formed, V_o, and that of the metal, V_m. The scale is continuous and protective to the substrate if $V_o \geq V_m$ and the opposite is the case when $V_o < V_m$. In the latter situation, the scale breaks as it forms.

Ingress of oxygen towards the substrate is facilitated and the weight w of the metal changes with time t as it oxidizes in a linear manner as follows:

$$w = C_1 t \qquad (3.16)$$

where C_1 is a constant which is dependent on temperature. A protective oxide film gives a parabolic relationship, viz.,

$$w^2 = C_2 t \qquad (3.17)$$

where C_2 is a constant at the particular ambient temperature. In certain metals such as aluminium, a tenacious oxide film forms and the weight, and hence the thickness, of the layer increases with time in a parabolic manner. The oxide layer of aluminium, however, breaks with time and heals, a process which continues throughout the period of oxidation. This gives rise to a stepped growth in the parabolic curve.

Most metal surfaces are covered with oxide layers and as long as they are coherent during sliding intermetallic contact will be hindered. An interesting experiment [5] is to slide metals in a vacuum. Even at a moderate vacuum of 10^{-6} Torr, the measured friction will be high. Introduction of oxygen in the system quickly lowers the friction, the time-scale being about two seconds. It will be necessary, of course, to examine the specimen so that it can be proved that a low coefficient of friction is due to oxidation and not a result of sorbed gases. If oxidation does occur, it is quite likely [6], that the film adjacent to the metal substrate will be the dominating factor in influencing the magnitude of friction and wear, always provided that it maintains its coherency.

If both the sliding surfaces are covered with oxides, the mode of failure to produce oxide wear debris is expected to be brittle, although plastic flow has been observed with certain non-metals [7]. A normal or tangential stress is expected to break the oxide layers. Although the cracks will heal with time as sliding continues, the oxide film will be damaged. The exposed metallic patches will adhere to their counterparts on the opposite surface so that the interactions will be both metallic and between oxide-covered junctions. The former will give rise to plastic shear whereas the oxides will probably fail by brittle tensile fracture as observed in materials like quartz [8].

Examination of wear debris produced at very light loads shows a predominance of oxides. We assume that only the oxide films fail due to a combined normal and tangential stresses and the metallic substrate is undamaged. Fresh oxides, of course, continue to form.

3.5.1 *Effect of temperature*
Sliding a 60/40 brass pin on a hard steel counterface initially shows a rise in the wear rate with increasing temperature [9]. The loss of wear resistance continues up to a temperature of 350 °C. However the difference in the wear rates between room temperature and 350 °C is very small, and the only explanation is that the oxide films are not effective and the rate of wear increases as a result of loss of hardness of the brass with rising temperature (Equation 3.7).

In the temperature range about 350–550 °C, there is a progressive rise in the contact resistance indicating a decrease of metallic contact. Obviously, this has happened due to the effectiveness of the oxide films, and the wear rate falls by about three orders of magnitude.

While evaluating the temperature dependence of the rate of wear, the effect of normal load should not be ignored. The obvious effect is that an increase in contact pressure will damage the oxide layer and expose the metal substrate. This will encourage the formation of metallic junctions and the possibility of increased surface damage and wear. Increasing the normal load, on the other hand, increases the interfacial temperature, but the positive effect of the protective oxide layer thus formed is not known.

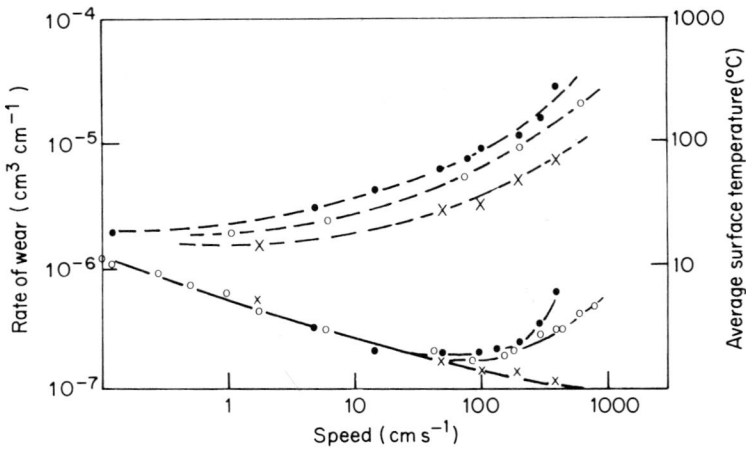

Fig. 3.4   Variation of wear rate and surface temperatures with speed. ——— wear rate; – – – – – surface temperature; ○ metal pin-holder; × water-cooled pin holder; • thermally insulated pin-holder. (Hirst, W. and Lancaster, J. K. (1960). *Proc. Roy. Soc. A* **259**, 228.)

3.5.2   *Effect of surface speed*

Undoubtedly, one way of increasing the interfacial temperature is to employ a high surface speed. Experiments [10] with brass pins on steel at heavy loads show a steady fall in the wear rate from a surface speed of 0.01 cm s^{-1} to about 100 cm s^{-1}. Beyond this speed, for a straightforward experiment, the wear rate increases. Some special experiments were carried out [10] to see if it was the rise in surface temperature due to increased surface speed which modified the wear rates (Fig. 3.4). If the pin is water cooled, the wear rate continues to fall. If the brass pin is thermally insulated, the wear rate is higher than when the pin is not insulated or is water cooled. Figure 3.4 also shows the variations in temperature as measured by a thermocouple placed 2 mm behind the brass interface. There is no abrupt change in temperature at the speed where the wear rate increases for the appropriate pins. The authors [10] argue that the fall and rise in the rate of wear is not altogether due to oxidation. We observe, however, that the temperature of the water cooled pin remains below 100 °C and it is known that any significant softening of 60/40 brass does not occur at that temperature.

The degree of oxidation depends on both the load and the nature of the metal. Apart from oxidation, very high sliding speeds can result in excessive heating and incipient fusion of the metal, especially if its melting point and thermal conductivity are low [11]. The nature of the friction couple with respect to the dependence of wear rates on surface speeds has been investigated at loads of up to 2 kg on an apparent contact area of 0·36 cm^2 and a velocity of 6500 cm s^{-1}. Sliding copper on steel or like combinations of copper and tungsten carbide respectively gives wear rates which fall at high speeds, although nickel on nickel behaves in an anomalous manner [12]. Addition of tin to copper generally has a beneficial effect on both friction and wear. The total amount of wear can be halved by doubling the amount of tin in certain composition ranges [13] and increased surface speeds appear to reduce wear further.

One way of estimating if the interface is oxidized is by electron microprobe analysis. Sliding bronze on steel will deposit some of the former on to the counterface. An increase in the sliding speed, particularly at high loads, gives rise to interfacial oxidation of the deposit, the oxide probably being of copper [14]. For a tin bronze with about 11% tin, the amount of material loss rises by a factor of four when the speed is reduced by a factor of about fourteen. Evidence of oxidation can also be obtained by X-ray analysis of the wear debris.

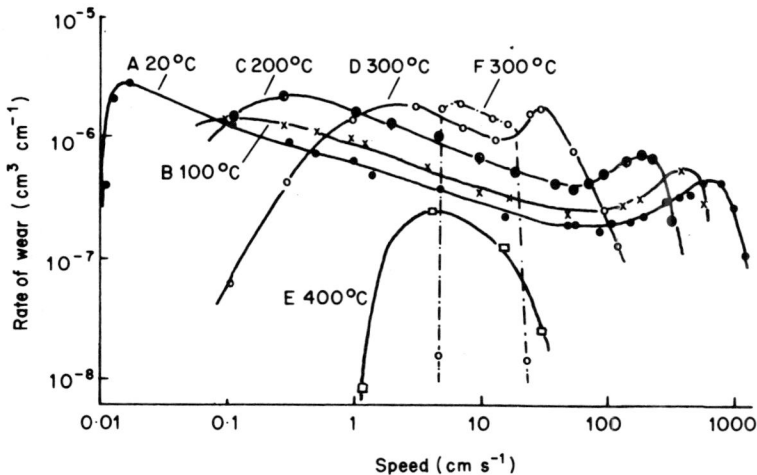

Fig. 3.5 Variation of wear rates of 2% leaded 60/40 brass pins at various surface speeds and ambient temperature. Hard steel counterface —— in air; – – – in oxygen. (Lancaster, J. K. (1963). *Proc. Roy. Soc. A* **273**, 466.)

Formation of oxide films and their protective action is a satisfying explanation until one encounters experimental results which show minor but confusing deviations. Figure 3.5 shows variations of wear rates with surface speeds for 2% leaded 60/40 brass pins rubbing on hard steel at various ambient temperatures [15]. To appreciate the range of speeds in these experiments, the bushes were run in the range 8 to 7700 r min^{-1} for a pin–bush machine with a bush diameter of 25 mm. The general pattern of relationships shown in Fig. 3.5 is that the wear rate first rises to a maximum with increasing speed. There is then a general fall followed by a rise and a second peak. A further fall is seen after this second peak. The first peak appears at higher surface speeds as the ambient temperature is increased. This is a complex example and simple explanations will not be possible. One can speculate that the phenomenon is probably tied up with strain rates at the interface.

3.6 Running-in wear

When a shaft begins its life on a new bearing, both the speed and the load are carefully controlled during what is termed the running-in stage. No doubt during the running-in period, any high spots on either the shaft or the bearing are removed and counterformality on an asperity scale is changed to a conformal engagement. Wear at this stage is beneficial as careful running-in of mechanisms ensures an optimum life. The secret of running-in is to induce conformity at gentle loads without increasing surface temperature excessively. In most engineering systems such as the car engine, the surfaces are flooded with a lubricant so that a high bulk temperature due to frictional heating is avoided.

Some surface heat is probably generated whenever one component slides over another under load. Even during gentle polishing such as preparing specimens for micro-examination, the surface layers flow and become physically different from the bulk of the metal. The layer was once known as the Beilby layer [16], and the nature of this was studied again by Cochrane [17] with the aid of electron diffraction techniques some three and a half decades after Beilby's report. It was also confirmed by Bowden and Hughes [18] around the same time that polishing causes surface flow in metals. The term Beilby layer is not fashionable nowadays and it has been shown to possess a high dislocation density and a disordered structure. The layer has considerable technological value and is regarded by most users as the *sine qua non* before an engine is used for its full duty.

A run-in surface is in a state of equilibrium topography [19]. This is obvious because the first effect of a load is to cause plastic flow of the asperities and, with time, the contact areas grow in size enough to support the applied load [20]. If both the components are hard, the necessary number of summits may be removed by some process other than the adhesive mode, but the junctions must shear progressively and assume a true area of contact commensurate with the applied load.

Apart from an equilibrium load-bearing area, the surface work hardens. In the case of grey cast iron piston rings or cylinder liners, electron diffraction has revealed [21] that the surfaces possess oxide layers and graphite flakes which orientate with their cleavage planes parallel to the sliding direction. Both these phenomena imply a situation of low friction and in all probability the rate of wear is diminished.

3.6.1 A wear law

A running-in wear regime shows a curvilinear pattern in the volume loss–sliding distance curve. The wear equations deduced previously are those for the steady state where the relationship between volume or weight loss and sliding distance is linear. A running-in wear law should, therefore, be different from that for the steady state and one has been derived [22].

Assume that two asperities interpenetrate as one or both of them yields plastically, holding between them a potential volume denoted by V. A wear particle will form from this volume so that we can say that, for a sliding distance S, the rate of wear dV/dS will be proportional to it. Thus,

$$\frac{dV}{dS} = -AV \qquad (3.18)$$

where A is a constant of proportionality and the negative sign is included to show that an amount of volume is being lost from the asperities under the sliding situation. Integrating Equation 3.18,

$$\ln V = -AS + B$$

where B is a constant of integration. Another way of expressing this is

$$V = \exp(-AS + B) \qquad (3.19)$$

To be more specific, if the potential volume is V_0 which will later give up an amount of material to produce wear of one or both of the components, the general Equation 3.19 can be used to provide an expression for V_0. That is, when $S = 0$,

$$V_0 = \exp{(B)} \qquad (3.20)$$

Therefore substituting for $\exp{(B)} = V_0$ in Equation 3.19,

$$V = V_0 \exp{(-AS)}$$

At this stage, it is wise to clarify again that V_0 is the volume which has undergone plastic yielding due to asperity interaction at $S = 0$. After the two asperities have slid a distance S, the residual volume of the junction is V, which is given by Equation 3.19. That is, the volumetric wear V_S for a sliding distance S, is,

$$V_S = V_0 - V \quad \text{or} \quad V_S = V_0 - V_0 \exp{(-AS)}$$

That is,

$$V_S = V_0\{1 - \exp{(-AS)}\} \qquad (3.21)$$

Suppose now the surfaces, say with hemispherical protuberances, first touch without any load as shown in Fig. 3.6(a). Upon applying a load, W, let the top surface move down by an amount z, making areas of contact shown within the shaded volumes in Fig. 3.6(b). For a particular asperity pair, the product of the area and the height of subsidence is the potential volume due to interpenetration. For real surfaces, the area is the sum of all the areas generated giving the true area of contact, A_t. Thus the total volume upon the application of load is,

$$V_0 = A_t z$$

Substituting this value in Equation 3.21,

$$V_S = A_t z \left[1 - \exp{(-AS)}\right]$$

Since $A_t = W/3\sigma_Y$, σ_Y being the yield stress of the soft material which is being worn,

$$V_S = \frac{Wz}{3\sigma_Y}[1 - \exp{(-AS)}] \quad \text{or} \quad V_S = \frac{Wz}{H}[1 - \exp{(-AS)}]$$

$$(3.22)$$

In a real situation, the centre line average by which the two bodies have moved together can be used to obtain a value of z, although it is not in

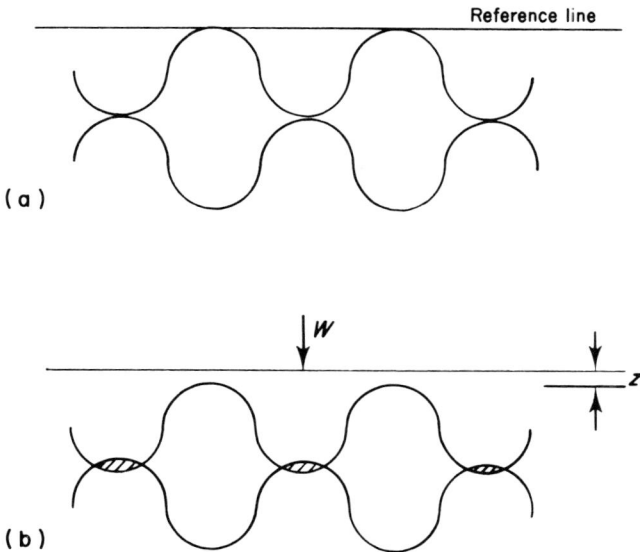

Fig. 3.6 Subsidence of a top surface by an amount, z, due to a normal load, W.

practice an easy exercise. The hardness of the metal is easily obtained and the sliding distance, S, will be recorded. The only unknown item in Equation 3.22 is the constant A.

If the cumulative volume loss, V, is plotted against the sliding distance, S, a curve is obtained which shows a curvilinear running-in stage followed by a steady state (AB, curve 1, Fig. 3.7). Unfortunately, replicate measurements under apparently identical conditions may provide a running-in wear, the magnitude of which is greater for curve 2 in Fig. 3.7. The steady state CD is parallel to AB so that the wear rate, V/S, is the same. The reason for this variation in running-in wear between replicate measurements is not known but one limitation of Equation 3.22 is that it predicts a constant value of V_s, other things being equal. Note that Equation 3.7 expresses the wear rate, V/S, and not the total volume lost which may vary between replicate measurements, while V/S is the same as predicted by the steady state wear law.

The reason for a lack of reproducibility in the amount of total wear at a particular sliding distance is not clear. Running-in, for the time being, is a bedding-in process for the two components to achieve microgeometric conformity. This is, perhaps, only a part of the picture

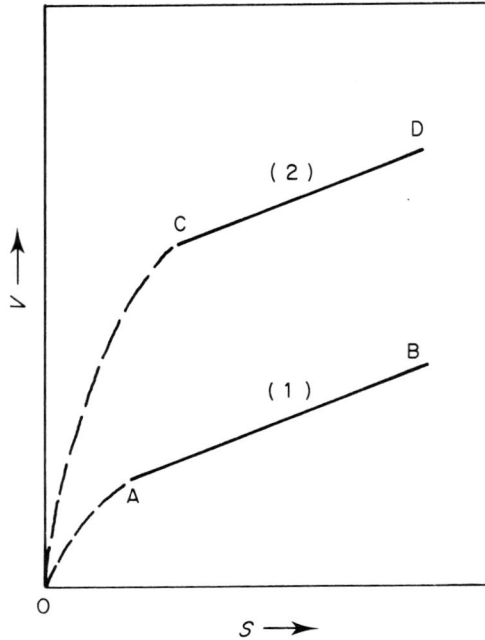

Fig. 3.7   Loss of volume, V, with sliding distance, S, in replicate measurements. OA and OC are running-in wear. The corresponding steady states are AB and CD.

because the regime probably also terminates in equilibrium metallurgical changes. It is thus possible that a distinct sublayer is produced which will have a characteristic metallurgical nature and there will be an equilibrium amount of interfacial strain. The amount of wear in the running-in stage is erratic, possibly because there may be microscopic structural variations between replicate samples although they may have been produced from the same bar stock, for example. All this is, unfortunately, speculative and a systematic experimental investigation of the phenomenon of running-in is long overdue.

3.7   Metal transfer

Whenever a soft slider is rubbed against a hard surface, the latter becomes covered with layers of the former. This is metal transfer which is visible to the unaided eye, but detail can only be observed by microscopy. The nature of the metal transfer is best established by observing it first at low magnification and then following up with an electron microscope for finer detail, if necessary. A scanning electron microscope provides detail in three dimensions, but microscopy cannot establish the amount of transferred material accurately. For a quantitative evaluation of the amount of material which is deposited, the soft metal is irradiated and then slid on the chosen counterface. The amount of soft metal which is deposited can be recorded by a suitably calibrated Geiger–Müller counter.

Quantitative evaluation of the amount of deposit provides valuable information about wear processes in metals and establishing the nature of the deposit confirms the theory of junction formation during metal–metal interaction. If an irradiated soft component is slid against an inactive hard surface, the transferred fragments will be radioactive. The autoradiographic technique places a photographic film on this deposit, which appears as black patches on the film. These can be measured to show [24] the discrete nature of the deposit. If an irradiated brass pin is held against a counterface under various loads and then separated from the counterface, autoradiography will provide information on the variation of junction size with load. It will also give a measure of the true area of contact. Junction growth due to creep or due to tangential stress can also be studied.

Unfortunately, the method is expensive and the feasibility of a project using the technique of autoradiography depends entirely on whether facilities are available for irradiating specimens. One alternative, albeit a less accurate method, is to weigh the counterface at frequent intervals during a wear run using a microbalance.

A rather striking observation is that the quantity of metal picked up is of the same order of magnitude irrespective of the hardness of the material which is being transferred [25]. If the amount of transfer is the same, the wear rate, as will be discussed in this section, will also be of the same order of magnitude irrespective of the metal under study. This is hard to accept but Rabinowicz and Tabor [25] come to that conclusion after carrying out experiments with a number of metal combinations. This adds an element of confusion to the established pattern of our knowledge and must be an area of further study.

However, similar studies have shown that if the interfaces are not chemically clean, the amount of metal transfer in a system can be reduced by a factor of 2×10^4 or more. This is to be expected since contaminants will obstruct metal to metal interaction. It is certain that

transfer occurs in discrete patches as the microscopically distributed
frictional bonds form. This is compatible with theory, as are the results
from experiments with metals such as copper and zinc which show that
the amount of metal transfer is proportional to the applied normal load.
So far, there does not appear to be any published information on the
nature and the amount of metal transfer *vis-à-vis* surface topography
and sliding speed.

Radioactive tracer techniques have been used in elegant experiments
to observe the phenomenology of wear processes of brass [26] and steel
[27, 28] sliding on steel. Sliding an irradiated 2% leaded 60/40 brass pin
on an inactive hard steel results in the brass being transferred at a very
rapid rate (Fig. 3.8). Measuring the volume of material lost with the
time of running the wear machine, the typical curve shows that during
the stage when brass is transferred at a fast rate on to the steel, the
volume–time relationship is curvilinear and is probably the running-in
stage. Steady state wear follows where the volume loss varies linearly
with time. At the end of the running-in time, the amount of deposit

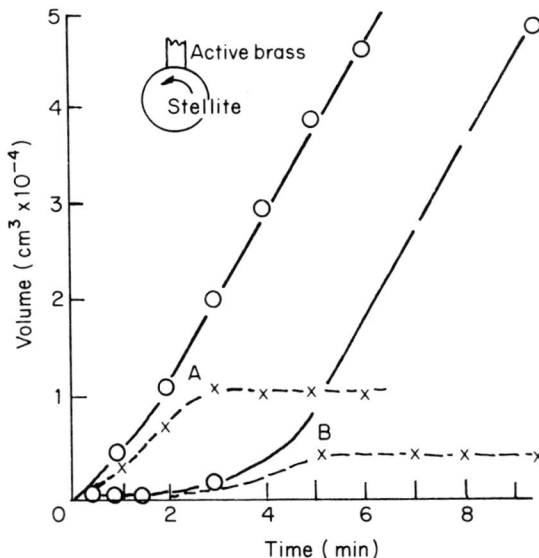

Fig. 3.8   Variation of wear and metal transfer with time of brass on stellite.
—o— wear; ---×--- transfer; *A* = unlubricated; *B* = lubricated. The load on *B* is 4·5 times that on *A*. (Kerridge, M. and
Lancaster, J. K. (1956). *Proc. Roy. Soc. A* **236**, 250.)

assumes a constant value with time and loose wear particles are detached from the interface. It is interesting to note that the amount of metal deposited onto the steel and hence the volume loss from the brass pin are generally low when the interface is lubricated, although the normal load is now greater by a factor of 4·5. The slopes of the loss of volume curves are the same for both the dry and the lubricated pins, which means that they have the same wear rates. This is coincidental, but note that, even at such a relatively heavy load, the wear rate is considerably reduced because of effective lubrication.

The mechanism of the detachment of wear particles is given as follows. In the first instance, a brass particle is deposited on the steel. As sliding continues, the deposit grows due to repeated encounters until it reaches a critical size when it detaches from the counterface. It is stated that the critical number of encounters is about 50. If each revolution means an encounter, which may not be the case at all, for a pin–disc machine running at 1500 r min^{-1} we should see wear particles in about two seconds. The interesting conclusion from the radioactive pins is that an equilibrium stage is reached when the rates of metal transfer and debris formation become equal.

The implication with brass sliding on steel is that oxide particles formed due to frictional heating do not take part in the wear process in any obvious manner. It is shown that this is not so with steel, and upon rubbing an annealed tool steel on itself in the hardened condition, metal transfer again occurs but the deposit oxidizes. Wear debris appears after a certain number of encounters but the rate of ejection from the counterface is determined by the rate of oxidation of the deposit.

The sequence of events in the early stages of the wear process in a laboratory test machine for tool steel sliding on itself applies, at least in part, to many combinations of metals. As the machine starts, the pin loses enough material to establish geometric conformity with the counterface. It is perhaps more correct to say that the pin loses mass to attain rapidly a true interfacial area, commensurate with the applied normal load and the tangential stress. After this initial weight loss, the pin rapidly transfers metal on to the disc and the deposit thickens with time of sliding. A transition stage follows in that there is a fall in the density of the deposit. During this stage, the pin wears rapidly followed by some disc wear. This is the beginning of the equilibrium state when both the pin and the disc wear at the same rate. We note that they are like metal couples and the equilibrium state is reached in about 2×10^4

revolutions. For a machine running at 1500 r min^{-1}, the time taken to reach the equilibrium state is about 13 min.

3.8 Energy model

We recall the observation that as sliding commences a brass particle is transferred and then detached from the interface once it reaches a critical size. There is always the possibility that two asperities interact elastically and are only detached from the main body after a number of interactions. This is probably a fatigue failure but the energy model [29] is equally admissible.

Suppose one asperity meets another and then moves on, the encounter being elastic. Let the first asperity which will detach receive an amount of energy, E. It is known that the energy received will decay with time t according to the following equation

$$E = E_0 \exp(-Kt) \tag{3.23}$$

where K is a constant and E_0 is the amount of energy the first asperity possessed before the collision.

Suppose at the first collision, the amount of energy received by the asperity is OA (Fig. 3.9). This energy decays exponentially with time,

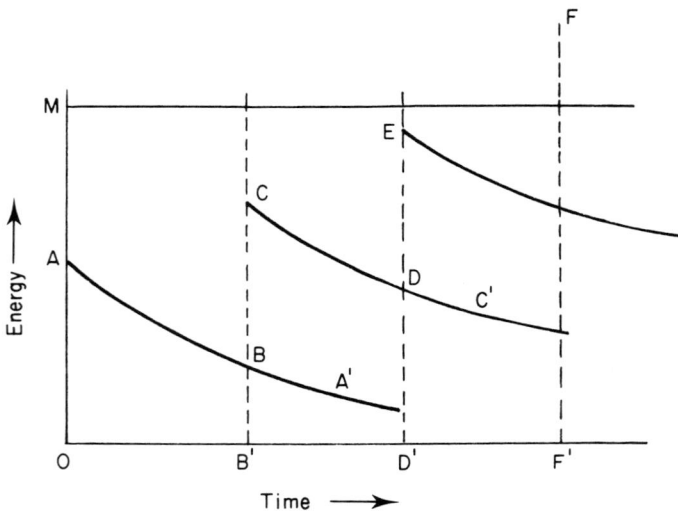

Fig. 3.9   Residual energy of an asperity at repeated encounters (from an idea of Davies [29]).

but suppose that the asperity encounters another quantum of energy at a time given by B' when it already had an amount of residual energy BB'. The total energy achieved by the asperity now is CB' and CB' > OA. At the third encounter at D', the energy attained by the asperity is D'E and D'E > B'C.

Fig. 3.9 also shows that a specific amount of adhesional energy holds the asperity to the substrate and the magnitude of this is represented by OM. The argument put forward by the energy model is that repeated collisions will increase the residual energy of the asperity. Eventually, it will be detached when its accumulated energy due to a specific number of collisions exceeds OM, e.g., that given by FF' in Fig. 3.9.

3.9 Wear debris

Examination of the nature of wear debris provides basic information regarding the degree of work hardening or the composition of the transferred layer. For the adhesive wear mode, junctions form as a result of plastic yielding. It is known that the many discrete areas thus formed are not equal to one another in size and the smaller junctions are the most numerous [30]. We have seen that when brass slides on steel, the discrete patches on the counterface grow by receiving fresh particles as the machine runs. The deposit becomes a wear fragment as soon as it reaches a critical size. In terms of energy, an order of magnitude of the residual energy due to collision with the opposite surface is about ten per cent of the energy of adhesion. We may assume that this is the energy stored even a long time after the intimacy, and the shorter the time interval between two successive encounters, the higher should be the residual energy.

To obtain an appreciation of the mean size of the wear particles, let V be the volume of a potential debris with σ_Y and E as the yield stress and Young's modulus respectively. The total energy of adhesion of this with the substrate is approximately $\sigma_Y^2 V/2E$. Therefore, the residual energy of the deposit after an encounter, designated as G_r, is

$$G_r = \sigma_Y^2 V/20E \qquad (3.24)$$

The adhesive energy G_a which holds the deposit to the substrate is $G_a = WA$, where A is the interfacial area between the deposit and the substrate and W is the work of adhesion. If the deposit is hemispherical with a radius a, $V = (2/3)\pi a^3$ and $A = \pi a^2$. The deposit will detach from the parent body when $G_r \geqslant G_a$. That is, the condition for detachment is

$$\sigma_Y^2 V/20E \geq WA$$

Substituting the appropriate values of V and A

$$2a \geq (60EW)/\sigma_Y^2 \qquad (3.25)$$

If H is the hardness, $\sigma_Y = H/3$ and, for most metals, $\sigma_Y/E = 3 \times 10^{-3}$. Substituting these in Equation 3.25,

$$2a \geq (W/H)6 \times 10^4 \qquad (3.26)$$

Experiments [31] with steel on steel have produced a mean diameter of a wear debris as 12×10^{-3} cm. This must be the critical diameter of the particle at the moment of its detachment from the substrate.

3.9.1 Surface–volume energy
The ratio W/H in Equation 3.26 can be regarded as the ratio of surface to volume energy of a wear particle. A high surface energy implies a large critical diameter before it detaches from the parent body. Soft metals have high surface energies and it is quite common for their surfaces to become rough during the sliding process. The reason should be that the deposits grow by repeated transfer to large sizes before they can be detached as wear debris. Hard metals, on the other hand, have high volume energies and usually run-in surfaces of these couples are smooth.

3.9.2 Effect of load
It has been shown [30] that the mass M of the largest wear fragment can be related to the applied load W in the following way:

$$M = W^\alpha \qquad (3.27)$$

where $\alpha < 1$ and has a value of 0·3 for copper sliding on steel. This suggests that the junction grows in size with load, albeit slowly. It is never easy to measure sizes of wear debris accurately and at best only ranges of sizes can be grouped together. There is scope for further investigation on particle size and there is emphasis on the fact that the predominant effect of increasing the applied load is to increase the number of junctions. We may speculate that initially the junctions increase in size by plastic flow as the load is increased. Since most metals work harden, there should be a limiting maximum size of junctions and

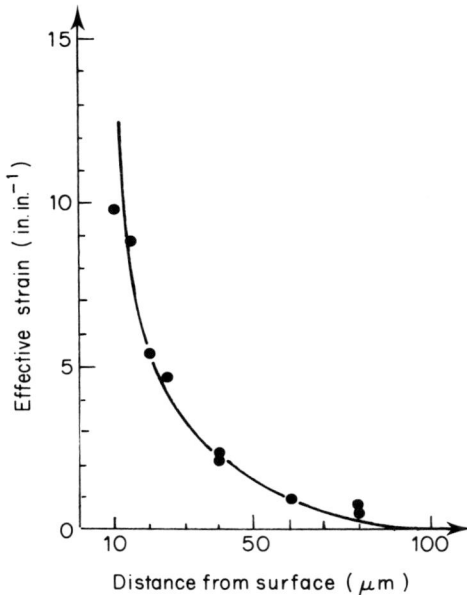

Fig. 3.10 Strain field at the subsurface of a worn 99·5% pure copper after being slid for 68 m in an atmosphere of argon. (Jahanmir, S., Suh, N. P. and Abrahamson, E. P. (1975). *Wear* **32**, 33.)

a further increase in the normal load may make contacts elsewhere if the compliance can change to allow this.

3.10 Delamination theory

The assumption that the bulk hardness of the material can predict the rate of wear (Equation 3.7) has been criticized often since the wear track and the layer supporting it are different in character to the parent metal. It is quite easy to obtain evidence of severe subsurface deformation by examining, say, the longitudinal section of a pin. The extrapolated plastic strain in the surface is very high and the actual values [33] show a decrease towards the interior (Fig. 3.10). The total depth to which the deformation extends varies between metals, being 40 μm and 100 μm for steel and copper respectively. Another criticism [33] aimed at the adhesive mechanism is that wear particles are in the form of sheets of metal and the adhesive wear model assumes that asperities are plucked after they adhere. This is unfounded because firstly, not all wear particles are in the form of sheets, and secondly, there never has been any explicit statement in the adhesive wear model that the asperities are plucked rather than smeared in the direction of motion. However, the variation of plastic strain in the deformed layer is a fact and any theory which takes this into account is worth examining.

The delamination theory has been proposed by Suh [34] and is only applicable to low speed sliding so that the bulk temperature of the interface does not rise appreciably. The theory suggests that as a rider moves over a surface and the latter is plastically deformed, dislocations are generated within the material. The dislocations from the surface region are eliminated by what is called the image force created by the free surface. As sliding proceeds, fresh dislocations are generated which move until they pile up at obstacles such as grain boundaries. This results in the formation of voids which coalesce in time by growth and also due to the applied shear stress at the interface. The next stage is for the agglomerate of voids to assume the form of a crack, parallel to the surface, which will propagate to form a wear particle.

An interesting set of experimental results [33] shows that the amount of wear of steel can be reduced considerably if both surfaces are coated with a layer of cadmium. Cadmium is a soft metal, and if it is thin any dislocations generated within this layer will escape through the surface. It is also known that if the shear moduli of the coating and the substrate differ greatly, dislocations generated by both regions will be repelled by the interface. Provided the applied stresses are low, the steel will probably escape deformation and plastic flow will be confined to the coating. A thick cadmium coating will produce voids somewhere below its own surface and wear by delamination will occur. The implication is that once it wears thin enough for the dislocations to be able to escape under the influence of the image force, the wear rate of the coating will be considerably minimized. The plating is ineffective in air because the oxidized layer of cadmium hinders the escape of dislocations and wear occurs as voids coalesce into cracks. The coating is probably also removed by the abrading oxide particles.

The importance of an optimum coating thickness has been demonstrated by others [35–37]. If a thick coating of lead is applied to a copper rider and then slid on a counterface made from the same metal, the lead film is lost rapidly until an optimum thickness is established when wear is low. The same thickness in turn is transferred onto the counterface and for a normal stress of $0 \cdot 1$ kg mm^{-2}, the optimum thickness is 2 μm. Other examples are coatings of gold, silver, copper and nickel on plain carbon and stainless steels. A practical situation which could be investigated further from the point of view of establishing an optimum thickness is the use of ball bearings and gears in vacuum, electroplated with precious metals [38].

3.11 Slip-line fields

The delamination theory assumes a plastic interaction, as do postulates of authors who advocate plane strain slip-line field solutions [39, 40]. The analyses are not always easy to comprehend and we shall only indicate the possibility of application of the theory in the field of wear.

Consider a hemispherical rider on a flat surface (Fig. 3.11); sliding under load causes a bulge in front of the rider [41]. Observations on paraffin wax and clay models with grids show the approximate slip lines (Fig. 3.11(a)), the distribution of which depends on the flow and shear fracture stresses of the material. A crack starts on the shear fracture line (Fig. 3.11(b)) and a wear particle is ejected (Fig. 3.11(c)). Bulges and

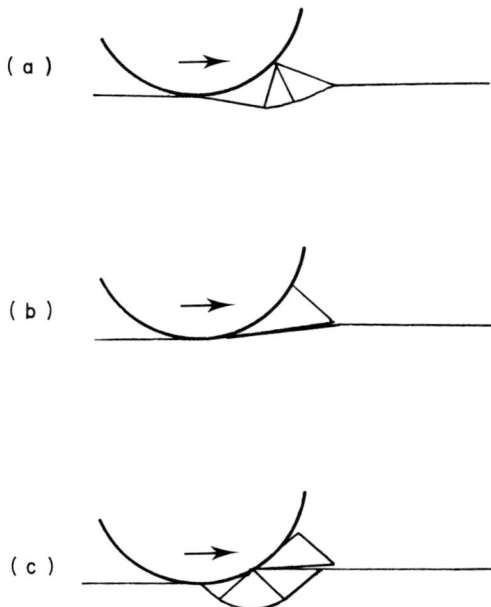

Fig. 3.11 Approximate slip line fields produced in a surface by a sliding hemispherical rider. (Bates, T. R., Ludema, K. C. and Brainard, W. A. (1974). *Wear* **30**, 365.)

cracks such as are shown in Fig. 3.11 can be found in large aluminium alloy bushes sliding against steel pins. Slip-line field analysis would be a welcome addition to the existing approaches to elucidate mechanisms. It has been suggested [41] that a quantitative analysis should include fracture toughness in shear.

References

1. Archard, J. F. (1953). *J. Appl. Phys.* **24**, 981.
2. Yoshimoto, G. and Tsukizoe, T. (1958). *Wear* **1**, 472.
3. Rowe, C. N. (1966). *Trans. ASLE* **9**, 101.
4. Guy, A. G. (1967). "Elements of Physical Metallurgy", Addison-Wesley, Reading, MA, USA.
5. Bowden, F. P. and Young, J. E. (1951). *Proc. Roy. Soc. A* **208**, 311.
6. Tingle, E. D. (1950). *Trans. Faraday Soc.* **46**, 93.
7. Finch, G. I. (1937). *Science Progress* **31**, 609.
8. Byerlee, J. D. (1967). *J. Appl. Phys.* **38**, 2928.
9. Lancaster, J. K. (1957). *Proc. Phys. Soc. B* **70**, 112.
10. Hirst, W. and Lancaster, J. K. (1960). *Proc. Roy. Soc. A* **259**, 228.
11. Bailey, A. I. (1961). *J. Appl. Phys.* **32**, 1430.
12. Cocks, M. (1957). American Society of Lubrication Engineers: Paper given at Lubrication Conference, Toronto, Canada.
13. DeGee, A. W. J., Vaessen, G. H. G. and Begelinger, A. (1969). *Trans. ASLE* **12**, 44.
14. Wellinger, K., Uetz, H. and Komai, T. (1969). *Wear* **14**, 3.
15. Lancaster, J. K. (1963). *Proc. Roy. Soc. A* **273**, 466.
16. Beilby, G. T. (1903). *Proc. Roy. Soc. A* **72**, 218.
17. Cochrane, W. (1938). *Proc. Roy. Soc. A* **166**, 228.
18. Bowden, F. P. and Hughes, I. P. (1937). *Proc. Roy. Soc. A* **160**, 575.
19. Burwell, J. T. (1957). *Wear* **1**, 119.
20. Barwell, F. T. (1967). *Proc. Inst. Mech. Eng.* **182**(3K) 1.
21. Finch, G. L. (1950). *Proc. Roy. Soc. A* **63**, 785.
22. Queener, C. A., Smith, T. C. and Mitchell, W. L. (1965). *Wear* **8**, 391.
23. Sakmann, B. W., Burwell, J. T. and Irvine, J. W. (1944). *J. Appl. Phys.* **15**, 459.
24. Gregory, J. N. (1946). *Nature* **157**, 443.
25. Rabinowicz, E. and Tabor, D. (1951). *Proc. Roy. Soc. A* **208**, 455.
26. Kerridge, M. and Lancaster, J. K. (1956). *Proc. Roy. Soc. A* **236**, 250.
27. Kerridge, M. (1955). *Proc. Phys. Soc. B* **68**, 400.
28. Archard, J. F. and Hirst, W. (1957). *Proc. Roy. Soc. A* **238**, 515.
29. Davies, R. (1959). "Friction and Wear", Elsevier, Amsterdam.
30. Rabinowicz, E. (1953). *Proc. Phys. Soc. B* **66**, 929.
31. Rabinowicz, E. (1958). *Wear* **2**, 4.
32. Rabinowicz, E. and Foster, R. G. (1963). American Society of Mechanical Engineers: Lubrication Symposium, Boston, MA, Paper no. 63-Lub S-1.

33. Jahanmir, S., Suh, N. P. and Abrahamson, E. P. (1975). *Wear* **32**, 33.
34. Suh, N. P. (1973). *Wear* **25**, 111.
35. Tsuya, Y. and Takagi, R. (1964). *Wear* **7**, 131.
36. Takagi, R. and Liu, T. (1967). *Trans. ASLE* **10**, 115.
37. Takagi, R. and Liu, T. (1968). *Trans. ASLE* **11**, 64.
38. Brown, P. E. (1970). *Trans. ASLE* **13**, 225.
39. Green, A. P. (1954). *J. Mech. Phys. Solids* **2**, 197.
40. Gupta, P. K. and Cook, N. H. (1972). *Wear* **20**, 73.
41. Bates, T. R., Ludema, K. C. and Brainard, W. A. (1974). *Wear* **30**, 365.

4 Abrasive Wear

We may recall that, while elucidating friction, there are two basic components, viz. shearing and ploughing, in tribological interactions. Ploughing of a soft metal is possible because the hard protuberances of the counterface first indent as the load is applied, and then cut a groove in the metal as the interface undergoes relative motion. The first traversal may simply cut a groove and the deformed material may just be heaped on either side of the indenter. Repeated traversals will, however, remove material from the soft body and loss of material by an indenter in this manner is classified under abrasive wear. Wear of this type is deliberate in certain cases, as in the use of grinding wheels for finishing operations. Another mode of abrasion is by particles of sand, rocks or ores flowing down a chute. Abrasive wear may occur due to trapped particles of wear debris and ingressed dust or grit at a bearing interface. The main source of wear of teeth is probably abrading media in food. There are many such examples and in certain cases abrasive wear can be minimized by a careful selection of materials. In other instances, attempts are made to avoid ingress of grits by efficient sealing devices. Although there are many examples where a low rate of abrasion will have obvious economic advantage, published information on planned laboratory studies of abrasion is not plentiful. The reason probably is that experimental techniques to carry out abrasive wear studies are not simple Typically, a pin on a grinding wheel will appear to be an obvious method of investigation but clogging of the spaces between the grits occurs readily. When this happens, rubbing is not solely between the metal and the abradants. The wheel can be dressed but very careful exploratory work is necessary before standardizing an experimental procedure. In this chapter, we shall examine the available theories and discuss results from laboratory exercises. Data are also available from field trials on ploughshares and a plethora of information has been compiled by a mining company.

4.1 Classification

It is helpful to recognize at the outset that an abrasive particle can be fixed to a rigid matrix or it can abrade while being essentially in loose form. The former may be regarded as a sharp cone or a pyramid which will plough material from the opposite face. If loose particles flow, for example over a chute, they may slide or roll and cutting by ploughing will depend on the probability of a sharp edge of a particle indenting the surface. In either case, we can expect, *ceteris paribus*, that a high contact pressure should increase the amount of wear. This means that we should identify two modes, viz:

(1) high stress abrasion;
(2) low stress abrasion.

The situation where there are only two bodies involved in the inter-action is known as two-body abrasion. A grinding wheel removing extraneous material from a metal surface or an emery paper polishing a solid provides a two-body abrasion mode, as does a ploughshare when it first dislodges the composite soil. Similarly, dragging a chain over the soil or drilling rocks may be mainly two-body abrasive processes where wear occurs by the ploughing action of the sharp abradants.

For grits trapped at the bearing interfaces, wear is said to occur by the three-body mode, the three bodies involved being the journal, the bearing and the grits. As we walk on the floor, which is covered with dust particles, our shoes and the floor undergo three-body abrasion.

Whereas the three-body mode is clearly identifiable in the above examples, there are situations where this is not so. These are the systems where loose particles flow over a surface causing wear. An example is an inclined chute along which sand is being conveyed or a worm mixer where solids are coated with binders. In the latter case, stresses involved are low but rapid wear of even a very hard facing such as stellite occurs. It is known that near laminar flow of water in a small stream causes pebbles to become smooth due to wear. All these cases could simply be classified under two-body abrasion but one could equally argue that a bed of sand in a chute produces three-body abrasion in the following way. One surface is provided by the chute, a second being the grains of sand immediately against the floor. The third body is the sand layer which does not directly abrade the chute but provides the normal load to the particles which do abrade the chute. Nevertheless, the whole sand stream or the solids in a worm mixer are loose so that they churn and the particles in the top layer may change position and find their way down to the surface of the chute. The third body, in the form of the bed of sand, is not rigid. Such systems will therefore be regarded as being subject to the two-body mode of abrasion. It is apparent that the majority of industrial situations where heavy wear of the abraded component occurs are subjected to two-body abrasion.

4.2   A simple theory

Whether the mode is two-body or three-body, it is fair to assume that significant wear will only occur if a sharp cutting edge is available. Consider therefore a model abrasive [1, 2] in the form of a cone indenting the component to a depth h under a load ΔW (see Fig. 2.11). The cone has a semi-apex angle θ, that is a base angle $(90 - \theta)$, and the half cone pushes a wall of metal ahead of it, producing a volume of wear debris. If the length traversed is dS, the volume dV which is removed is

$$dV = \tfrac{1}{2}2ah \; dS$$

where $2a$ is the diameter of indentation.

Since $h = a \cot \theta$,

$$dV = a^2 \; dS \cot \theta \qquad (4.1)$$

The load is supported by only half the basal area πa^2. If H is the hardness of the indented metal,

$$\tfrac{1}{2}\pi a^2 = \Delta W / H \quad \text{or} \quad a^2 = 2\Delta W / \pi H$$

Substituting this value of a^2 in Equation 4.1,

$$dV = \frac{2\Delta W \cot \theta}{\pi H} dS$$

The rate at which metal is removed is dV/dS and

$$\frac{dV}{dS} = \left(\frac{\Delta W}{H}\right)(0 \cdot 63 \cot \theta)$$

By analogy, for the whole surface being attacked by many cones, supporting a total load W,

$$\frac{V}{S} = (0{\cdot}63 \cot \theta)(W/H) \qquad (4.2)$$

According to this analysis, the harder a material is, the more wear resistant it will be. This aspect will be discussed again later. Equation 4.2 shows that an increase in the normal load will increase the rate of wear. For a cone, the larger the semi-angle θ, the smaller the rate of wear. This is plausible, since a cone with a large value of θ is blunt and we know that a sharp cutting edge is essential for improving the ploughing efficiency.

4.3   Angle of attack

Although the sharp edge of an abrasive is expected to plough the body of a soft material, it does not follow that the body will show wear if defined by loss of weight. As stated earlier, the displaced metal may simply be heaped as ridges on either side of the groove. Wear, that is complete removal of matter from the main body under attack, will only be that amount which has adhered to the abrasive particle by plastic flow [3]. This situation is normally termed rubbing but significant wear by two-body abrasion occurs only by chip formation as in turning a component on a lathe. The efficacy of chip formation will depend on the angle of attack by the abrasive in relation to the surface being abraded and this has been investigated [4] by using macromodels of pyramidal tools representing the abrasives. The tools were carefully prepared from tungsten carbide and the counterface was made from lead because of its non-work hardening nature. If the pyramid is brought down slowly onto the lead surface under a normal load W, that is at an attack angle $\alpha_{90} = 90°$, the first effect is indentation of the soft surface to a depth dictated by its static yield strength. Applying a tangential pull at right angles to the face ABC of the pyramid (Fig. 4.1) causes it to sink first in the lead before moving in the horizontal direction. This is the onset of chip formation and, as the horizontal pull is maintained, the frictional force at the interface opposes the motion. This results in the pyramid climbing out of the indented groove slightly and establishing a steady depth during the rest of the sliding process. Sliding, however, is not steady so that adhesion and disruption of the junctions must occur. This results in friction fluctuating with time about a mean value and material continues to be removed by a machining action.

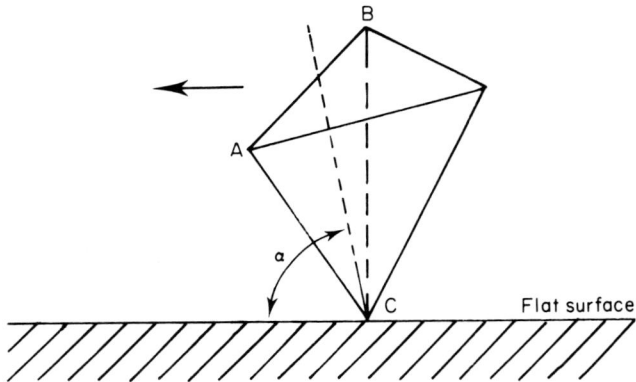

Fig. 4.1   A pyramidal tool contacting a flat surface at an attack angle α. The direction of sliding is indicated by the arrow (after Sedriks and Mulhearn [4]).

For a small value of $\alpha = 50°$, a groove is again produced and the displaced lead is deposited on either side of the channel. The tool again digs in initially but the motion is not jerky. The amount of material lost by the lead block, however, is very small, being in the form of a small button scooped out by the tungsten carbide tool. Experiments such as these where the attack angle is varied establish that there is a critical value of α for maximum wear of the counterface. If the grits on a grinding wheel were disposed favourably, one would get a maximum output from the dressing operation, say, on castings in a steel foundry. Emery papers are also used for material removal on a fine scale but experiments [6] show that only 10% of the grits are arranged in the most favourable manner. On the other hand, wear on a dredger bucket could be minimized if the abrading particles which are scooped out could be arranged at an unfavourable angle. This, of course, is an impossibility in practice.

4.4   Particle characteristics

When used as an abrasive an initially sharp grit will become blunt because it will wear. This will decrease its efficiency from the point of view of metal removal. It is commonly believed that a hard abrasive such as diamond (Table 4.1) will retain its sharpness longer than, say, quartz. Unfortunately, hardness is not always an accurate guide and quartz is known to be very aggressive in certain situations. However, a

Table 4.1 Typical Knoop microhardness of
materials

Material	Knoop hardness
Diamond	7575
Titanium carbide	2955
Silicon carbide	2585
Corundum	2020
Iron carbide	1025
Quartz	840
Martensite (0·6% C)	800
Work-hardened manganese steel	645
Plastics	25

ratio H_m/H_a is used [7] to predict the efficiency of grits where H_m is the hardness of the metal and H_a is that of the abrasive. If the ratio is high, the rate of wear of the metal will be low but not nil. This is because the metal may have soft micro-constituents, and also repeated attacks will probably remove surface layers of the metal by a fatigue mechanism. The ratio H_m/H_a for a martensitic steel, being abraded by quartz (Table 4.1), is 0·95 and the latter is known to wear the former successfully.

Experience with certain combinations suggests that when the ratio $(H_m/H_a) < 0·8$, the victim undergoes hard abrasive wear. That is, the metal wears rapidly. A situation of soft abrasive wear is experienced when H_m/H_a is greater than 0·8 and the metal should not wear at all if $(H_m/H_a) \geq 1$. Zero wear is an unlikely proposition in practice and we can only say that there is no wear by the abrasive mode. Abrasive wear is minimized in certain circumstances by imparting hardness to the subject and there is a suggestion [8] that the modulus of elasticity of the material which is being abraded is a more appropriate parameter than hardness in ranking materials from the point of view of abrasive wear resistance.

The question is, should abrasion occur for a while and then cease because the life of the sharp edges has ended due to work? This should be so but, in both two-body and three-body abrasion, the particles fracture at intervals, thus exposing new sharp edges for cutting. One cannot predict how frequently this will happen but a statistically predictable pattern should be established for a particular system. Similarly, an equilibrium stage will be reached when the wear rate will slow down to a small value as the particle size gets smaller due to repeated fracture. We expect the particles eventually to assume a constant size and become rounded.

Whereas much emphasis is placed on the hardness of the component and of the abrasives together with the sharpness of their cutting edges, the effect of particle size does not become obvious in theoretical treatments of abrasive wear processes in metals and non-metals. Laboratory studies [2] (Fig. 4.2) show that up to a certain size, the wear rate rises in a linear manner with increasing diameter of the abrading particles. Characteristically, for a particular friction couple, there is a critical particle diameter beyond which the size of the abradants is ineffective. For bronze on bronze (Fig. 4.2), increasing the size of the abrasive by a factor of six raises the rate of wear by an order of magnitude. This is a significant change but the reason for the constant wear rate when the bronze is abraded by large particles is not clear.

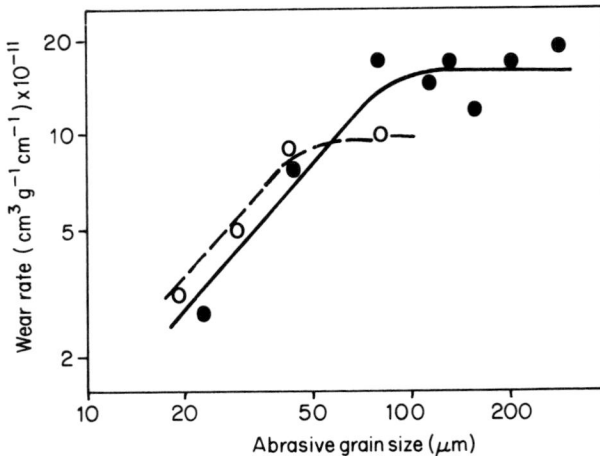

Fig. 4.2   Abrasive wear rate as a function of abrasive grain size. Like couples of bronze (•) and steel (o). (Rabinowicz, E., Dunn, L. A. and Russel, P. G. (1961). *Wear* **4**, 345.)

An interesting facet of the wear rate in the presence of loose abrasives is that it is usually an order of magnitude lower than that in a two-body situation. The suggested reason is that loose abrasives spend about 90%

of their life in rolling, the remainder being spent in a state favourable to incurring material loss by some direct action such as cutting. It is also known that if one of the components is soft, abrasives will be buried in it, and if they protrude, wear of the opposite face by a two-body mode will also occur.

4.5 Component hardness

Rather than carry out an experiment on a single material as required and express its wear rate by noting the volume or weight loss per unit sliding distance, a ratio called the relative wear resistance, R, has been used [9]. This is the ratio of the wear resistance of the material under investigation to that of a reference sample, which could be a soft polymer. An advantage with this approach is that variables such as load and speed can be arbitrary provided that they are identical for both materials. A plot of relative abrasive wear resistance of materials against their corresponding values of hardness produces a linear relationship from which an empirical relationship can be obtained:

$$R = K_1 H + K_2 \qquad (4.3)$$

where H is the hardness of the material under investigation and K_1 and K_2 are empirical constants.

Denoting the applied contact stress by σ and using a range of commercially pure metals and some alloys, all in their fully annealed state, a plot of R against the annealed hardness H produces a linear curve which passes through the origin. Values of R for a range of metals from molybdenum to aluminium have been obtained and the empirical law changes as follows where K is a constant:

$$R = K\frac{H}{\sigma} \qquad (4.4)$$

Equation 4.4 is really very similar to that deduced for the wear rate by the adhesive mode.

Wear resistance of a number of heat treated steels plotted as a function of hardness, gives a linear relationship, but the results suggest that even at zero hardness a metal will have a finite amount of wear resistance. This is illogical but it has been suggested that a plot of wear resistance against hardness will show the line to pass through the origin provided R is plotted against the hardness of the steel under investigation, and for each steel R is obtained by dividing the wear resistance of the hard steel with that of the same steel in the fully annealed condition.

An overall increase of the abrasive wear resistance of certain steels with hardness has been reported [2] but this is not in proportion to the hardness values as dictated by wear laws. There are many possible factors we could speculate upon to explain this but wear processes in materials are complex phenomena and neat trends with variables are not always attainable. As far as the hardness of the material which is being abraded is concerned, it can be stated that there is a good chance that wear resistance will increase if this is high. The unresolved question is whether we should consider the as-manufactured hardness or the increased hardness of the abraded surface.

4.6   Energy criteria

The hardness of a metal can be affected in a number of ways so that one single chemical composition can be made to provide a range of hardnesses. Although the design engineer is compelled to use hardness as the criterion for wear resistance, we have seen that the parameter is not altogether reliable. Attempts have therefore been made to see if more intrinsic properties of metals can be used to correlate with the resistance to abrasion.

One probable criterion is the solid state cohesion of metals and non-metals [10]. The total bond energy, G, in a solid metal can be calculated from a knowledge of H_s, the heat of sublimation and the coordination number, N. Thus,

$$G = 2H_s/N \tag{4.5}$$

The relative abrasive wear resistance of a range of metals shows a linear relationship with the total bond energy, G. A similar pattern is also obtained when G is plotted against T_m, the melting point of a metal. Empirically, for every kcal increase in the bond energy in metals, there is an approximate improvement in the abrasive wear resistance by 1·5 units. That is,

$$R = 1·5G. \tag{4.6}$$

Since both the bond energy and the melting point of metals are measures of cohesive forces in their solid state, the evidence is there to show that abrasive wear resistance can be correlated with strengths of metals at an atomic level. This also seems to be true with a range of non-metals such as quartz and gypsum where the relative abrasive wear resistance is a linear function of the melting point. The reason for such encouraging correlations probably lies in the non-dependence of the bond energy on the presence of very small quantities of second elements. Second elements even in very small quantities may influence the hardness which will also be affected by cold work and residual stresses. On the other hand, these factors should not alter the melting points of metals and alloys.

Another energy criterion has been used in connection with metal removal by a grinding wheel. This has been termed the specific energy [11] which is the energy expended per unit volume of material removed by two body abrasion. The energy involved during grinding comprises three terms, viz:

(1) energy required to produce a chip;
(2) energy required in the ploughing process which forms ridges of metal on either side of the indenter;
(3) energy expended due to sliding of a blunt abrasive against the workpiece.

The energy required to produce a unit volume of chip remains constant but it is held that the respective energies for processes (2) and (3) above can be decreased if a thick chip is produced by increasing the depth of cut. By progressively increasing the depth of cut to produce thicker and

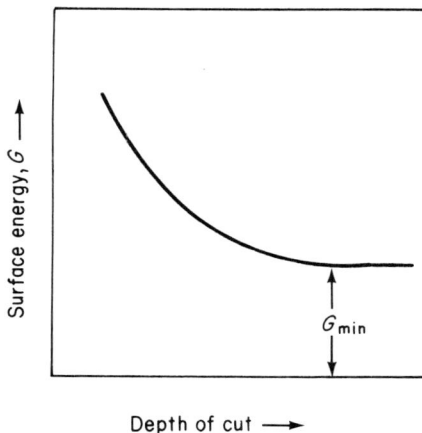

Fig. 4.3 Variation of surface energy with depth of cut in two body abrasion (schematic—after Malkin and Joseph [11]).

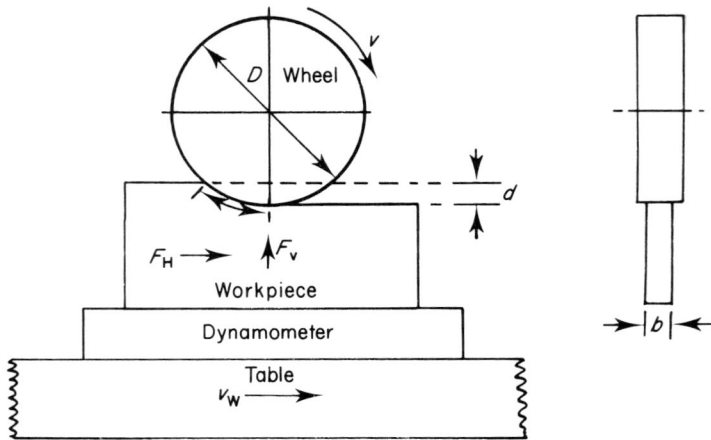

Fig. 4.4   A grinding wheel removing material by two-body abrasion. (Malkin, S. and Joseph, N. (1975). *Wear* **32**, 15.)

thicker chip, a stage is reached when at or beyond a characteristic depth of cut the system requires a minimum specific energy (Fig. 4.3).

To carry out an experimental investigation, the grinding wheel is rotated at a fixed surface speed, v, to remove material from a flat workpiece fed to the wheel at a fixed speed of v_w (Fig. 4.4). The flat workpiece is, of course, the material under investigation. A semiconductor strain gauge dynamometer is used to record grinding forces. If b and d are the width and the depth of cut respectively, for a wheel exerting a horizontal grinding force, F_H, the specific energy, G, of the grinding operation is given by,

$$G = F_H v / b d v_w \qquad (4.7)$$

From all these measurable parameters, the specific energy of this two-body abrasion process can be obtained and plotted against various depths of cut to obtain a relationship similar to that shown in Fig. 4.3. This is the energy required in producing a chip and a plot of this against the melting energies of the respective metals gives excellent correlation. The melting energy is obtained [11] from published data [12] and is the enthalpy difference of the metal between its melting temperature and 25 °C.

Examination of data shows that G_{min} is about 1·4 times the melting energy. We note that in Equation 4·6, a similar factor is 1·5. The

Table 4.2 Wear resistance and minimum grinding energies

Relative wear resistance, R [10]		Minimum grinding energy, G_{min} [11]	
R	Metal	G_{min} (lb in.$^{-3}$)	Metal
4·0	Aluminium	0·60	Aluminium
9·3	Copper	0·99	Bronze
20·5	Iron	1·60	Steel
39·0	Molybdenum	2·00	Molybdenum

correlation is again good, possibly because neither the minimum energy of grinding nor the melting energy is affected by heat treatment or due to the presence of small amounts of alloying elements.

Taking the relative wear resistance from a different source [10] and those for the corresponding values of G_{min} from the results of Malkin and Joseph [11], Table 4.2 has been constructed. Values of R for bronze

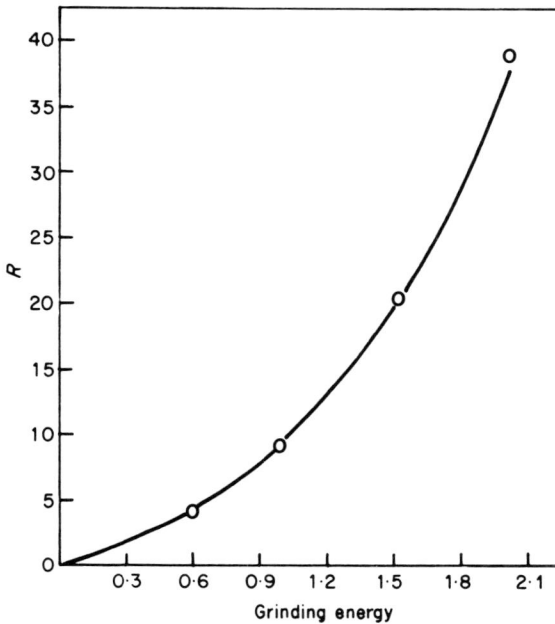

Fig. 4.5 Variation of relative wear resistance, R, with the minimum grinding energy for various metals (see Table 4.2).

and steel are not available but those for copper and iron respectively have been taken. Plotting the values of R against G_{min} (Fig. 4.5), a pattern is evident. This will be a fruitful approach for further study.

4.7   Two-body abrasion

The two-body abrasion process has been investigated in detail from the metallurgical point of view. Friction and wear experiments were carried out very simply on pure metals and binary alloys by moving specimens, often by hand, over emery papers or files. The latter are manufactured with a deliberate macrotopography, but it is not so easy with emery papers because of the difficulty inherent in the manufacturing process. The cutting grits in emery papers are crushed and graded silicon carbide particles which are deposited electrostatically on a paper covered with a layer of uncured resin. Once deposited, the resin is cured so that the silicon carbide particles are held strongly in a resin matrix. The particles are then given a thin coating of lacquer. The object of the electrostatic method is to align the silicon carbide particles with their long axes at right angles to the paper. The success, however, is limited.

Experiments with specimens $2 \cdot 5$ cm $\times 2$ cm $\times 1$ cm deep, traversed once only, show that of the total groove volume only about 15% is removed as wear debris and this must be related to the silicon carbide grits which have favourable attack angles [3, 14, 15]. There is sufficient evidence now [16] to conclude that the frictional resistance is decided by all the abrasives which are in contact with the workpiece. However, wear can only occur if the sharpness and the orientations of the grits are favourable. It has been shown [13] that if the abrasives are spherical in shape, M, the mass of metal worn per unit sliding distance is given by

$$1/(M/\rho) \propto (H_D/W)^{3/2} \tag{4.8}$$

where ρ is the density of the metal under investigation in g cm^{-3}, W is the applied normal load and H_D is the hardness of the metal.

The ratio M/ρ is the volume of material lost per unit sliding distance expressed as cm^3 cm^{-1} so that its reciprocal in Equation 4.8 is the wear resistance of the material. A plot of wear resistance measured in this manner against $(H_D)^{3/2}$ shows that wear resistance increases with increasing hardness of metals. The relationship between $\{1/(M/\rho)\}$ and $(H_D)^{3/2}$ is not, however, linear and probably the reason is that Equation 4.8 is deduced assuming hemispherical asperities whereas the grits and the files will have near pyramidal shapes.

For a copper–zinc system, the coefficient of friction decreases as the zinc content is increased up to 63%. The reason for this is attributed to the increased hardness of the alloys with a high zinc content. The hardness of the copper–zinc alloys decreases as the zinc is increased beyond 63% and there is a rise in friction. An alternative explanation for the variation in the frictional resistance is attributed to the change in the metallurgical structure of the surface layers. Up to 63% zinc, although the body of the metal retains the b.c.c. structure, the surface layer assumes the f.c.c. form. With a further increase in the zinc content, the structure of the surface layer becomes h.c.p. The structural evidence, unfortunately, is contrary to our understanding of the effect of crystallographic structure on friction. An f.c.c. structure normally gives a high value of friction, whereas h.c.p. crystals slip readily along the basal planes. The shapes of the wear curves as plotted by Alison, Stroud and Wilman [13] do not follow a pattern which can be given an explanation consistent with our ideas about mechanisms of wear. We must observe, without ignoring some useful information emanating from work of this nature, that to draw conclusions about quantitative wear rates from a sample run for a few centimetres only is subject to criticism. For the copper–nickel system, however, the mass wear follows the expected pattern as the hardness of the alloys increases with the amount of nickel.

Further work [17–23] along these lines has produced enough information regarding the crystallographic nature of the surface layers of metals abraded against emery papers and files. Not unexpectedly, the grains of the metals assume preferred orientations due to sliding. The h.c.p. and the f.c.c. metals develop orientations in the $\langle 0\,0\,1 \rangle$ and $\langle 1\,1\,0 \rangle$ directions. The surface layers, about 1 μm thick, develop preferred orientations by slip and plastic flow and the directions are independent of the crystallographic size or orientation of the underlying metal which has not been deformed.

If rubbing is continued long enough, useful information can be obtained. Thus, with emery papers incorporating angular abrasives, metal is removed by ploughing but an interesting feature is that the coefficient of friction increases as some of the abrasives are detached and embedded in the metal. It seems that the embedded abrasives are removed with continued sliding and fresh ones are picked up until an equilibrium amount of abrasives remains in the metal. Thus the nature of the friction couple changes from one of metal on abrasives to one of metal plus abrasives on abrasives. The effect of this equilibrium amount of grits in the metal is that the coefficient of friction increases and the amount of wear of the metal decreases, both parameters becoming constant at the equilibrium stage of pick-up of abrasives. This is reasonable since it suggests that the frictional resistance is high if the interaction is between the non-metallic constituents and the reduced amount of wear can be understood if we consider the protective effect of the embedded abrasives.

Experiments show that if the size of the abrasives is small, the interstices are readily blocked by metal wear debris, and friction, measured after an equilibrium sliding distance is low. The low friction has been attributed to the effective sliding between metal plus oxide and the oxide of the metal in the friction couple. This is so because the wear debris is a mixture of metal and oxide, and the metal sample under investigation must have embedded grits and a layer of oxide on its sliding surface as a result of frictional heat.

Provided the abrasive particles are sharp, the coefficient of friction becomes higher with increasing particle size and assumes a constant value beyond a certain critical size. The suggestion again is that with increasing particle size of the grits, the spaces are not necessarily clogged by metal wear debris so that the frictional interaction is that expected of a metal sliding on a hard abrasive.

The effect of increased particle size of the abrasives is initially to cause an increase in the amount of wear, which reaches a constant value beyond a characteristic particle size. The explanation probably lies again with the clogging of the interstices which reduces the cutting efficiency of the emery paper. The constant value of wear at large particle sizes should be associated [3] with an equilibrium amount of embedded particles in the metal so that its direct contact with the abrasives is not total. This argument as that given for friction goes against the postulate that both the frictional resistance and the amount of wear are independent of the apparent area of contact.

The experiments of the authors generally show that at any applied load W, the loss of metal per unit sliding distance or the wear rate, q, is proportional to the coefficient of friction, μ. Thus,

$$q = KW(\mu - \mu_0)$$

where K and μ_0 are constants.

We have noted that as a result of sliding, a thin surface layer achieves a maximum hardness due to deformation and this influences both friction and wear. For a wide range of metals, μ decreases with increased surface hardness as does the rate of wear. However it is worth noting that the hardness is not that of the annealed metals as reported by Khruschov [9].

We have noted that at a first glance the explanation of the authors [13–19] working with emery paper contradicts our understanding of friction and wear in relation to the apparent area of the interface and the crystal structure. However, electron diffraction studies by the authors show that if brass samples undergo three-body abrasion with carborundum, the surfaces appear black to the unaided eye. At a slightly higher magnification, the impression is that the carborundum particles have lodged in the valleys of the undulating surfaces. Electron diffraction studies, however, give minute detail which is not obtainable from microscopic studies. For example, the former technique shows that minerals such as γ-Al_2O_3 embed to a finite depth in the metal. They exhibit a preferential orientation and the planes involved are those of easy cleavage or fracture, lying parallel to the metal surface. We conclude that the true area of contact is shared between metal asperities and the non-metallic embedded particles in the metallic component. The easy cleavage planes, if that is how the embedded abrasives are aligned, will contribute to low friction. Similarly, since the true metallic contact is diluted, the amount of wear should be low.

4.7.1 *Stored energy*

There is ample evidence to show that as abrasion starts, the surface strain hardens with time to the maximum possible amount. Wear occurs possibly by the removal of part of this surface layer and further work hardening continues with sliding so that a certain depth of hard layer is maintained.

If a material, prior to tribological deformation, has an amount of strain energy, G_0, which increases to a maximum value G_m due to sliding under load, the amount of stored energy per cm^3 is given by $(M/\rho)(G_m - G_0)$. For annealed metals, $G_0 \simeq 0$. For a total sliding distance of 1 cm, the work done per unit sliding distance is μWg, where g is the acceleration due to gravity. Therefore,

$$\mu Wg = (M/\rho)(G_m - G_0) + Q + AS \qquad (4.9)$$

where Q = amount of frictional heat, A = total surfaces comprising those of the created grooves and the surfaces of the wear product and S = surface energy per unit area.

If experiments are conducted at slow speeds and light loads, Q can be neglected and since S is small compared with the volume energies, G_0 and G_m, Equation 4.9 can be written as

$$(M/\rho)(G_m - G_0) = \alpha\mu Wg \qquad (4.10)$$

where α is the fraction of the energy stored.

For annealed metals, $G_0 \simeq 0$, therefore,

$$G_m = \alpha g\mu/(M/\rho W) \qquad (4.11)$$

Note that the significance of α is really that only a fraction of the work expended is stored in the metal. α is usually taken as $0\cdot1\%$ for cubic metals but has a different value for those with hexagonal structure. The latter show a linear relationship when G_m is plotted against the hardness of the fully abraded metals. The corresponding relationship is near parabolic with cubic metals. Similar calculations show that, for the Cu–Zn system, the G_m–hardness relationship is parabolic like that for cubic metals. However, the change in energy with hardness is not continuous and at both the copper-rich and zinc-rich ends of the alloys, G_m increases as the amounts of zinc and copper are increased respectively.

One problem with experiments on emery papers is that the backing resin is elastic which introduces what is known as the hysteresis effect in the friction term. The other drawback has already been mentioned, viz. the very short duration of the experiments of these authors [13–19]. Nevertheless, the work has added a much needed metallurgical background to the study of abraded surfaces.

From a knowledge of G_m, even the dislocation density in the fully work hardened surface layers can be calculated. This is done by dividing G_m by the energy per atom spacing along a dislocation [22]. This is justified by the unreasonable assumption that stored energy in pure metals is mostly in the form of dislocations. That is, contributions from point defects, stacking faults and twin boundaries are negligible.

4.8 High stress abrasion

A typical example of high stress abrasion is in the mining and beneficiation of rocks containing minerals. Abrasive wear problems have been studied in detail by investigators involved in obtaining molybdenum from low grade ore body [24]. The minerals are disseminated in quartz and feldspar in the form of hard rocks. The principal mineral is MoS_2, molybdenite, in amounts less than 0·4%. Other minor amounts of byproducts from this ore body are wolframite-hubnerite, cassiterite, pyrite and monazite, from which tungsten, tin, sulphur and rare earth elements respectively, are eventually extracted. The major steps beginning with mining and recovering the minerals are blasting, drilling, crushing, grinding, flotation etc., and they all incorporate components which are subjected to abrasion. The most important ones from the point of view of loss of materials as a result of high stress abrasion are crushing and grinding operations. The economic disadvantage can be appreciated when one is told that the liner of a cone crusher weighing about seven tons can lose five tons of metal in six weeks' service.

Before proceeding to summarize the philosophy of selecting materials for high stress abrasion, it is useful to look at the historical perspective of materials of one particular industry [25]. As reported in 1974, the mining industry, excepting the one mining for fuels, handles over 4000 000 000 tons of material to produce over 2600 000 000 tons of crude ore. The value of the annual output of metals and non-metals is about $10 000 000 000. As has been pointed out [25] the future statistics will be decided on the availability of energy and technology in extracting minerals from low grade ores. The phenomenon of high stress abrasion is related to the energy available and the corresponding duty that is imposed on machinery by the advance in technology to utilize the full potential offered by the available energy. Up to about the middle of the nineteenth century, mining was underground and ores were removed by hand drilling, blasting with black powder and transporting the material by horse-drawn carts. Evidently, abrasion resistance did not assume a great deal of importance because of the non-arduous nature of the drilling operation. The Chilian-mell (Fig. 4.6), a mill to grind gold ore and fireclay, used to be first driven by horses and later by steam power. In Fig. 4.6, the tribological components are a cast iron plate placed on the millstone A and the rollers C, presumably made of stone. The metallic plates were made of chill-cast iron for improving wear resistance. These were low duty mills and vertical stamps were used for crushing ores, the surfaces subject to abrasion being made from cast iron. The progressive development of equipment as increased duty was imposed led to the roll crusher mill and the ball mill, among others.

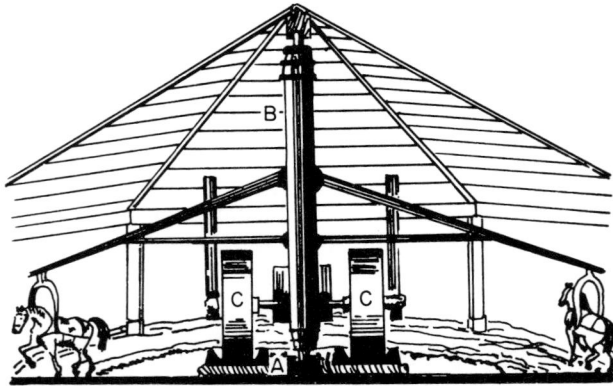

Fig. 4.6   A Chilian-mell to grind soft ores (courtesy of Climax Molybdenum Company).

Towards the latter part of the nineteenth century, pearlitic chill-cast iron was thought to be the most wear resistant material.

The year 1865 probably saw the application of power-driven drills. A steam shovel was first used in 1877 in the mining industry in the USA, although an excavator, powered by steam, was developed in England in 1796. We have seen how industrial progress accelerated in European society with the advent of power-driven equipment. The demands for metals increased which provided the impetus for increased output from mines. This led to heavier and faster cutting and crushing devices which demanded superior materials to withstand the relentless attack on interfaces under two-body and three-body abrasion. We shall now discuss some of the processes in mining and beneficiation operations of ores.

4.8.1   Crushing
Crushing is carried out in three stages. In the primary stage, lumps of ore about 1·5 m in diameter are reduced to about 200 mm in diameter in gyratory or large jaw crushers. In the secondary stage, large cone crushers reduce the product from the primary stage to about 38 mm in diameter followed by the tertiary stage to reduce the diameters further to 9 mm. Further reduction of these particles is then carried out in ball mills.

The rubbing interfaces of all the three types of crushers undergo impact and high stress three-body abrasion. The velocity is very low and typical metal losses of the sliding surfaces due to wear in kg per 1000 tons of ores crushed are 7·2, 13 and 24 for the primary, secondary and tertiary stages respectively. Loss of material from rubbing surfaces during crushing of large size rocks has been termed gouging wear [26]. As the name implies, the sharp edges of the rocks cause deep gouges and large particles of metal are removed from the surface. Laboratory evaluation of candidate materials for liners or other surfaces acting as interfaces for three-body abrasion utilizes a small jaw crusher [27, 28], but correlation is also sought with field trials to establish acceptance tests. The laboratory test prepares the stationary plate from the candidate material and the movable plate is a reference material. Duplicate runs are made for each material and a complete test involves crushing about 900 kg of rocks. A wear ratio, R, is then obtained as,

$$R = \frac{\text{Weight loss of the test material}}{\text{Weight loss of the reference material}} \qquad (4.12)$$

An arithmetic average of the two tests is taken for the value of R. The test is reproducible and can be used for comparative studies of materials.

Performance of materials in actual service is also observed, but because of the unavoidably variable nature of operation, wide scatter between the values of R is observed. Nevertheless, methodical observations on actual components in service help in the development of new materials and of manufacture of components. Although expensive to produce, a 2% Mo-containing austenitic manganese steel gives good wear resistance. The manufacturing problem in austenitic manganese steel is obvious when one considers heat treatment of castings weighing about 8 tons with an average section thickness of 250 mm.

4.8.2 Austenitic manganese steel

The first patent on this steel was granted to Robert Abbot Hadfield, a steel manufacturer in England. The Hadfield manganese steel, as it is also known, contains about 11% manganese and 1·1% carbon. If the cast component is austenitized at a high temperature and then quenched in water, it remains in its austenitic form. The structure is f.c.c. and the austenitic component has good fracture resistance. During abrasion, a thin layer of the component transforms to a wear resistant martensitic structure. As this hard layer wears with use, fresh martensitic surfaces are produced due to strain hardening by abrasion. In this way, the wearing interface is frequently replenished by a martensitic layer. As well as this, the austenitic core makes the component resistant to shock loads. However, there appears to be some evidence [29] to show that the role of manganese in steels is to reduce wear resistance. If the abraded surface is not fully martensitic, that is if it is austenitic in patches, wear resistance will be impaired. This may be the reason why, for abrasion under very high stress, an austenitic manganese steel is used as a backing which is tough under impact load. A wear resistant surface is produced by weld-depositing a high carbon, highly alloyed white iron.

4.8.3 Ball and rod mills

Ball and rod mills are essentially rotating cylinders containing a charge of balls or rods which pulverize the solid ore. The input ore has an average diameter of about 9 mm which is ground to a fine powder. The inside diameters of the mills vary from about 2·7 to 3·9 m. Steel balls, about 75 mm in diameter, are used as the charge to carry out the pulverizing action by impact and sliding. The mill is kept half full with balls and the net weight of the balls charged daily could be between 1 and 2 tons. The inside surface of the mills is built with shells of liners cast in steel. The liners can wear out in a period ranging from two months to a few years and the amount of wear of both these and the balls depends mainly on two factors:

   (1) operating conditions which include the size and hardness of the input ore;
   (2) the metallurgical nature of the liners and the balls.

A ball mill makes use of three-body abrasion. The two bodies are the liner and the balls which trap the third body and crush it under impact loads. The most abrasive constituents are quartz and the silicate minerals in the ore, and localized stresses have been estimated [30] at 220 kg mm^{-2}. Loss of liner and ball occurs by microspalling or fracture of the brittle constituents such as carbides. Wear of both components is considerable and the rate of ball consumption alone is in the range 2·2 to 3·3 kg per ton of ore ground.

Considering the reported tonnage [24] of ore mined in the USA in 1969 for extracting molybdenum, given as 15 million tons per annum, even a marginal reduction in the wear of balls and liners will result in huge financial savings. Efforts continue, therefore, to devise suitable tests to rank materials from the wear resistance point of view.

To assess materials in actual mills, test samples in the form of balls are used to evaluate both liner and ball materials. For the liner material, the balls are large to ensure a similar solidification rate to the actual liner and the dimension of the ball diameter is standardized at 125 mm. For each candidate material, duplicate balls are tested. The procedure is to charge the pair of test samples in the mill and allow them to wear until about 3 mm of the surface layer has been removed. This takes about a week and the object of this is to remove any casting defects or decarburized layers due to heat treatment. Each ball is weighed and charged with another weighed ball made from the reference material into the mill. In fact a group of test samples, paired with an equal number of reference balls, are given identification marks and charged into the mill. They are run for a sufficient length of time to remove a further 3–6 mm of surface material. The balls are then weighed and the surface hardness measured. From a knowledge of the weight and density, the volume lost and hence the loss of surface area by wear can be calculated.

Measurement of the surface area is important because it has been shown [31–33] that balls of different sizes and weights but of identical chemical composition and metallurgical structure give reproducible wear rates if wear is expressed as loss of material per unit surface area. The amount of relative wear is expressed as an abrasive factor, defined as,

$$\text{Abrasive factor} = \frac{\text{Wear rate of material under test}}{\text{Wear rate of the standard material}} \times 100$$

(4.13)

This is the same as that expressed by Equation 4.12. Equation 4.13 shows that the standard material will have an abrasion factor of 100 and the higher the abrasion factor, the more inferior is the wear resistance of the candidate material.

4.8.3.1 *Metallurgy*
Studies indicate that the microstructure of the steel is important in deciding the wear resistance of balls and liners. There are three types to consider, viz. martensitic, pearlitic and austenitic steels. Before drawing any conclusions from a set of field trials it is important to emphasise that the results may only be applicable to that particular situation. Considering high stress abrasion in a ball mill, for steels heat treated to

produce a martensitic structure, the component is cast in steel containing 1% C with 6% Cr and 1% Mo. If the steel is austenitized and cooled in air, the microstructure comprises spheroidized carbides in a fine-grained martensitic matrix. If low alloy steels of varying carbon contents are fully hardened so that the structure is fully martensitic without any retained austenite, the wear resistance is seen to improve with increasing carbon content (Fig. 4.7) up to a limiting amount of carbon of 0·7%. As shown in Fig. 4.7, the wear resistance is affected by the degree to which the pro-eutectoid carbon goes into solution during austenitizing of the steel.

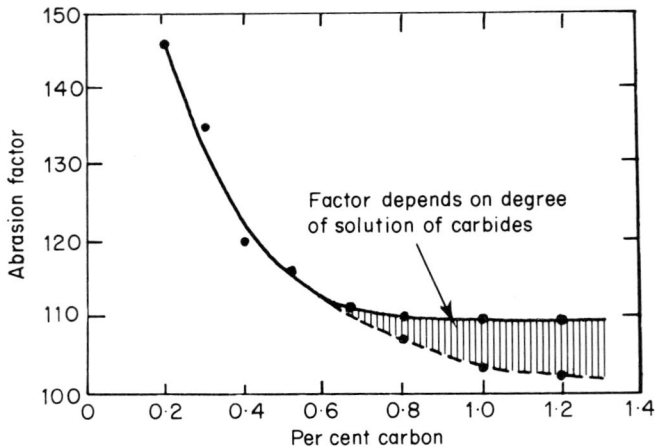

Fig. 4.7   Influence of carbon content on the wear rates of martensitic cast steels. (Norman, T. E. (1958). *Modern Castings* May, p. 89.)

The obvious conclusion is that a high austenitizing temperature will improve the abrasion factor (Equation 4.13) and tests show this to be so (Fig. 4.8). However, the danger remains that a high austenitizing temperature will result in a large amount of retained austenite and grain growth will occur because the spheroidized carbides will dissolve at high temperatures. The combined effect will be to lower the impact resistance of the steel, resulting in spalling of balls and liners on a macro-scale. Retained austenite, of course, will lower the abrasive wear resistance.

Toughness in martensite can be imparted by tempering. Tempering a high carbon martensitic steel at 204 °C reduces the propensity to

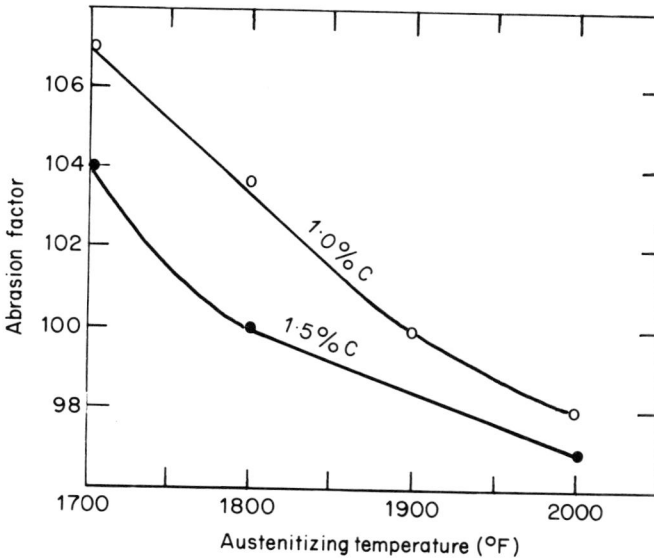

Fig. 4.8 Influence of austenitizing temperature on the wear rates of two martensitic steels. (Norman, T. E. (1958). *Modern Castings* May, p. 89.)

spalling without reducing the wear resistance appreciably. A further improvement in toughness is possible by tempering above 426 °C but considerable wear resistance is sacrificed. In fact, pearlitic steels with comparable hardness to tempered martensite show higher wear resistance. Alloying elements are necessary to improve hardenability in thick sections but selection of elements should be such that retained austenite can be avoided.

While comparing sliding interfaces, it is important to consider the hardness and the nature of the abrading particles as well. Quartz has comparable hardness to martensite and the latter is worn readily by the former. Unlike hard martensitic steels tempered at a low temperature, pearlitic steels wear readily if abraded against feldspar.

4.9 Low stress abrasion

Conditions of low stress abrasion can be found in the mining industry, two examples of its occurrence being in pumps and impellers for agitators in the flotation cells. Experience shows that high chromium or chromium–molybdenum irons provide three times as much wear resistance as the plain carbon or low chromium white irons. Improvement in wear resistance is offered by moulded natural rubber and pumps and impellers make use of this. It should be noted, however, that rubber degrades in regions where oil is present and a substitute material is 15/3 Cr–Mo alloy. The alloy is also used in the spirals which separate out heavy minerals and a life of 20 years is expected.

Dry abrasion is probably not as severe as when it is combined with certain chemicals. Worm mixers mixing carbon black or asbestos before shaping into blocks are typically weld deposited with hard facing alloys. These deteriorate very rapidly, often after two weeks, necessitating costly replacement. Carbon particles oriented suitably can act as abrasives but the mode for the loss of the hard deposit would appear to be chemico-mechanical in nature. It is not easy to devise an accelerated low stress abrasion machine. One approach seems to be the use of a small impeller weighing about 50 g which will rotate in slurries of controlled acidity. The wear is expressed in a number of ways:

(1) weight loss per unit sliding distance;
(2) weight loss per unit surface area;
(3) the use of the abrasion factor (Equation 4.13).

Whatever the choice, reproducibility of results is essential.

Published information suggests that the rubber wheel test (Fig. 4.9) simulates low stress abrasion and gives reproducible experimental

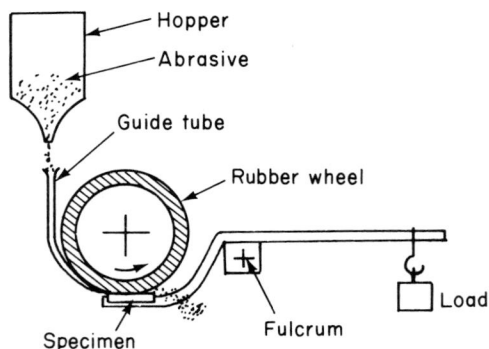

Fig. 4.9   The rubber wheel test. The loaded specimen wears against the slurry from the hopper. The rubber wheel rotates slowly and cushions the abrasives, thus reducing the pressure at the interface. ("Abrasive Wear—J965" (1966). Information Report by the Society of Automotive Engineers, New York.)

results. According to this test, low stress abrasion resistance of metals is improved by increasing their hardness. Other things being equal, hardness of a steel will increase with increasing carbon content. One could therefore correlate abrasion resistance of steels with their carbon contents and the mechanism is probably associated with the presence of increased amount of carbides. The role of carbides in wear resistance is discussed later but a rubber wheel test shows clearly the positive value of martensite from the view point of abrasion. The rubber wheel test also provides the logical answer that the higher the temperature of tempering, the greater is the loss in both the hardness and wear resistance of a steel. This is encouraging but the unhappy practical difficulty is that the presence of solvents in a slurry will probably attack the rubber. The surface finish of the rubber in this test is important from the point of view of consistency of results. The solvents will make the experimental results unreliable and the rubber wheel is therefore unsuitable for studying low stress abrasion in the presence of unsuitable solvents.

4.10 Carbides

There may be two possible reasons why the hardness of a metal does not correlate with wear resistance for all situations. These are:
(1) The Vickers or the Brinell hardness gives the overall hardness of a metal and the tests cannot distinguish between, say, solid solution hardening or that imparted to an alloy by a phase change.
(2) Wear in certain situations may be dictated by a soft micro-constituent only and the Vickers or the Brinell hardness will not identify separate micro-constituents.

Considering high chromium white irons with a typical composition (by weight) of 20% Cr, 2% Mo, 1% Ni and 2·7% C, maximum strength, toughness and abrasion resistance are obtained with a fully martensitic matrix [34]. The microstructure shows (Fig. 4.10) chromium carbides, Cr_7C_3, in a matrix largely transformed to martensite. Used with crushers, the surface is severely deformed and shows wear debris forming in the work hardened layer (Fig. 4.11).

Experiments have shown that the phenomenology of wear can be followed on a microscale. The abrasives, especially if they are fine, remove micro-constituents by a cutting action or form grooves and scratches without causing any loss of material. Microscopic observation reveals the particles on either side of a groove as swarf and it is evident

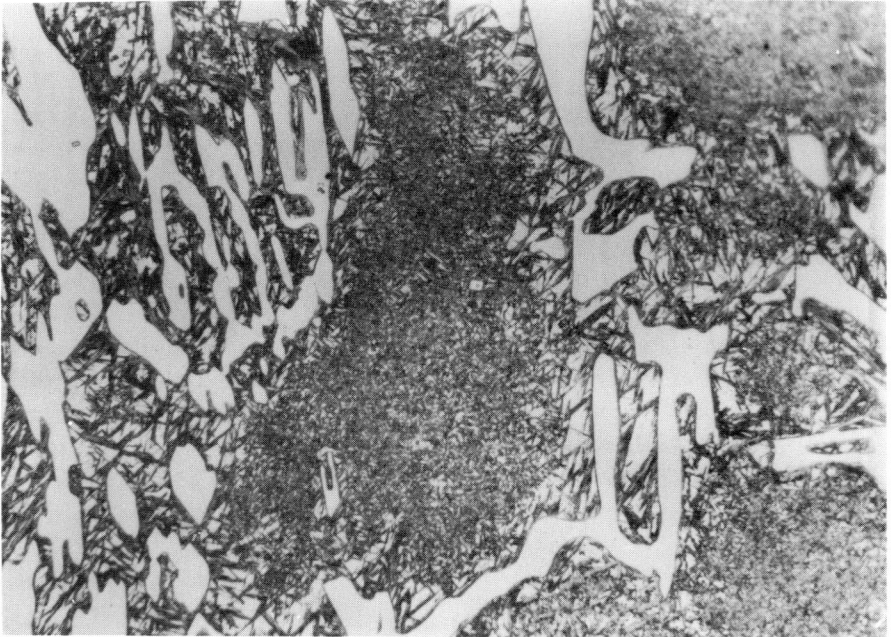

Fig. 4.10   Photomicrograph of the inside of a gyratory crusher segment showing Cr_7C_3-type carbides in a matrix of austenite, largely transformed to martensite ($\times500$). Composition (in %): C 2·47; Mn 0·75; Si 0·6; Cr 19·2; Mo 1·2; Ni 0·7 (courtesy of Climax Molybdenum Company).

that the micro-constituent which is removed is softer than the abrasive particle. Quartz, often the major constituent in rocks with a Knoop hardness of 840, is harder than any of the possible matrix constituents in white cast iron (Table 4.3). In low alloy white cast irons produced by chilling the casting, the carbide is in the form of cementite and is easily abraded by quartz. The alloy carbides shown in Table 4.3 are certainly much harder than quartz and they resist abrasion. There are, unfortunately, two difficulties. One of them is that the harder carbides such as WC, TiC and VC are expensive to produce. Secondly, since even a martensitic matrix is softer than the quartz abradant, the carbides are useless in offering protection, being only 50% of the total volume of the iron. Thus once the matrix is weakened sufficiently, the carbides are dug out.

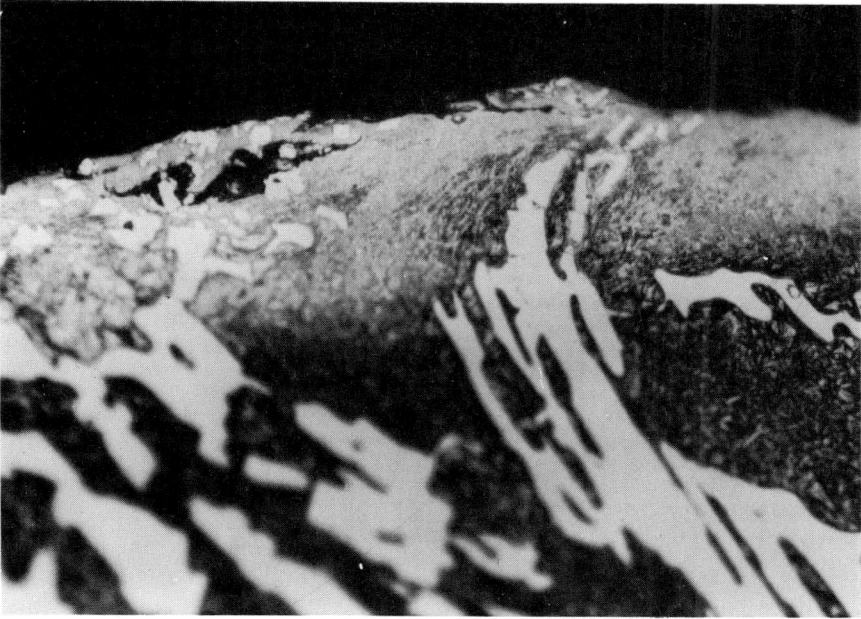

Fig. 4.11   Photomicrograph through the abraded surface of the crusher in Fig. 4.10. Note severely distorted and work hardened metal at the surface and the appearance of wear debris (×520) (courtesy of Climax Molybdenum Company).

Table 4.3   Knoop microhardness of micro-constituents in alloy cast iron

Constituent	Hardness
Ferrite	235
12% Mn, austenite	305
Martensite	500–800
Cementite	1025
Cr_7C_3	1735
Mo_2C	1800
WC	1800
TiC	2470
VC	2660
B_4C	2800

The solution obviously lies in the production of a matrix which is hard but tough enough to withstand applied impact stresses. One way of achieving this is to have a high carbon martensite with further reinforcement in the form of dispersion hardened secondary carbides. If the chromium content is allowed to fall below 12%, an eutectic or a slightly hypoeutectic iron will tend to form cementite with a Knoop hardness higher than quartz but lower than that of any of the alloy carbides such as TiC or WC. The problem remains that cementite may be worn by abrasives such as topaz, emery and corundum, although these are softer than the stable chromium carbide. If the chromium is increased to about 20% and the carbon is left unaltered at 3·2%, the eutectic composition, most of the carbon is unavailable for any other reaction because it combines with the chromium only to form carbides. Consequently, the martensite will be impoverished in carbon and will have diminished wear resistance because of relatively low hardness. Increasing the carbon content beyond the eutectic composition is unhelpful because, rather than increasing the proportion of carbon in the austenite and hence in the martensite, hypereutectic coarse carbides are formed. These are brittle and will readily fracture under impact stresses and loss of material by wear will be aggravated.

The use of Hadfield steel has already been described. It has limitations but a general statement regarding the role of microstructure in the wear rates of alloy cast iron can be made which is that a high carbon martensitic matrix with well-dispersed alloy carbides will give a wear resistant material in three-body abrasion. The actual production of such components, however, is very involved and while selecting a specific material for a particular application, the cost of production must be balanced with the expense of replacement and the irretrievable loss of material by wear.

4.11 Ploughshares

It is not uncommon to use near eutectoid steels as ploughshare materials, but these wear readily due to abrasion by soil. Abrasion by soil is by no means as ruthless as gouging caused in the mining industry but conclusions from experience gained in that industry should be of interest in the agricultural field. Alloyed steels and white irons have thus been the logical alternatives to plain carbon steels for longer life under abrasion by soil.

Laboratory tests such as the use of a pin on an abrasive disc are carried out but the most useful results have derived from field trials [35]. The effect of interaction with soil is, in common with most abrasive situations, to produce a work hardened layer and a fair correlation with the hardness of the deformed layer and the wear resistance has been found. Dry soils are known to cause a high amount of wear but most agricultural soils contain about 40% of gas together with water. The ploughing action compresses this soil ahead of the ploughshare so that force is exerted on the cutting face. Agricultural soils always contain stones which help to consolidate the soil and thus offer extra resistance to cutting. This is the indirect role which stones play in aggravating wear of ploughshares but the direct effect of rubbing against hard stones also contributes to the wear. An analysis of the nature of the interaction between a stone and the cutter has been carried out [35].

Experiments with initially parallel-sided plates undergoing abrasion against soil show that the interface becomes nearly parabolic and the total wear is proportional to the initial plate thickness. To appreciate the reason for this dependence, it is necessary to recognize that the total wear path, ds, is made up of two components:

(1) a distance ds_1 is involved if the stone skids round the parabolic edge;

(2) a distance ds_2 which is slid by the edge over the stone surface when it is held stationary by the soil.

Therefore,

$$ds = ds_1 + ds_2 \qquad (4.14)$$

The parabolic edge is represented in Fig. 4.12 and has a radius of curvature, R_1, whose centre is O_1. The stone moves about an instantaneous centre, O, as shown at a radius R measured to its circumference. In Fig. 4.12, $AB = ds_1$ and $BC = ds_2$. For small angles $d\theta$ and $d\phi$,

$$ds_1 = AB = R_1\, d\theta$$

Therefore,

$$\frac{ds_1}{d\theta} = R_1 \qquad (4.15)$$

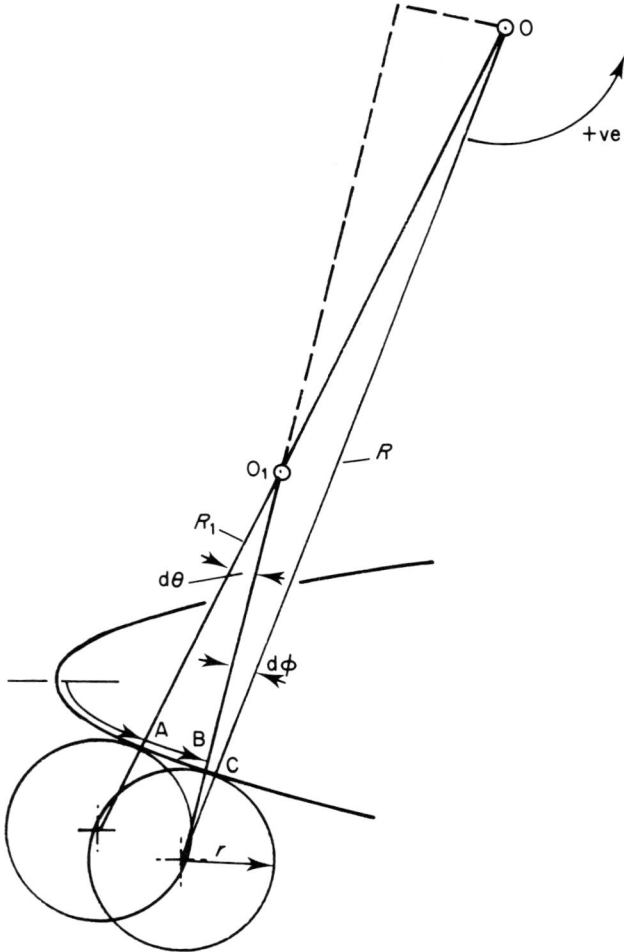

Fig. 4.12 A diagram showing the cutter and a stone in agricultural soil. (Richardson, R. C. D. (1967). *J. Agric. Engg. Res.* **12**(1), 22.)

and $(R + r)\, d\phi = (R - R_1)\, d\theta$ or

$$\frac{d\phi}{d\theta} = \frac{R - R_1}{R + r} \tag{4.16}$$

Since, $ds_2 = r\, d\phi$, the total friction path is

$$ds = AB + BC = ds_1 + r\, d\phi$$

Therefore,

$$\frac{ds}{ds_1} = 1 + r \frac{d\phi}{ds_1}$$

Substituting appropriate values in the above relationship from Equations 4.15 and 4.16,

$$\frac{ds}{ds_1} = \frac{1 + (r/R_1)}{1 + (r/R)} \tag{4.17}$$

The factor ds/ds_1 is the expression for the intensity of wear at a point on the metal edge. If the instantaneous centre O is on the surface of the sphere, i.e., at $R = 0$, $(ds/ds_1) = 0$, since the denominator in Equation 4.17 is infinitely high. This is a situation of rolling by the sphere and there is no wear. When R is not zero, ds/ds_1 will have various values resulting in skidding and wear.

The performance of a material used for ploughing can vary depending upon the service condition. Abrasion by flint is soft, meaning low, if it is distributed in a loose sandy soil. The same particles embedded in a strong clay will have a severe abrasive effect. The most abrasive constituent again is quartz or other forms of silica. Silica is also present as a network in many plant tissues and can be abrasive when the plants are sheared. Apart from quartz and flint, there may be present other types of solids, e.g., ironstone. Impact is involved during cutting a soil. We conclude that the philosophy of materal selection and of manufacturing techniques will be analogous to that described for high stress abrasion, although the scale of events may be less aggressive in certain respects.

References

1. Rabinowicz, E. (1966). "Friction and Wear of Materials", John Wiley and Sons, New York.
2. Rabinowicz, E., Dunn, L. A. and Russel, P. G. (1961). *Wear* **4**, 345.
3. Avient, B. W. E., Goddard, J. and Wilman, H. (1960). *Proc. Roy. Soc. A* **258**, 159.
4. Sedriks, A. J. and Mulhearn, T. O. (1963). *Wear* **6**, 457.
5. Tabor, D. (1954). *Proc. Phys. Soc. B* **67**, 249.
6. Mulhearn, T. O. and Samuels, L. E. (1962). *Wear* **5**, 478.
7. Richardson, R. C. D. (1968). *Wear* **11**, 245.
8. Oberle, T. L. (1951). *J. Metals* **3**, 438.

9. Khruschov, M. M. (1957). Institute of Mechanical Engineers: Conference on Lubrication and Wear, Paper 46.
10. Vigh, A. K. (1975). *Wear* **35**, 205.
11. Malkin, S. and Joseph, N. (1975). *Wear* **32**, 15.
12. Kelly, K. K. (1967). US Bureau of Mines, Bull 584.
13. Alison, P. J., Stroud, M. F. and Wilman, H. (1964). *Proc. Inst. Mech. Eng.* **179**, (3J) 246.
14. Goddard, J., Harker, H. J. and Wilman, H. (1959). *Nature* **184**, 333.
15. Stroud, M. F. and Wilman, H. (1962). *Brit. J. Appl. Phys.* **13**, 173.
16. Goddard, J. and Wilman, H. (1962). *Wear* **5**, 114.
17. Evans, D. M., Layton, D. N. and Wilman, H. (1951). *Proc. Roy. Soc. A* **205**, 17.
18. Agarwala, R. P. and Wilman, H. (1953). *Proc. Phys. Soc. B* **66**, 717.
19. Wilman, H. (1957). *Acta Cryst.* **10**, 824.
20. Scott, V. D. and Wilman, H. (1958). *Proc. Roy. Soc. A* **247**, 353.
21. Porgess, P. V. K. and Wilman, H. (1960). *Proc. Phys. Soc. B* **76**, 513.
22. Wilman, H. (1960). *Acta Cryst.* **13**, 1062.
23. Kerr, I. S. and Wilman, H. (1955). *J. Inst. Metals* **84**, 379.
24. Norman, T. E. and Hall, E. R. (1969). American Society for Testing and Materials, Special Technical Publication 446.
25. Climax Molybdenum Company Symposium: Materials for the Mining Industry (1974). Vail, Colorado, USA.
26. "Abrasive Wear—J965" (1966). Information Report by the Society of Automotive Engineers, New York.
27. Borik, F. and Sponseller, D. L. (1971). *J. Mater.* **6** (3), 576.
28. Borik, F. and Scholz, W. G. (1971). *J. Mater.* **6** (3), 590.
29. Maratray, F. J. and Norman, T. E. (1961). *Revue de Métallurgie* **58**, 489.
30. Norman, T. E. (1958). *Trans. Amer. Found. Soc.* **66**, 187.
31. Norman, T. E. and Loeb, C. M. (1948) *Metals Technol.* April, TP 2319.
32. Garms, W. I. and Stevens, J. L. (1946). *Mining Technol.* March, TP 1984.
33. Norquist, D. E. and Moeller, J. E. (1950). *Mining Eng.* **187**, 712.
34. Fairhurst, W. and Röhrig, K. (1974). *Foundry Trade Journal* May 30.
35. Richardson, R. C. D. (1967). *J. Agric. Eng. Res.* **12** (1), 22.

5 Erosive Wear

Erosive wear is defined as the process of material removal from a surface subjected to impingement attack by solid or liquid media, being particulate in nature for the former and in the form of droplets for the latter. It is not known if, in areas of heavy rainfall, buildings undergo erosion to a significant extent but sand storms have caused destructive loss of material from structures left unmaintained such as temples erected centuries ago. In the world of engineering, an example of solid particle erosion is the impingement by atmospheric grit in aero gas turbines as used in helicopters, hovercraft and civil aircraft. Observations under service conditions show that gas turbine blades undergo wear from quartz, which is the main erosive constituent in the impinging solid particles which are sucked into the vicinity of the blades by the air intake. Impact takes place in the velocity range 250–450 m s^{-1} and the particles are known to hit the leading edges of the blades, generally at right angles to the surfaces. Experiments show that normal impact is inefficient from the point of view of removal of surface material and the propensity to wear increases at small angles of attack. The latter is the case for the trailing edges of the blades which encounter the abradants at glancing angles. This results in wear, and hence loss of efficiency because of a change in the aerodynamic profile. Attempts are made to obstruct the entry of sand particles by filtering the air intake but small grits can still be taken in if filtration is inadequate. Filtration, however, incurs a weight penalty on engines and power losses due to the induced pressure drop. An important approach in obtaining better life from blades for engines operating under conditions conducive to erosive wear is to develop new materials. Whatever the approach, we must understand, firstly, the mechanisms of erosive wear, and secondly, establish the variables which aggravate the damage. In this chapter we shall collate information available for both solid and liquid particle erosion

5.1 Impact

Erosion studies of components involve, essentially, an analysis of impact forces acting on a surface excited by a projectile at a reasonably high velocity. As discussed in Chapter 4, in certain worm-type mixers producing blends of carbon black, asbestos, etc., the particulate matter tends to traverse the surfaces when the worm or impeller rotates. This has been loosely termed erosion in certain quarters, but in this text the phenomenon is described as a low stress abrasion process. As defined in the introductory paragraph, erosion is that process which embodies impingement at a reasonably high velocity.

Basic studies of impingement attack have been carried out by striking plates of the material under study with single particles such as a hard steel ball or, as an example, with a stream of sand. The general subject of experimental study of impact can cover a range of velocities up to 10^5 m s^{-1}, and apart from speed, the mode of damage to the surface is dependent on the mechanical and certain physical properties such as specific heat of the plate material. Of no less importance are the characteristics of the offending particles, chief among which are shape, size and bulk hardness.

5.1.1 Damage number

A criterion used to categorize the impact phenomenon is the *damage number*, defined [1] as $\rho v^2 / \sigma_Y g$, where ρ = density of the target material, v = relative velocity of normal impact, σ_Y = flow stress of the target material, and g = acceleration due to gravity.

A cursory glance at the parameters comprising a damage number shows that, at identical velocities, a material with a high value of ρ / σ_Y will have a high damage number. A high value of ρ / σ_Y is not automatic if a material with a high density is used. For example, steel has a high density but the difference in the values of ρ / σ_Y between steel and aluminium may not be great. This is because the flow stress σ_Y is also higher than that of aluminium. Obviously, the velocity component has a decisive effect and it has been shown that, for mild steel, using the criterion of damage number, the interaction due to impact is quasi-static and elastic at a speed of $0 \cdot 762$ m s^{-1}. Plastic yielding of the steel target begins at $7 \cdot 62$ m s^{-1}, giving way to extensive plastic deformation at a velocity of 762 m s^{-1}, the speed of ordinary bullets.

It is certainly important to establish the nature of deformation for all tribological interactions to understand the mechanisms. It is not easy to ascertain whether the deformation is elastic or plastic, and the damage number will be of value if it can identify the nature of the interface since the quantities ρ, v and σ_Y are measurable. However, the damage number does not take the shape of the projectile into account and it is difficult to quantify σ_Y when both the velocity and strain rate are high with an attendant rise in the bulk temperature.

In the same manner as in the model proposed in Chapter 3 for adhesive wear, a shape factor could probably be introduced to account for variations in the geometry of the colliding particles. It is unlikely that it will be easy to predict changes in mechanical properties in terms of the heat gained by the target due to collision, but the loss of yield strength of steels with rising temperatures has been measured experimentally. It should not be difficult to bury thermocouples in the target and measure subsurface temperature to compare this with the theoretical value. This value can be calculated as follows:

Assume that the total kinetic energy of a projectile given by $\rho v^2/2g$ is completely converted into heat due to collision giving a mean adiabatic rise in temperature, ΔT. In that case, if C is the specific heat of the target and is independent of temperature,

$$\rho v^2/2g = \rho C \ \Delta T \quad \text{or} \quad \Delta T = v^2/2gC \qquad (5.1)$$

Knowing the change in temperature in this manner and selecting the appropriate yield strength of the target a more realistic damage number can be obtained for predicting the mode of deformation.

5.2 Crater volume

If a simple experiment is carried out where a steel ball is allowed to fall normally on a lead plate, say, from a height of about 3 m giving a striking velocity of over 7 m s^{-1}, only a circular indentation is observed. Most of this indented volume is produced by plastic accommodation and there is little loss of material from the target apart from a small amount which may adhere to the steel ball. If the impact is at an angle, the geometry of the crater changes, being ellipsoidal, and examination under the microscope shows evidence of slip. At all angles of impact, the striker rebounds and spinning of the projectile has been observed at high velocities.

We could speculate that using an initially ductile material such as brass, the dimensions of the crater would increase under repeated impact to a limiting value when the region in the vicinity of the crater had fully work hardened. Material loss from the target would then occur, possibly by a fatigue process if the same part of the target was impacted repeatedly. Theoretical considerations show that erosive wear is a function of both the particle velocity and the yield strength of the target material and both these parameters influence the volume of the crater produced. Since it is not difficult to obtain the crater volume experimentally, it may be possible to correlate wear rates with crater volumes from laboratory studies. Before collating available information on experiments on erosive wear, we should attempt to establish the relationship between the crater volume V_c, the impact velocity v and the yield strength σ_Y of the target.

5.2.1 Conical projectile

Consider a conical projectile with a semi-apex angle θ and mass m striking a plate with an initial velocity v_i at a normal angle of incidence

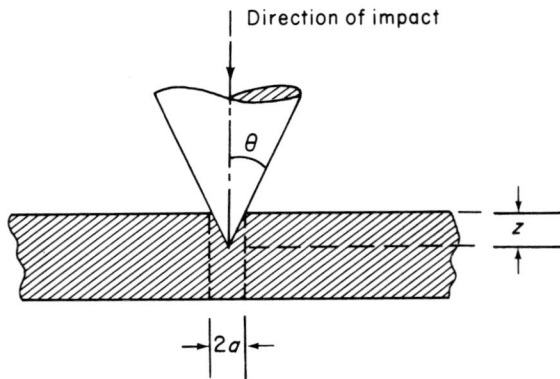

Fig. 5.1   A conical projectile striking a flat surface.

(Fig. 5.1). Let the indentation diameter be $2a$ with a depth of penetration z from the free surface of the plate which has a flow stress, σ_Y. Assume that the striking velocity is low and that of rebound, v_r, is only a small fraction of v_i. Then

$$mv \frac{dv}{dz} = -\sigma_Y \pi a^2 g \qquad (5.2)$$

where v is the speed of the projectile in the normal direction from the surface of the plate towards its centre. From Fig. 5.1, $a = z \tan \theta$; substituting this value of a in Equation 5.2

$$-\sigma_Y (\pi z^2 \tan^2 \theta) g = mv \frac{dv}{dz} \qquad (5.3)$$

when $z = 0$, velocity $= v_i$ and at a distance z, velocity $= v$. Thus integrating Equation 5.3.

$$-\sigma_Y \pi g \tan^2 \theta \int_0^z z^2 \, dz = m \int_{v_i}^v v \, dv$$

or

$$\sigma_Y \pi g \tan^2 \theta \frac{z^3}{3} = \frac{m}{2} (v_i^2 - v^2) \qquad (5.4)$$

Note that in Equations 5.2 and 5.3, the crater diameter, $2a$, and the depth of penetration, z, change with time. To avoid clumsiness by using too many suffixes, denote the final crater diameter (Fig. 5.1) by $2a$ and the depth of penetration again as z when $v = 0$ in Equation 5.4.

Therefore, from Equation 5.4,

$$z = [(\tfrac{3}{2}m)/\sigma_Y \pi g \tan^2 \theta]^{1/3} v_i^{2/3} \qquad (5.5)$$

Since $2a = 2z \tan \theta$, from Equation 5.5,

$$2a = (2 \tan \theta) \left(\frac{3m}{2\sigma_Y \pi g \tan^2 \theta} \right)^{1/3} v_i^{2/3}$$

The indenting cone generates a crater volume V_c, and

$$V_c = \tfrac{1}{3}\pi a^2 z$$

Substituting for $a = z \tan \theta$ and using the value of z from Equation 5.5,

$$V_c = (\tfrac{1}{2}mv_i^2)/\sigma_Y g \qquad (5.6)$$

Equation 5.6 shows that a material with a high yield strength will resist indentation and hence the relative success of the nickel-based alloys for turbine blades. We note also that V_c is a function of v_i^2, the square of the striking velocity. For experimental work, the striking velocity at high impact is obtained by high speed photography.

5.2.1.1   Shape factor

We posed a question earlier as to whether the crater volume is affected by the geometry of the projectile. It is clear that the geometry of the crater will depend on the shape of the indenter but we need to deduce the volumes as for a conical projectile. Let us consider a spherical and a rectangular-sectioned indenter, respectively. As before, the yield stress of the target is σ_Y and the penetration depth is z. Consider a sphere of diameter D and mass m. From Fig. 5.2(a),

$$mv\frac{dv}{dz} = -\sigma_Y \pi a^2 g \tag{5.7}$$

Now using a suitable triangle,

$$a^2 = \left(\frac{D}{2}\right)^2 - \left(\frac{D}{2} - z\right)^2$$

Since z is small, the higher powers of z can be neglected so that

$$a^2 = zD$$

Substituting this in Equation 5.7,

$$mv\frac{dv}{dz} = -\sigma_Y \pi z Dg$$

The initial velocity is v_i. At the point of impact, the velocity is zero. The corresponding penetration depths are 0 and z. Therefore, integrating the above expression,

$$m\int_{v_i}^{0} v\,dv = -\sigma_Y \pi Dg \int_{0}^{z} z\,dz$$

It can be shown that the volume of the spherical cap, which is the crater volume, V_c is

$$V_c = \frac{\pi z^2}{3}\left(\frac{3D}{2} - z\right) \tag{5.8}$$

The solution of the above integral gives,

$$\tfrac{1}{2}mv_i^2 = \frac{g}{2}\sigma_Y \pi D z^2 \quad \text{or} \quad z^2 = mv_i^2 / \sigma_Y \pi Dg$$

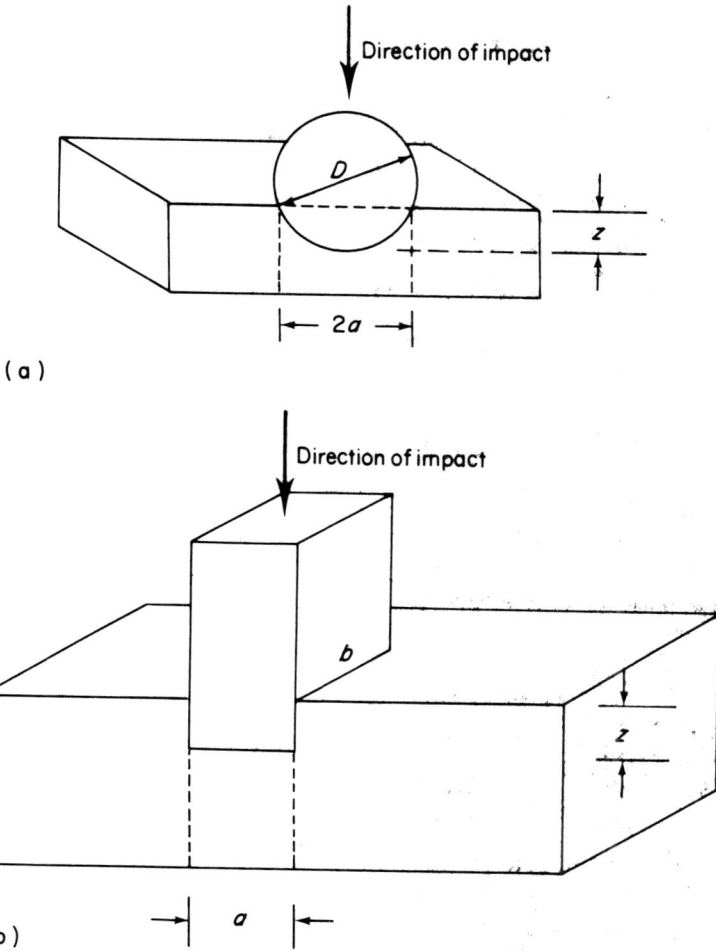

Fig. 5.2   A spherical (a) and a rectangular sectioned (b) projectile striking a
surface respectively.

Substituting this value of z^2 in Equation 5.8,

$$V_c = \frac{1}{6}\left(3 - \frac{2z}{D}\right)\left(\frac{mv_i^2}{\sigma_Y g}\right)$$

Neglecting $2z/D$,

$$V_c = (\tfrac{1}{2}mv_i^2)/\sigma_Y g \qquad\qquad (5.9)$$

Consider a rectangular section of sides a and b (Fig. 5.2(b)), penetrating a depth z. With the same boundary conditions as for the sphere,

$$m \int_{v_i}^{0} v \, dv = -\sigma_Y abg \int_{0}^{z} dz \quad \text{or} \quad \tfrac{1}{2}mv_i^2 = \sigma_Y abzg$$

Since $abz = $ volume of the crater $ = V_c$,

$$V_c = (\tfrac{1}{2}mv_i^2)/\sigma_Y g \tag{5.10}$$

Comparing Equations 5.6, 5.9 and 5.10, we see that there is no shape factor in the volume of a crater, which is decided entirely by the kinetic energy of the projectile and the yield stress of the target material.

5.3 Deformation mode

In typical practical situations, such as impingement of buildings by sand, the projectiles are compound shaped. That is to say a striking particle may be regarded as incorporating two or more of the basic shapes, viz., sphere, cone, cube, etc. Experimental data using quartz particles or other grits provide information on the attack of a target by compound shapes. For theoretical analysis, however, unpredictable non-uniformity in shapes makes compound particles unmanagable for mathematical manipulation and theoretical treatments of erosive wear classify the impinging particles into one of the following shapes:

(1) conical;
(2) ogival;
(3) spherical;
(4) square-ended.

Many targets will be indented even at low speeds because of the possibility of plastic flow as a result of the kinetic energy received from the projectile. However, the size of the indentation will be small if the target is hard with a large yield stress (Equation 5.6). If the projectile maintains its shape, a coronet of the target material is thrown around the particle if the target undergoes plastic deformation. If it does not and is brittle, shattered material is seen to be removed by the unaffected projectile during an encounter, leaving a crater in the target as the colliding particle rebounds off the surface it attacked.

If the head of the indenter is ogival, the deformation process displaces the target material outwards. This results in a situation of radial compression in the medium which is being impacted.

Experiments [2] with hard steel projectiles having conical heads show strong adhesion at the interface when the target material is mild steel. If the projectile surface is rough, the body of the target appears to undergo ploughing by the protruding asperities so that the mode of wear is that of two-body abrasion. In that case, the situation is analogous to a plough scraping metal from the indented solid and the projectile does not show any significant amount of transfer of mild steel on to its surface. On the other hand, large scale metal transfer on to the hard projectile is observed if it has a smooth surface so that the target can be said to have undergone adhesive wear. It is speculated that a projectile with a rough surface will provide a larger area for the frictional heat to be dissipated. For the opposite reason, a smooth surface will allow a build-up of temperature at the interface which will result in a lowering of the flow stress of the mild steel, facilitating adhesion.

It can be expected that if one particular spot of a ductile medium is impacted normally, the diameter of the indentation will continue to increase to a limiting value decided by the total kinetic energy needed to work harden the area of the target fully. Some material may be removed by adhesion to the projectile, particularly if it has a smooth surface, but there is a strong possibility of transverse cracks and spalling of layers of the target by fatigue occurring. As far as it is known, such studies have not been undertaken possibly because of the difficulty of hitting the same spot each time under repeated impact.

The onset and the nature of wear, however, have been studied [3] by striking mild steel plates with steel spheres, 9·5 mm in diameter, at various angles of impingement. The impact process was photographed by using a high speed cine camera and subsequent examination of the still frames provided information about the sequence of the pattern of deformation and velocity including that of rebound, and spinning of the projectile if any.

The loss of material from the target, that is wear, was assessed by weighing it before and after impact. At low striking velocities of 175 m s^{-1} and less, the sphere creates a crater by displacing material which remains attached to the target in the form of a lip at one end of the depression (Fig. 5.3(a)). When the velocity is increased a lip still forms, as examination of the frames shows, but it is detached, the amount being recorded as the weight loss of the target. The angle of impact is important and up to about 30°, lips form and are removed at the experimental velocity of 270 m s^{-1}. A lip is still formed but is folded

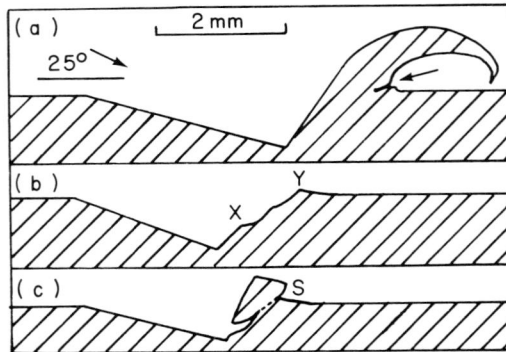

Fig. 5.3   Section through impact craters. Impact angle 25° by flat ended steel cylinders. Targets (a) mild steel at 194 m s^{-1}; (b) titanium at 175 m s^{-1}; (c) titanium at 120 m s^{-1}. The position of shear concentration is shown by the arrow in (a) and the shear band by the dotted line in (c). (Winter, R. E. and Hutchings, I. M. (1975). *Wear* **34**, 141.)

over the underlying metal at the same velocity but when the striking angle is 40°.

A comparative study [4] between mild steel and titanium targets was made by impacting these with 4 mm diameter × 5 mm long hardened steel rods striking at an angle of 25°. There were interesting differences between the two target materials as far as the mode of deformation was concerned. With mild steel, a lip of metal was extruded ahead of the hard steel rod but it curved in a manner similar to that shown in Fig. 5.3(a). The projectile then rotated and left the crater, leaving a region of intense shear at the base of the lip which remains attached to the body of the mild steel. A lip also forms with titanium but it leaves the parent body with an escape velocity much higher than the speed of impact of the hardened steel rod.

Angular impact appears to remove material by a process of micro-cutting and in both mild steel and titanium, a chip possesses bands which show deformation whose magnitude is much greater than the surrounding matrix which has been left behind. This localized deformation is much higher in titanium than in mild steel and the intense strain produced probably causes the titanium lips to escape as they form. The suggested mechanism [4] of detachment is that a combination of tensile stresses arising as a result of the inertia of a chip causes a tensile fracture within the deformed band. The forward motion of a chip will, of course, be as a result of induced shear forces. The possibility exists that this process of ductile deformation, which detaches the lip with an efficiency depending on the intensity of the strain induced, is influenced by the thermochemical properties of the material. The process of ductile deformation will be facilitated by a fall in the flow stress of the target material as a result of thermal softening due to collision, the deformation being rapid enough for conditions to be nearly adiabatic. Apart from a temperature dependent flow stress, softening due to impact will be favoured by a low heat capacity per unit volume of the target [5]. The heat capacity is defined as the product of the specific heat C and the density ρ of the material and it is assumed that the propensity to thermal softening is also a function of the melting point T_m of the target. The term used in fact, is $(T_m - T_a)$, where T_a is the ambient temperature at which collision occurs, and to a first approximation, the lower the value of $(T_m - T_a)$ the greater is the probability of thermal softening.

5.3.1 Orientation of impact

A particle and a target undergoing impact can be envisaged as a situation analogous to two-body abrasion where material is removed from a solid by a cutting action of the abrasive grit. Cutting and ploughing are also expected of striking projectiles depending on the angle of attack [6]. The orientations of a compound sand grain can be described in terms of rake angles in a similar manner to a cutting tool removing material by machining. Depending on the orientation of the impacting particle, there will be two distinct modes of deformation. The modes are ploughing and cutting, the former occurring at large negative rake angles, situations where the angle between the leading edge of an indenting particle and the target is small. At positive rake angles material is removed from the body of the target by cutting.

5.4 Brittle target

It is not an implausible proposition that a solid, initially capable of plastic deformation, will strain harden under repeated attacks and become a brittle material when subsequently undergoing collision. Instead of being micro-machined or ploughed by the projectile, the mode of failure of the target should then be similar to that experienced with glass when impacted by a harder but ductile metal such as steel. The effect of such a collision is to form a ring crack [7] as shown schematically in Fig. 5.4 which shows the failure pattern of a glass plate struck by a steel sphere at normal angle of incidence. The crack forms at right angles to the impact direction and around the circumference of the contact area, flaring out below the surface to form a truncated cone.

Fig. 5.4   A ring crack in glass produced by a steel sphere striking normally (schematic). (Finnie, I. (1960). *Wear* **3**, 87.)

Experiments with glass show that, as impact commences, the first effect is to form these ring cracks on locations where the plate encounters the blasting particles. As the attack continues, several ring cracks join up and material is removed by erosive wear. Ring cracks occur as a result of Hertzian stress induced by a non-plastic sphere striking, at a normal angle of incidence, a brittle material which remains elastic as it fractures. The maximum tensile stress acts radially at the surface around the periphery of the contact area. The magnitude of this maximum radial stress, σ_r, is

$$\sigma_r = 0 \cdot 187 (1 - 2\nu_2) \rho^{1/5} v^{1/5} \left(\frac{1 - \nu_1^2}{E_1} + \frac{1 - \nu_2^2}{E_2} \right)^{-4/5} \tag{5.11}$$

where ρ, v, ν_1 and E_1 are the density, velocity, Poisson's ratio and modulus of elasticity respectively of the impacting sphere and ν_2 and E_2 are the corresponding parameters for the flat brittle surface.

Provided impact is normal, the Hertzian equation relating the radial diameter, $2a$, of the crater to the striking velocity v_i is

$$2a = Av_i^{0.4} \qquad (5.12)$$

where A is a constant.

The difference between the contact diameter and that of the ring crack is small so that $2a$ is, in effect, the diameter of the latter. It has been suggested [7] that when the sphere is incident at an angle α, the diameter should be expressed by the following relationship:

$$2a = B(v_i \sin \alpha)^{0.4} \qquad (5.13)$$

where B is a constant. However, experiments do not confirm the above relationship.

5.5 Velocity

As a particle strikes a surface, a crater is produced if conditions are conducive and the projectile rebounds with an angular impulse $I\omega$ where I is its moment of inertia and ω, the angular velocity. A knowledge of the incident and rebound velocities allows the calculation of the loss of kinetic energy due to collision [3]. Thus if v_i and α_i are the velocity and the angle of impact respectively and v_R and α_R are the corresponding parameters at rebound, the impulse received by the ball of mass m, normal to the surface, is

$$m(v_i \sin \alpha_i + v_R \sin \alpha_R)$$

For oblique collision, the ball slides throughout contact and the tangential impulse is $\mu m(v_i \sin \alpha_i + v_R \sin \alpha_R)$, where μ is the coefficient of friction. For a ball of diameter $2r$, the angular impulse is $I\omega = (2/5)mr^2\omega$. Equating the angular impulse with the tangential impulse,

$$\mu = \frac{2r^2\omega}{5(v_i \sin \alpha_i + v_R \sin \alpha_R)} \qquad (5.14)$$

Rather than use a single ballistic, abrasive particles such as quartz are propelled on to a solid with the aid of a gas gun [8]. A variant is to use a whirling arm rig which holds the target material under study and rotates at a predetermined speed in an abrasive medium such as sand. Erosive wear is expressed as weight loss of the target per unit weight of abrasive particles used in both these experiments.

As shown in Equation 5.6, kinetic energy considerations suggest a proportionality of the volume of the crater produced to the square of the velocity of impact. It has been predicted [9] that if the quantity ρv_i^2, ρ being the density, is adjusted to a constant value for projectiles of the same size the crater volume for targets whose composition or properties are not varied will be the same. That is a quartz particle of the same size and a similar particle of steel will produce the same crater volume V_c in annealed mild steel provided the velocity for quartz is increased to a certain value. The dependence of the crater volume on the term ρv_i^2 has been verified by experiments [3].

Figure 5.5 shows a plot of log (wear rate) against the log of impact velocity. The wear rate is that of mild steel targets assessed by weight loss technique [3]. The velocities chosen, however, were only those which removed material completely; that is, which did not leave even a small amount of lip clinging to the crater formed by cutting or ploughing. Figure 5.5 produces a relationship of the form $q = (5 \cdot 82 \times 10^{-10}) v_i^{2 \cdot 9}$, an aspect which is discussed later (q is the wear rate and v_i is the impact velocity).

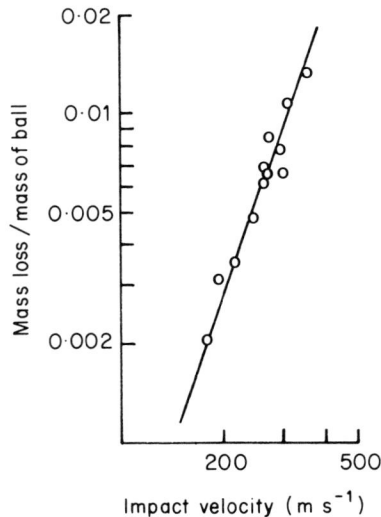

Fig. 5.5   A plot of log (wear rate) against log of impact velocity. Mild steel targets. (Hutchings, I. M., Winter, R. E. and Field, J. E. (1976). *Proc. Roy. Soc. A* **348**, 379.)

Experiments such as this suggest that the erosive wear rate can be generalized in the form

$$q = Kv^n \qquad (5.15)$$

where K and n are constants.

We have noted that the crater volume $V_c = \phi(v_i^2)$. Since Fig. 5.5 represents only those velocities which removed material from the body of the target to form a crater, the volume of material lost equals that of the crater. The wear rate in that case should be proportional to the square of the velocity.

A relationship of the form given by Equation 5.15 has been obtained in controlled experiments [8] in a vacuum of 10^{-3} Torr. The apparatus was the whirling arm rig and the abradant was diamond dust. Whereas kinetic energy considerations predict a value of $n = 2$ and the experiments with steel spheres [3] gave n as 2·9, the diamond dust particles produced a value of $n = 2·3$.

5.6 Fragmentation of projectiles

A probable reason for a high value of the exponent n is found in the fragmentation of brittle projectiles such as sand upon impact. An explanation for steel spheres which do not shatter due to collision is given at the end of this section.

The nature of particle fragmentation has been observed experimentally [8] by photographic techniques when glass spheres were made to strike metal targets. It was observed that the glass particles shattered upon impact and flowed across the surface of the target in a radial wash at a speed of about 1·7 times the striking velocity. A circular indentation was produced at normal impact, the diameter of which was $2D/3$ where D was the diameter of a glass sphere. Disintegration of the particles was completed in 20 μs and the radial wash lasted about 70 μs. It was observed that radial scarring to a diameter of $3D$ occurred around the indentation formed by the glass sphere when it first struck. Firing the target with irregularly shaped quartz particles showed the radial scarring to be unsymmetrical. The degree of fragmentation was found to depend on the prior particle size and the speed of impact. The average velocity of rebound of the fragments was about a third of the striking velocity. Experiments [10] with grits of varying prior particle size have established that the average diameter of a chip formed by collision with the target material is several times smaller than the striking particles but becomes proportionately larger with increase in the prior size of the grits. An important observation was that there was a threshold size of the eroding particles below which they did not fragment.

Not all the kinetic energy of impact is therefore expended in erosion but some is consumed in disintegrating the particles into smaller pieces. Since the crater volume and hence wear is dependent for a given target material upon the kinetic energy, one would expect a low value of n (Equation 5.15). As experiments show, this is not the case and the reason why $n > 2$ is that, for the same amount of kinetic energy, erosion occurs in two sequential stages. The first stage is direct cutting or ploughing when the angle is not normal. Upon impact the particles disperse as they fragment in a radial wash at a higher speed than the initial velocity of impact. That is, the numerous particles moving out radially from the original crater simply plough material from the target and cause further wear. Because of these two stages of wear and since the velocity of attack of the radial wash is high, for the same mass of grits the total amount of erosion should be higher giving a value of n greater than 2.

It is important to note, however, that although the total mass of the particles constituting the radial wash is the same as the original particles making the first strike, n depends on the prior particle size [11]. Experiments show that, with quartz particles, $n = 2$ for a particle diameter of 25 μm but increases to 2·3 when the impacting grits have an average particle size of 125 μm. This is tantamount to saying that the kinetic energy laws are successfully applied for situations where the grit size is kept at an optimum value and complications set in when this is exceeded. The role of particle size in erosion has been examined by a few authors but we should note two points. One is that from a mass of particulate matter of varying sizes, it is not easy to assign a particular size for the average particle diameter. The second point to note is that for a group of particles with a certain size there will be a characteristic degree of fragmentation. It can be argued that the kinetic energy wasted in fragmentation will tend to reduce the value of n whereas the radial wash will increase it. So that the algebraic effect, for a particular degree of fragmentation, will be for n to assume a value of 2. All this, of course, is speculation but there is a threshold diameter D_0 of quartz particles below which a steel target will not undergo erosive wear (Fig. 5.6) unless the collision experiment is unduly prolonged. The rate of erosion then increases with the particle diameter up to a limiting size and becomes

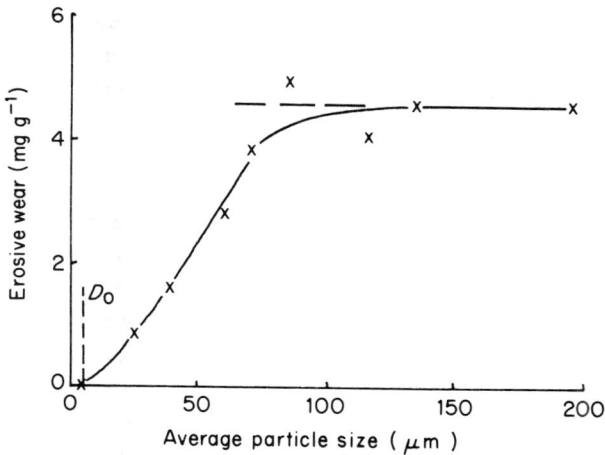

Fig. 5.6   Influence of particle size on erosion rates of steel being struck by quartz. (Tilly, G. P. (1973). *Wear* **23**, 87.)

asymptotic thereafter. This maximum diameter for quartz particles striking an 11% chromium steel is about 100 μm. It has been experienced [10] that particles below a diameter D_0 do not fragment and prolonged attack is necessary to cause a small amount of wear. Large particles fragment so that the two stages of erosion prevail resulting in a steady increase in wear with particle size. The reason why the curve becomes asymptotic beyond a particular size of the particles is not clear.

According to the foregoing discussion, there is no cutting or ploughing when a projectile hits a target normally and wear probably occurs by a fatigue process due to repeated contacts. It has been suggested, however, that once erosion sets in, the crater is roughened and subsequent collisions will, in effect, become angular impingement. That is, even normally directed projectiles will strike a target at glancing angles most of the time.

The idea of the disintegration of particles giving a high value of the velocity exponent would be a neat explanation were there not the frustrating anomaly that steel balls, which do not break up, give $n > 2$. Whether a high value is the result of sliding and rotation of the ball after impact has not been investigated. It has been suggested, however, that, even with steel balls, the craters become rough and further collision occurs at glancing angles, resulting in a high rate of wear. That is, a form of secondary erosion takes place again during rebound even without fragmentation. The idea is probably not very convincing with steel balls but the theories based on radial wash for brittle particles are prevalent at the moment and are discussed fully in Section 5.8.

5.7   Indentation theory

The explanations offered for the value of the velocity exponent obtained from kinetic energy consideration being different from experimental facts are as a result of speculation based upon observations on the phenomenology at the projectile–target interface after intimacy. A relationship between the wear rate and the striking velocity has been provided by Sheldon and Kanhere [12] from an analysis of the interaction based on indentation hardness theory with the main proviso that a projectile strikes a target normally. The authors [12] state that when the steel balls are of similar size to the ones used for, say, Brinell hardness measurements, the effects of strain hardening, inertia of the displaced material and the deformation of the projectile can be neglected provided the velocity of impact does not exceed 500 m s^{-1}. It is postulated that, as a projectile strikes the surface, a lip of the target material forms immediately but becomes detached from its anchored position after an appropriate number of collisions with the striking particle. It is further assumed that the amount of material removed from the target is related to the depth of penetration reached by the projectile during the normal impact.

The Meyer hardness test is quoted for this analysis which gives the following empirical formula:

$$W = W_0 \, d^p \tag{5.16}$$

where $W =$ applied normal load, $d =$ diameter produced by the penetration of the ball, $W_0 =$ the normal load at which $d = 1$, and $p =$ logarithmic index. Consider a spherical particle of diameter D and mass density ρ making a normal impact on a surface at a striking velocity v_i (Fig. 5.2(a). Note that in Fig. 5.2(a), the diameter of indentation is $2a$ but we shall denote it here as d as the authors [12] did.

The associated kinetic energy of the particle is

$$\frac{1}{2}\left[\frac{4}{3}\,\pi\left(\frac{D}{2}\right)^3\right]\rho v_i^2$$

where ρ is the density of the striking particle. Let the direction of indentation normal to the surface be z and let the penetration cease at a depth z from the surface of the plate. Then, the work done during the indentation process is

$$\int_0^z W \, \mathrm{d}z$$

Substituting for W in the above term from Equation 5.16, but changing d to δ, the instantaneous diameter of indentation at any distance z,

$$\text{work done} = \int_0^z W_0 \delta^p \, \mathrm{d}z \qquad (5.17)$$

From geometry (Fig. 5.2(a)),

$$2z = D - (D^2 - \delta^2)^{1/2}$$

Therefore, differentiating,

$$2 \, \mathrm{d}z = (D^2 - \delta^2)^{-1/2} \delta \, \mathrm{d}\delta$$

Substituting for $\mathrm{d}z$ in Equation 5.17, for the final indentation d,

$$\text{work done} = \tfrac{1}{2} \int_0^d \frac{W_0 \delta^{(p+1)}}{(D^2 - \delta^2)^{1/2}} \, \mathrm{d}\delta$$

Equating the above expression with the kinetic energy of the sphere,

$$\frac{\pi}{12} D^3 \rho v_i^2 = \tfrac{1}{2} \int_0^d \frac{W_0 \delta^{(p+1)}}{(D^2 - \delta^2)^{1/2}} \, \mathrm{d}\delta \qquad (5.18)$$

To solve the integral, put $\delta = D \sin \theta$ and $d = D \sin \theta_1$. Therefore,

$$\mathrm{d}\delta = D \cos \theta \, \mathrm{d}\theta$$

and

$$(D^2 - \delta^2)^{1/2} = D \cos \theta$$

Therefore, from Equation 5.18,

$$\frac{\pi}{12} D^3 \rho v_i^2 = \tfrac{1}{2} \int_0^{\theta_1} W_0 D^{(p+1)} \sin^{(p+1)} \theta \, \mathrm{d}\theta \qquad (5.19)$$

To simplify the mathematics, take $p = 2$ as is normally found from hardness tests. Also, Meyers hardness HM is given by

$$\text{HM} = W/(\pi d^2/4)$$

That is $W/d^2 = (\pi/4)\text{HM}$, and at $p = 2$, $W = W_0 d^2$, therefore

$$W_0 = \frac{\pi}{4}\text{HM} = \frac{\pi}{4}\text{HV}$$

where HV is the Vickers pyramid number. Also since, $2z = D - (D^2 - \delta^2)^{1/2}$, at the maximum indentation depth z,

$$\frac{2z}{D} = 1 - \cos\theta_1 \qquad (5.20)$$

Rearranging Equation 5.19 and substituting for $W_0 = (\pi/4)\text{HV}$

$$\frac{\pi}{12}\frac{\rho v_i^2}{D^{(p-2)}} = \frac{1}{2}\frac{\pi}{4}\text{HV}\int_0^{\theta_1}\sin^{(p+1)}\theta\,d\theta$$

or putting $p = 2$ in the left-hand side of the above relationship,

$$\frac{2}{3}\frac{\rho v_i^2}{\text{HV}} = \int_0^{\theta_1}\sin^{(p+1)}\theta\,d\theta$$

or

$$\left(\frac{\rho}{\text{HV}}\right)^{1/2}v_i = 1\cdot23\left[\int_0^{\theta_1}\sin^{(p+1)}\theta\,d\theta\right]^{1/2}$$

Solving the above integral,

$$v_i\left(\frac{\rho}{\text{HV}}\right)^{1/2} = 1\cdot23[\tfrac{1}{3}(2 - 3\cos\theta_1 + \cos^3\theta_1)]^{1/2}$$

Putting $\cos\theta_1 = 1 - (2z/D)$ from Equation 5.20, and neglecting z^3/D^3,

$$v_i\left(\frac{\rho}{\text{HV}}\right)^{1/2} = K_1\frac{z}{D}$$

where $K_1 = \text{constant} = 2\cdot46$. We can state that

$$z = K_2 v_i \quad\text{or}\quad z^3 = K v_i^3 \qquad (5.21)$$

where $K_2 = (1/K_1)D(\rho/\text{HV})^{1/2} = \text{constant}$ for a ball of fixed diameter and for a target whose hardness is assumed to remain constant during the interaction, and $K = K_2^3 = \text{constant}$. It is expected that q, the mass of material removed from the target per unit mass of abrasives, should be proportional to z^3. Comparing this with Equation 5.21,

$$q \propto v_i^3 \qquad (5.22)$$

This agrees fairly with experimental results, particularly when a single hard projectile is used. The velocity exponent as obtained from experiments is always lower than 3, and this is so because according to Sheldon and Kanhere [12], not all the crater volume generated by impact is removed as wear product.

5.8   Primary and secondary erosion

The indentation theory, as described in the previous section, would appear to be confirmed by experiments for steel spheres but results from non-metallic brittle projectiles show a range of values for the velocity exponent. Disintegration of blasting grits falling on hard obstructions is a fact, so that a theory should take cognizance of this aspect. Since in practice there is a size effect in terms of rates of erosion, the particle diameter should also be considered. An attempt has been made to incorporate both these aspects in a theory of erosive wear and a study of the subject would be incomplete without an appreciation of this. The premise is that, as a particle such as quartz attacks a surface, a crater is created and a wear particle is formed if the extruded lip as a result of impingement is severed from the body of the target. This is called *primary erosion*. Upon impact, a particle disintegrates in accordance with its prior size and the kinetic energy received and further erosion occurs due to the radial wash which follows at high velocity. Material removal by the radial wash of the fragmented particles is termed *secondary erosion*. It is appreciated that a normal impact should result in the radial wash being over the total target surface, giving rise to a situation where the fragmented particles will slide over the metal. This is a potential situation for maximum wear. Wear should be proportionately low when the sliding distance by the fragments is small at oblique impact.

5.8.1   Primary erosion

The energy balance proposed by Bitter [13] is that the initial kinetic energy is expended in producing erosion and in causing elastic deformation of the target, so that

$$G_p = (G^{1/2} - G_e^{1/2})^2 \qquad (5.23)$$

where G_p = energy used up to produce erosion, G_e = energy required for elastic deformation, and G = initial kinetic energy.

If v_i is the impact velocity and m, the mass of the particle,

$$G_p = \tfrac{1}{2}m(v_i - v_0)^2 \qquad (5.24)$$

where v_0 is the threshold velocity below which the encounter is entirely elastic and the colliding couple remains undamaged. Tilly's experiments (Fig. 5.6) show that there is a critical diameter D_0 of the abradant below which there is no wear according to the experimental technique used [10]. This shows that the impact stresses generated by small particles of quartz are not high enough for plastic yielding and wear. That is, there are two modes of interaction, viz., elastic and plastic. In the former case, a crater volume is still expected but this should disappear leaving an undented surface, once the impressive force discontinues.

Comparing Equations 5.23 and 5.24, $G_e = \tfrac{1}{2}mv_0^2$ and assuming a spherical particle of diameter D_0 and a mass density ρ,

$$G_e = \frac{\pi}{12}\rho D_0^3 v_0^2 \qquad (5.25)$$

Consider now the general case when a particle of diameter D and a mass density ρ is striking a metal surface at an impact velocity v_i. Examining Equation 5.25, Equation 5.23 can be used to express the amount of energy used to produce erosion as

$$G_p = \frac{\pi\rho}{12}(D^{3/2}v_i - D_0^{3/2}v_0)^2 \qquad (5.26)$$

Now, if G_0 = energy required to remove unit mass of material by primary erosion, and q_p = primary erosive wear rate expressed as target material removed per unit mass of grit, then,

$$G_p = G_0 q_p m_p$$

where m_p = total mass of the abrading particles. This can be written as

$$G_p = G_0 q_p(\pi\rho D^3/12)$$

Substituting this value of G_p in Equation 5.26,

$$G_0 q_p = [v_i - (D_0/D)^{3/2}v_0]^2$$

Dividing both sides of the above expression by v_i^2,

$$q_p = \frac{v_i^2}{G_0}[1 - (D_0/D)^{3/2}(v_0/v_i)]^2 \qquad (5.27)$$

It would be useful if the energy term in Equation 5.26 could be eliminated and the erosive wear rate expressed in terms of measurable parameters such as velocity of impingement and grit diameter. To do this, consider the case where both the striking velocity and the diameter of the indenting particle are large and experimental conditions are adjusted to give a maximum rate of erosive wear, q_m. In that case, the terms D_0 and v_0 in Equation 5.27 can be neglected so that it can be expressed in a modified form as follows:

$$G_0 = v_i^2/q_p$$

Suppose for the experimental conditions chosen, there is a striking velocity v_m which gives a maximum amount of wear, q_m. We can then express G_0 as $G_0 = v_m^2/q_m$ and substituting this value in Equation 5.27,

$$q_p = q_m \left(\frac{v_i}{v_m}\right)^2 [1 - (D_0/D)^{3/2}(v_0/v_i)]^2 \qquad (5.28)$$

One problem with Equation 5.28 is the statement of a maximum velocity v_m which gives a maximum amount of wear q_m. Kinetic energy considerations show that the higher the velocity of impingement, the greater is the propensity to wear. Undoubtedly, experiments show that the erosive wear rate q can be expressed in terms of the impacting velocity v_i and the impingement angle α as

$$q = \phi(v^n)\phi_1(\alpha)$$

n has the usual value of about 2·4 and the role of α is fully described in Section 5.9. It is sufficient to mention here that q is zero at an attack angle of zero and rises to a maximum at $\alpha \simeq 30°$. The wear rate then falls progressively with increasing α until at normal impact it is about 30% of the maximum wear rate obtained.

Equation 5.28 is probably of practical value if experiments establish the values of v_0 and D_0 for a specific type of eroding particle or target material. A particular experiment should then be conducted to obtain q_m at the velocity v_m. Provided the target material was the same, the wear rate q_p could be computed for a particular velocity v_i or for any particle diameter D from Equation 5.28, for example, for design information.

5.8.2   Secondary erosion

The force causing secondary erosion is assumed to be directly proportional to the prior kinetic energy and the degree of fragmentation of the grits upon collision. Observations on indented targets under the microscope suggest that the shattered particles either rebound from or give rise to radial indentation of the primary crater. The conditions of a threshold velocity or a particle diameter below which there is no wear are not relevant to the analysis of the secondary erosion process.

Weight loss q_s of the target material per unit mass of grit due to secondary erosion at a velocity of impact v can be expressed as,

$$q_s = (v^2/\psi)(f) \tag{5.29}$$

where ψ is the secondary erosion factor and f is the degree of fragmentation which is known to depend on impact velocity, particle size and the incident angle. The degree of fragmentation can be quantified as

$$f = \frac{W_0 - W}{W_0} \tag{5.30}$$

where W_0 = the proportion of the indenting particles by weight within a specified size range before impact, and W = the proportion after impact. If the particles are completely fragmented, $W = 0$ and, therefore, $f = 1$ and a maximum amount of secondary erosion q_{sm} will be expected, other things being equal. Thus from Equation 5.29,

$$\psi = \frac{v^2}{q_{sm}} \tag{5.31}$$

If for any striking velocity v_i with an initial particle diameter D, f_i is the degree of fragmentation, from Equation 5.29, the erosive wear rate q_{si} is

$$q_{si} = \frac{v_i^2}{\psi} f_i \tag{5.32}$$

Substituting for ψ from Equation 5.31 in Equation 5.32,

$$q_{si} = \left(\frac{v_i}{v}\right)^2 q_{sm} f_i \tag{5.33}$$

If the total erosion rate per unit mass of grit is q and is as a result of both q_p and q_{si}, the primary and the secondary components respectively,

$$q = q_p + q_{si}$$

Taking the expression for q_p from Equation 5.28 and that for q_{si} from

Equation 5.33

$$q = q_m\left(\frac{v_i}{v}\right)^2 [1 - (D_0/D)^{3/2}(v_0/v_i)]^2 + \left(\frac{v_i}{v}\right)^2 q_{sm}f_i \qquad (5.34)$$

Note that the term v_m in Equation 5.28 has been replaced by an experimental velocity v, and q_m is the wear at that velocity. We also note that Equation 5.34 does not include the impact angle as such but f_i, the degree of fragmentation will be maximum at normal impact.

5.9 Impact angle
Neilson and Gilchrist [14] suggest that when a projectile attacks a ductile material, the normal force causes failure of the target by spalling of surface layers due to work hardening and this mode has been termed the *deformation wear*. Angular impact, on the other hand, removes a chip from the target material by a cutting or by a ploughing action. The authors [14] ignore the ploughing action and consider the wear phenomenon comprising one or both of the following modes, viz.,

(1) deformation wear, the mode expected due to normal impact;
(2) cutting wear when impingement is at an angle.

A partial analysis of the erosive process is then given as follows.

Let q_t be the amount of material removed from a target by the total mass m of an impacting particle impinging the surface at an angle α, the striking velocity being v (Fig. 5.7). Assume that the mode is cutting

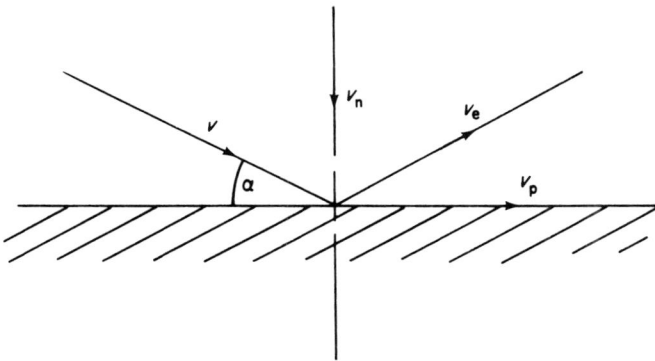

Fig. 5.7 Impingement angle of a particle showing escape velocity v_e; v_p and v_n are residual and normal velocity respectively. (Neilson, J. H. and Gilchrist, A. (1968). *Wear* **11**, 111.)

wear by the non-normal component of the projectile, and at small angles of incidence, the particle leaves, simply grazing the surface at a residual velocity v_p. Another probable speed of escape is v_e (Fig. 5.7) and v_n is the velocity component normal to the surface.

The modes of wear can then be quantified as follows. If G_a is the part of the total kinetic energy expended during a non-normal impact to produce a unit mass of wear debris,

$$\text{Cutting wear} = \frac{\frac{1}{2}m(v^2 \cos^2 \alpha - v_p^2)}{G_a}$$

If G_n is the part of the total kinetic energy expended to produce a unit mass of wear debris by the normal component,

$$\text{Deformation wear} = \frac{\frac{1}{2}m(v \sin \alpha - v_n)^2}{G_n}$$

If α_0 is the angle of attack at which $v_p = 0$,

$$q_t = \frac{\frac{1}{2}m(v^2 \cos^2 \alpha - v_p^2)}{G_a} + \frac{\frac{1}{2}m(v \sin \alpha - v_n)^2}{G_n}, \qquad \alpha < \alpha_0 \quad (5.35)$$

and

$$q_t = \frac{\frac{1}{2}mv^2 \cos^2 \alpha}{G_a} + \frac{\frac{1}{2}m(v \sin \alpha - v_n)^2}{G_n}, \qquad \alpha > \alpha_0 \quad (5.36)$$

At an impacting angle $\alpha = 0$, the impacting velocity $v = v_p$, the component parallel to the surface of the target (Fig. 5.7). Putting an arbitrary symbol $\gamma = (v_p/v \cos \alpha)^2$, the component v_p increasing with decreasing α, at $\alpha = 0$, $\gamma = 1$ and at $\alpha = \alpha_0$, $\gamma = 0$.

Experiments with ductile metal targets show that as α is increased from zero, the amount of erosion increases at a rapid rate but decreases again when α is large. The role of the impinging angle in the erosive wear rate expressed as mass of target removed per unit mass of particles is shown in Fig. 5.8. The target is aluminium, being attacked by hard particles of aluminium oxide. The erosive wear rate is seen to reach a maximum at an angle of about 25°, the value being about an order of magnitude higher than that at $\alpha \approx 5°$. At normal impact, the wear rate is considerably lower than the maximum value.

Equations 5.35 and 5.36 can be expressed in a more generalized manner by assuming,

$$\gamma = 1 - \sin(n\alpha) \quad (5.37)$$

where n is a constant.

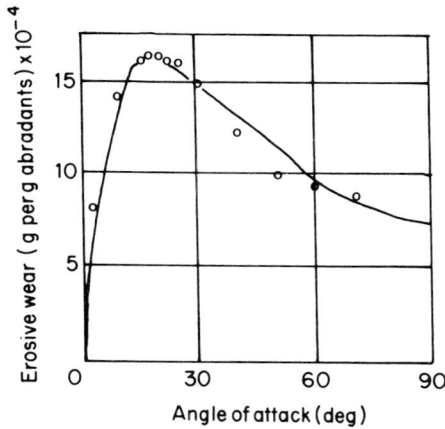

Fig. 5.8   Erosion as a function of the angle of attack. Aluminium targets attacked by 210 μm aluminium oxide particles at 424 ft s^{-1}. (Neilson, J. H. and Gilchrist, A. (1968). *Wear* **11**, 111.)

Comparing Equation 5.37 with Equations 5.35 and 5.36 respectively,

$$q_t = \frac{\frac{1}{2}mv^2 \cos^2 \alpha \sin(n\alpha)}{G_a} + \frac{\frac{1}{2}m(v \sin \alpha - v_n)^2}{G_n}, \qquad \alpha < \alpha_0 \quad (5.38)$$

and

$$q_t = \frac{\frac{1}{2}mv^2 \cos^2 \alpha}{G_a} + \frac{\frac{1}{2}m(v \sin \alpha - v_n)^2}{G_n}, \qquad \alpha > \alpha_0 \quad (5.39)$$

When $\alpha = \alpha_0$, $\sin n\alpha_0 = 1$ or $n\alpha_0 = \pi/2$, i.e. $\alpha_0 = \pi/2n$. Equations 5.38 and 5.39 can be applied to any situation and the erosive wear rates calculated provided G_a and G_n are known. These, unfortunately are not easily obtainable parameters and hence the solution by the authors [14] remains only partial as far as applications of these two equations are concerned. Equations 5.38 and 5.39 can, however, be simplified if v_n can be neglected since this is usually very small at oblique impact. For brittle materials, the cutting component is redundant in these two equations but is the dominating mode when a metal such as aluminium undergoes impact at oblique angles.

5.10   Material properties

It is clear that the rate of erosive wear depends largely on the velocity of attack, the angle of impingement and the prior diameter of the particles.

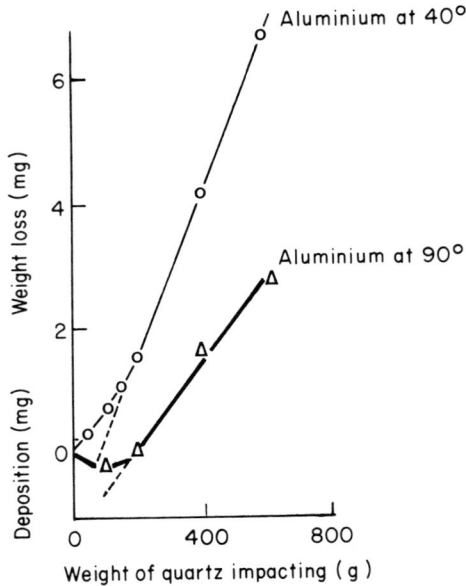

Fig. 5.9   Variation of weight loss of an aluminium alloy target with the quantity of quartz as the abrasive particles. (Tilly, G. P. (1969). *Wear* **14**, 63.)

Although there is now a body of knowledge on the mechanism of erosive wear, there is no universal agreement among investigators regarding the role of certain variables.

Experiments [15] with quartz particles show that, at normal impact, the projectiles may embed in the soft surface giving an incubation period such that there is an actual gain in weight of the target (Fig. 5.9). As collision continues, however, the embedded particles are freed and the amount of erosion increases with the flux of projectiles blasting the surface. Figure 5.9 shows clearly the presence of an incubation period for aluminium at normal impact. An incubation period showing a weight gain is absent when the impact angle is 40° in this experiment.

The combined effect of impact angle and ductility of the target material has also been reported (Fig. 5.10) [15]. A ductile aluminium alloy shows a typical rise to a maximum erosion rate at an impacting angle of about 15° and then falling to a low value at normal impact. A brittle material such as glass continues to wear with an increase in the angle of impingement of the projectiles and erosion is the highest at

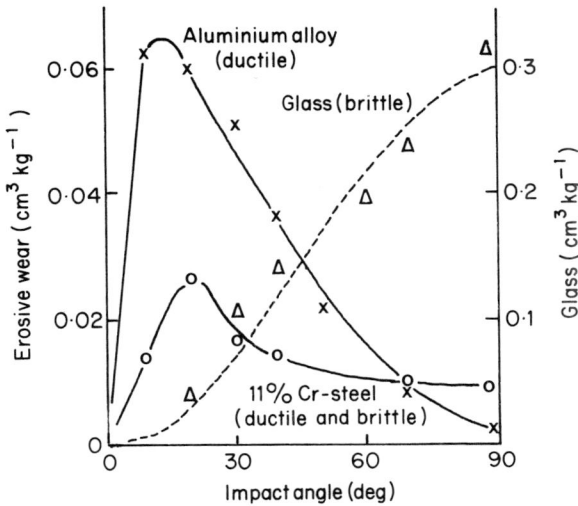

Fig. 5.10   Erosion of ductile and brittle materials. (Tilly, G. P. (1969). *Wear* **14**, 63.)

normal impact, unlike that for a ductile metal. An 11% chromium steel with relatively low ductility gives the same general pattern as the aluminium alloy, but the overall rate of erosion is low. However, the amount of wear at normal impact is greater than that of the aluminium alloy. Experiments with polymers show a similar pattern to metals when erosive wear is plotted against impact angle. Crystalline polymers such as polypropylene and nylons simulate the behaviour of ductile metals with a peak wear at an angle of $\alpha = 30°$ (Fig. 5.11). A fibreglass shows the characteristics of a brittle material, and being soft, the crystalline polymers show that the grits embed in them as the impact tends towards $\alpha = 90°$.

The evidence on the hardness of the target *vis-à-vis* erosive wear is by no means conclusive. Various metals such as aluminium and copper have been blasted with silicon carbide grits, the metals being in both annealed and cold worked conditions [16]. A fall in the wear rates is noted with increasing hardness of the target material. Since an initially annealed metal will work harden under repeated impact, hard targets produced by cold work should not show any difference in the rate of wear from annealed targets. This is confirmed by experiments but targets of steels, non-ferrous metals and non-metallic materials subjected to attack by grits show no obvious relationship between hardness

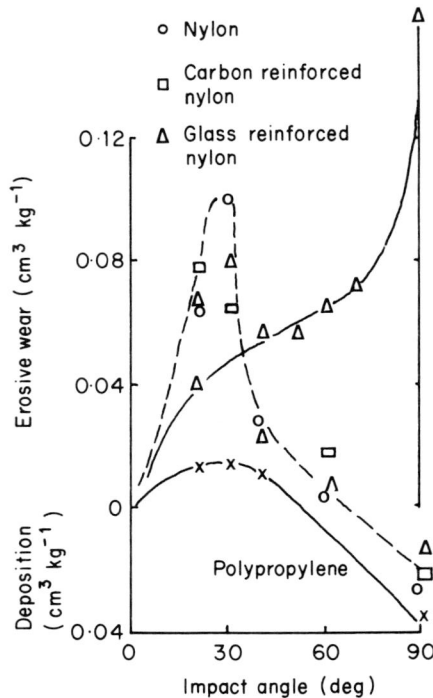

Fig. 5.11 Brittle and ductile type of erosion of polymeric materials. (Tilly, G. P. (1969). *Wear* **14**, 63.)

and erosion rate. Similar conclusions are arrived at by propelling quartz particles normally on soft and hard steels [17].

The nature of the impacting particles has a decided effect on the erosive wear rates of targets. Dusts become increasingly aggressive as their quartz content increases. This is probably the reason why in sandy terrains erosion of compressor blades is severe enough to reduce their life to one tenth of that in a non-dusty environment [15]. Erosion rate is known to increase with the size, hardness and the sharpness of the indenting particles. Soft organic materials such as walnut shells and rice husks are used successfully to erode scales and rust from steel surfaces, while the underlying metal is left undamaged.

Erosive situations are often encountered at a high temperature, an example being a helicopter engine whose compressor blades may reach a temperature of about 300 °C. If combustion is not perfect, carbon particles form in the turbine section giving rise to a potential situation of erosive wear at temperatures of about 900 °C. Even higher temperatures are experienced in more sophisticated engines.

It has been reported [18] that the effect of ambient temperature on erosion rates of components is small. On the other hand, experiments with such materials as copper and aluminium alloys, mild steel and a plastic show a reduction in the rate of wear with a rise in temperature and this is possibly tied up with the increased ductility of the material [15]. That is, unlike the case where a ductile metal is blasted at room temperature, the metal probably maintains its ductility all the time at elevated temperatures by a continuous annealing effect. There is scope for further research on the erosion of materials at high temperatures.

Our understanding to date suggests that the critical parameter is the angle of attack. This is not easy to manipulate in practice. The most successful treatment for metals from the point of view of a high resistance to erosion in room temperature is known to be an application of a protective coating of rubber. Of course, unlike some other forms of wear, erosive wear is encouraged in many industrial applications. The attractive method of removing rusts from steel by using rice husks has already been mentioned. The phenomenon of erosion is also utilized in cleaning of parts by shot peening and sand blasting or in shaping of components as in erosive drilling of hard materials.

5.11 Liquid particle erosion

A paper [19] in 1932 describes in some detail the damage observed on the exhaust end blades of steam turbines as a result of impact forces from the water of condensation. The loss of metal was localized, the severity of which increased as turbines were designed to run at high speeds to meet higher duty. The author [19] observed that the steel used had a yield stress of about 165 kg mm^{-2}. The estimated pressure at the impinging liquid–solid interface was 65 kg mm^{-2}. Since the stress was elastic, it was suggested that yielding occurred at the summits of asperities and surface damage and wear were presumed to begin on a microscale.

5.11.1 Phenomenology

On a macroscopic basis, the assault of a surface by a liquid drop is regarded [19] to be analogous to an elastic collision between two solids. Consider a rod whose end strikes a rigid body at a certain velocity (Fig. 5.12). Assuming normal strike, the pressure σ at the interface rises

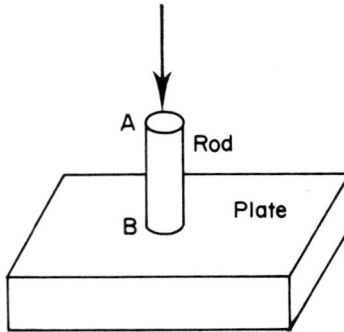

Fig. 5.12   A rod striking a flat plate. The length AB = *l*.

instantaneously to,

$$\sigma = v(\rho E/g)^{1/2} \tag{5.40}$$

where v is the striking velocity, g is the acceleration due to gravity, E is the modulus of elasticity of the rod and ρ its density. It will be shown later that, strictly, the pressure does not rise instantaneously but there is a time interval, albeit very short, between the moment when collision just occurs and the time when the peak pressure develops. The effect of collision is the generation of a compressive wave at B (Fig. 5.12), travelling upwards. For a rod of length l, the interval Δt for the compression wave to reach A, assuming that the wave front remains plane throughout, is given by

$$\Delta t = l(\rho/Eg)^{1/2} \tag{5.41}$$

Once the wave reaches A, it is reflected back to B as a tension wave. As this reaches the interface, the rod rebounds from the plate and the pressure at the interface is released instantly. The time taken for this negative pulse is assumed to be Δt so that the total time in which the rod strikes the plate, maintains an amount of pressure and then detaches from the surface is $2\Delta t$. For a liquid drop striking a rigid surface, similar expressions are used to arrive at the interfacial pressure but the particle now is a compressible drop of liquid.

Mapping of the liquid drop as it strikes an obstacle and then spreads has provided much needed information. Two papers published more than a century ago [20, 21] describe the kinetic pattern of a liquid drop after it strikes a flat plate. Experiments with drops of mercury and milk simply confirm the intuitive knowledge that the head of the drop after striking a flat plate disappears progressively in the form of an expanding liquid with radial arms, the tendency increasing with the size of the liquid particles and the height of fall. The obvious technique for a study of the sequence of flow is to obtain stills from high speed cinephotography of liquid drops at impact. This has been done to confirm that a drop spreads out laterally as its head subsides.

The process of liquid impact produces acoustic waves like a solid striking a solid, and at the first strike the wave travelling to the head of the liquid drop is compressional in nature. It then rebounds as a tension wave followed again by a compression wave and this continues until the head of the liquid drop disappears completely. Typically, the load exerted by a jet of water may rise to a peak within a microsecond of

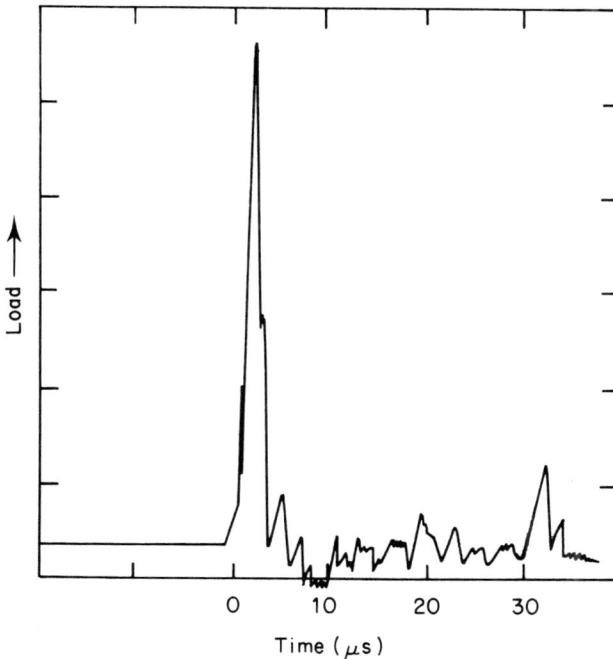

Fig. 5.13   Variation of load with times as exerted by a water jet impacting a steel target at a velocity of 720 m s^{-1}. The peak load for a 1·3 mm diameter jet was 630 kg. (Bowden, F. P. and Brunton, J. H. (1961). *Proc. Roy. Soc. A* **263**, 433.)

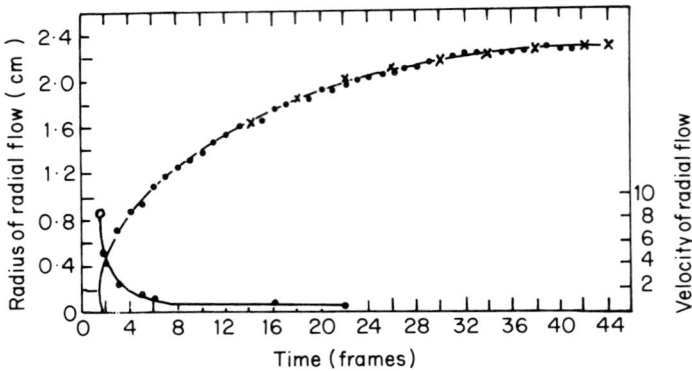

Fig. 5.14   Variation of the radius and of velocity of flow with time (multiply
the ordinate by 820 to convert velocity into cm s^{-1}). (Engel, O. G.
(1955). *J. Res. Nat. Bur. Stand.* **54**(5), 281.)

strike (Fig. 5.13) and fall rapidly to a low value in two or three
microseconds.

Experiments with liquid drops on glass at a striking velocity of
8·2 m s^{-1} show that the radial flow does not commence until about
152 μs after the collision (Fig. 5.14). The radial wash then continues to
spread laterally with time. The radial velocity is high initially as it
starts by being about 9 times the velocity of impact. This rapid spread,
however, slows down and equals the velocity of strike in less than 1 ms
when there is still some head in the liquid drop. The head disappears
completely in 1·2 ms, and this is about the time when the force of
impact decays to zero. Clearly, the speed of the radial flow is derived
from the energy of collision.

The first effect of liquid impact on solids is to promote surface
roughening. These rough surfaces have been regarded as centres which
grow to produce erosive wear debris. Roughening of surfaces occurs not
as a result of direct impact forces but due to cavitation. Cavitation is the
terminology used when gas bubbles generate in a liquid. The bubbles
form from gas nuclei already present in a liquid which grow when
$p_0 < p_c$ where, at the operating temperature, p_0 = pressure on the liquid,
and p_c = vapour pressure of the liquid.

It is evident that within the total volume of a drop, the fluid across the
liquid–vapour interface will vaporize to facilitate the growth of the
nuclei into bubbles. Only certain parts of a liquid drop will have

conditions conducive to nucleation and growth of bubbles so that ambient pressure will prevail elsewhere. Since the drop of liquid is in a dynamic state some of the bubbles will move into the high pressure zone and collapse. It is known that this gives rise to very high localized pressure, being, for example, greater than the yield stress of steels. This should cause surface roughening.

Cavitation conditions in liquid particle impact can arise in at least two ways. The first explanation is connected with the fast flowing liquid in the radial direction. As the head of the liquid disappears, the liquid still continues to flow outwards, which should result in a lowering of pressure at the centre of the liquid, which is now shaped like a disc. This will produce bubbles if the pressure is lower than the vapour pressure of the liquid. Incidentally, the pressure in the centre falls sufficiently to result in a break in the disc so that eventually a ring of liquid forms.

The second explanation is obtained from the presence of alternate compression and tension waves in the column of the liquid throughout its life after collision. The first negative pressure is large, and it is argued [22] that since the compression waves get smaller as the head subsides, the negative pressure remains larger than the compression wave when the head vanishes. A negative pressure will encourage bubbles. The higher the impact force, the greater will be the final negative pressure. This means that at flight velocities of aircraft of the order of 270 m s^{-1}, a high negative pressure in the impinging liquid drop should give rise to cavitation.

The characteristics of the radial wash depend as much on the impact velocity as on the bulk property of the solid and the topographical features of the recipient surface. Collision of a liquid drop with an elastic surface produces a rebound spray of fluid particles in the general direction reverse to that of impact. Natural rubber, being much more resilient than certain synthetic elastomers, offers an interface which promotes near vertical sprays when the incident angle is 90°. The elastic property of the rubber plays a part in deciding the nature of rebound but rough surfaces also produce a similar effect. For a liquid drop to impinge with a minimum amount of upward jetting and a laminar radial wash, a smooth surface is quite important.

5.11.2 Radial flow
The water that flows out radially exerts a shear stress at the solid–liquid interface and the stress patterns have been described [23] with the aid of

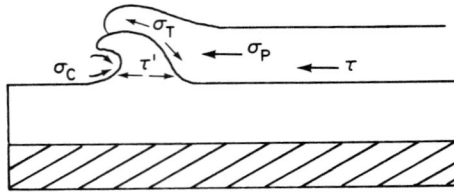

Fig. 5.15   Stress distribution in the surface of rubber supported on a metal substrate due to radial wash of an impinging water droplet. (Engel, O. G. (1958). *J. Res. Nat. Bur. Stand.* **61**(1), 47.)

a model comprising a liquid drop impinging upon an elastic rubber coating mounted on a hard metal substrate. The effect of impingement will be to indent the rubber, but suppose in the first instance we ignore this and only consider the situation where a surface asperity opposes the radial wash (Fig. 5.15). A stress σ_P at right angles to the vertical axis of the surface presses against the protrusion which is bent radially as a result. This gives rise to a shear stress τ' at the base which is different from τ acting in the solid surface because of liquid flow. The elastic bending produces a tensile stress σ_T and a compressive stress σ_C at the protrusion–air interface as shown. The results of all these forces alone will be failure of the rubber coating by fracture of the protrusion or by partial detachment from the substrate. Both these failure modes are aggravated by the indentation caused by the impinging liquid drop.

5.11.3   *Experimental studies*
There are three broad experimental techniques for studying the effect of impingement attack by liquids. These can be grouped as follows:
   (1) low velocity impact by a single water drop;
   (2) high velocity impact by a single water drop;
   (3) a whirling rig in water droplets.

5.11.3.1   *Low velocity*
Work has been carried out on milk and mercury dropped from a small height of 275 mm [22]. It is unlikely that at such low speeds, of the order of $2\,\mathrm{m\,s^{-1}}$, damage to targets will be appreciable unless impacted continuously for months even. The information from such experiments becomes useful in another direction. This is the nature of the pattern of radial wash.

The technique is to map the pattern of a drop of water as it strikes a plate either by photography or chemically. The latter is simple to carry out as an experiment and there is one [22] which uses a pipette to produce a drop of water which is allowed to fall on to a horizontal surface through a vertical height of about 0·5 m. A small crystal of sodium dichromate is held at the bottom of the water drop. The horizontal surface consists of a filter paper which is covered with an aqueous solution of acidified starch and potassium iodide. As the water drop strikes, the dichromate produces a starch iodide print. It is important that the dichromate crystal is placed in position immediately before the drop is released from the pipette and this can be done by holding a small crystal of dichromate at the end of a needle and placing it at the bottom of the drop. As the dichromate reacts with the flat surface it can be deduced from the changing nature of the print that the bottom of the drop flows out radially but the remaining head of water just subsides without causing a radial wash. The structure of the radial

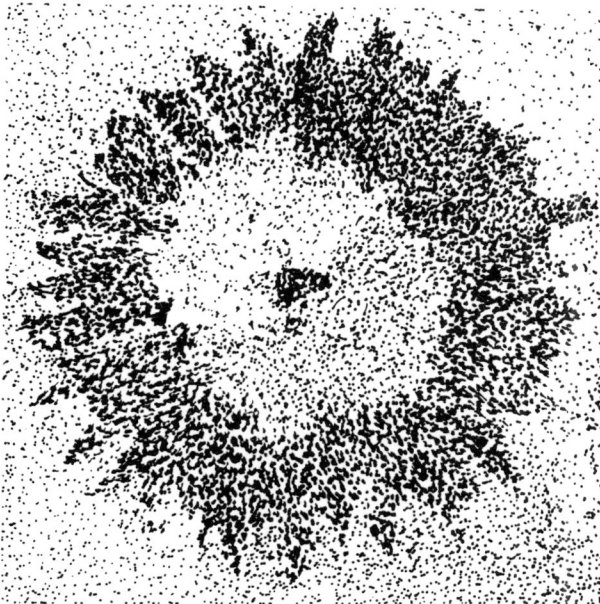

Fig. 5.16 Radial wash created by a water drop impinging normally. (Redrawn from Engel, O. G. (1955). *J. Res. Nat. Bur. Stand.* **54**(5), 281.)

arms of the water drop (Fig. 5.16) shows channels. The pattern of the liquid drop is followed by what are known as *schlieren pictures*.

The first published information on the schlieren technique appeared in 1866 [24] and there are many variations of this. A schlieren object is one in which there is a variation in thickness. For our case the water drop shows a height variation with time as it disappears progressively in the form of the radial wash. The schlieren effect can be used to obtain information about the progress of this wash [22]. What is needed is to provide a drop of liquid through a pipette into a tube (Fig. 5.17). A light source is collimated by a parabolic mirror and then directed into the drop tube with the aid of an inclined plane mirror, the inside of the tube being dyed black to reduce light scattering. A glass plate receives the falling water drop and the collimated beam of light passing through the drop is focused onto an opaque spot. The opaque spot is contained in a glass plate and should be the same size as the sharp ring of light focused by the lens. A high speed cine camera is mounted at the bottom and is focused on the upper surface of the glass plate. The speed of the camera can vary between 15×10^3 and 10^6 frames per second. Interpretation of

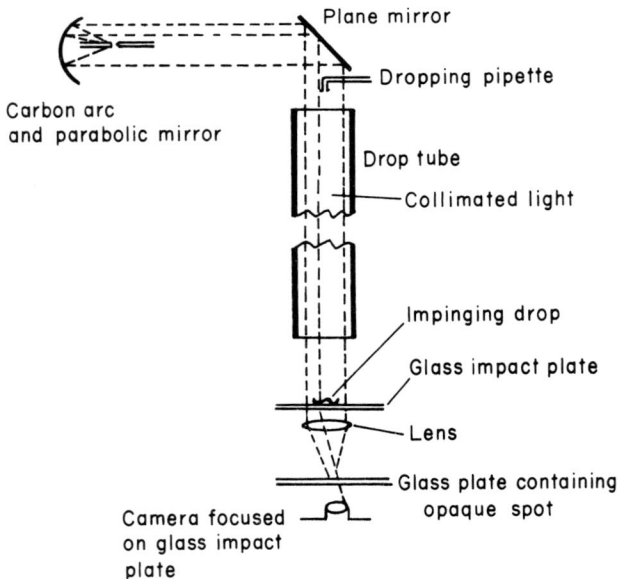

Fig. 5.17   An apparatus to study the schlieren effect. (Engel, O. G. (1955). *J. Res. Nat. Bur. Stand.* **54**(5), 281.)

the stills from the high speed camera entails careful calibration and experience. Glass models have been produced [22] to represent various stages of the radial wash. A flat area appears light in the schlieren patterns and an inclined region appears dark. Thus for a drop undergoing radial flow, the flat wash will appear light and the head of liquid above this will show up as a dark circle. Knowing the frame sequence, the radial velocity, among others, can be evaluated. Unfortunately, the radial wash can appear dark if it is inclined and the interpretation of the stills is not straightforward.

5.11.3.2 High velocity
A drop of liquid accelerated to a speed beyond a critical velocity breaks up into small droplets. To obtain an idea of the scale, the critical velocity for a drop 2 mm in diameter is 17 m s^{-1}. This makes studies with aqueous single drops at high speeds nearly impossible, but an apparatus has been designed [25] where a drop of water has been propelled onto a solid stationary surface at high velocities successfully (Fig. 5.18).

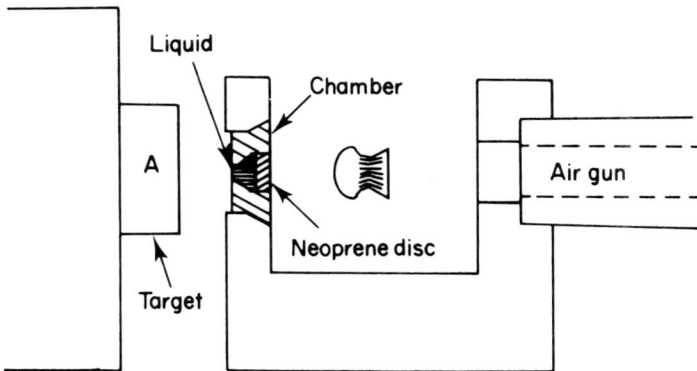

Fig. 5.18   An apparatus to study impingement attack by a single liquid drop propelled at high velocity. (Bowden, F. P. and Brunton, J. H. (1961). *Proc. Roy. Soc. A* **263**, 433.)

The target A is mounted a variable distance 4–12 mm from the orifice which holds the liquid. The orifice diameter can also be varied between 0·5 and 3 mm and the entrance angle is 120° to minimize turbulence. To avoid turbulence, the parallel portion of the orifice is kept small. For the same objective, the interior surfaces are polished and all sharp corners are given appropriate radii. A slug is fired onto the neoprene disc which extrudes the liquid drop through the orifice. The disc spreads during compression and prevents back leakage. The velocity of the water drop can be as high as 1200 m s^{-1} and is varied by the speed of the slug. The experiments use high speed cinephotography at 10^6 frames per second to observe the impact process.

High velocity interaction has also been attempted [26] by firing solid specimens at liquid drops.

5.11.3.3   Whirling arm
Whirling arm rigs have used a device where metal specimens rotate at speeds of about 340 m s^{-1} in streams of water [19, 27, 28]. These studies have shown that even hard steels undergo erosion damage under the speed range obtained in this rig. The rig supports two or more specimens at the periphery of a disc which impinge against continuously running liquid streams. This, of course, is a case of repeated impact, but the size and the velocity of the drops are more controlled in the apparatus shown in Fig. 5.18.

5.11.4   Impact forces
Detailed knowledge of the nature of interfacial and subsurface stresses is still not available, but certain parameters have been isolated which provide an insight into the deformation processes in materials under liquid particle impact. We have seen that, as a liquid drop strikes a surface, compression and tension waves are generated at speeds of ultrasound. The area of the wash expands with time and the radius, $a(t)$, of it at an instantaneous time t is shown to be governed by v_i, the velocity of impact and R, the prior radius of the spherical drop [29]. Thus,

$$a(t) = [2Rv_i t - (v_i t)^2]^{1/2} \tag{5.42}$$

An approximate form of this equation is,

$$a(t) = Kt^{1/2} \tag{5.43}$$

where

$$K = (2Rv_i)^{1/2}$$

The relationship in Equation 5.43 suggests that $a^2(t) \propto t$, but the area will not continue to increase indefinitely. Thus at characteristic values of R and v_i, there will be a limit on t at which the radial wash ceases to flow. This probably happens when only a monomolecular layer of the fluid covers the target surface but it is innocuous in that condition. Putting $R = 1$ mm and $v_i = 3$ m s^{-1} the radius is $0 \cdot 134$ mm, 3 μs after impact. The value is not an unrealistic one.

The stresses at the interface are transient and their exact spatial and temporal distribution pose a formidable problem to deduce exactly, but a number of expressions are available. Equation 5.40 is one. Strictly, if the liquid drop attacks a homogeneous isotropic solid, it will cause some elastic deformation of the medium. Taking this into account, the stress σ is expressed [28] as,

$$\sigma = \left[\rho_1 v_i u_1 \bigg/ \left(1 + \frac{\rho_1 u_1}{\rho_2 u_2} \right) \right] \qquad (5.44)$$

where ρ_1 and ρ_2 are the densities of the liquid and the solid respectively; $u_1 =$ velocity of compression wave in the liquid; $u_2 =$ velocity of the compression wave in the solid.

The stress has also been expressed [22] in terms of the velocity c of the compression wave in the liquid, so that

$$\sigma = \tfrac{1}{2} \alpha \, (\rho c v_i) \qquad (5.45)$$

where ρ is the density of the liquid and α tends to unity at high velocities.

We note that most of the expressions are similar to the water hammer equation. This is derived by assuming the liquid to be a compressible fluid and the pressure is that which exists at the first instant of collision. Lateral flow of the fluid is prevented and the water hammer equation states that,

$$\sigma = \rho c v_i \qquad (5.46)$$

5.11.5 *Target damage*
There is always a stagnant point at the centre of a liquid drop as it strikes the solid, and at a characteristic distance from this centre, circular damage marks are clearly seen. The intensity of this radial scarring increases with the density of the liquid and the velocity of impact, as is to be expected, but the mechanical property of the target plays a significant role. A highly elastic material damps the impact force. This reduces the velocity of the radial wash and hence damage to the target.

However, the target must be able to resist the considerable radial tensile stresses which result due to elastic depressions of surfaces. Hard materials with negligible ductility produce a high velocity lateral flow of the liquid. They sustain heavy impact forces but fail at their fracture stress by shattering.

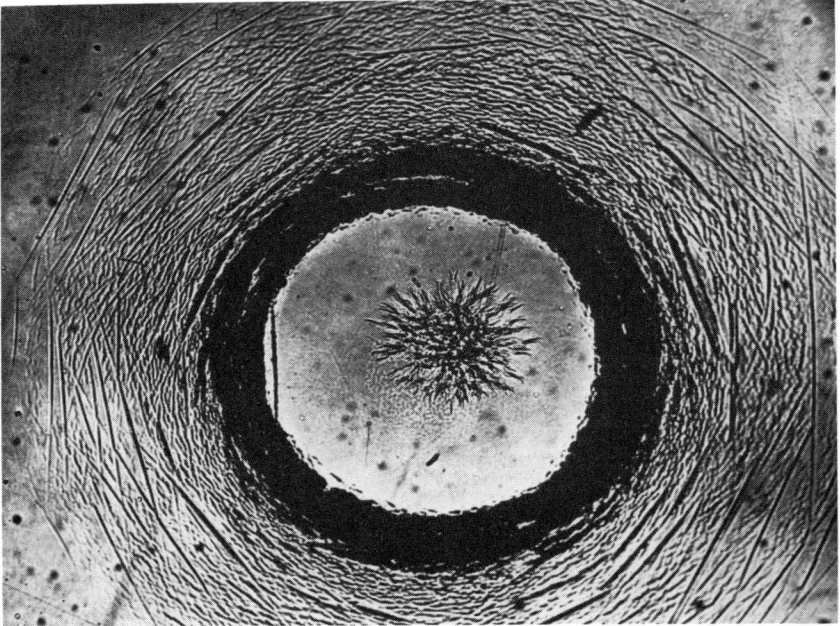

Fig. 5.19   Ring crack produced due to impingement attack by liquid drops. (Bowden, F. P. and Field, J. E. (1964). *Proc. Roy. Soc. A* **282**, 331.)

A typical failure pattern is shown in Fig. 5.19 where the central undamaged part is surrounded by an annulus area showing cracks [30]. This annulus region has been termed the ring fracture and is similar to that produced by static indentation test. There are, however, some differences in that rather than a continuous crack, the ring shows many small cracks. Outside this ring are additional circumferential fractures which are short and discrete in nature. A ring crack is also observed with a solid particle impinging on glass but the circumferential discrete cracks are absent. The reason is that, unlike liquid particle impact, in the case of collision between solids, the duration of the compressive stress pulse is about 100 μs which is about 2 orders of magnitude larger than that in a liquid drop striking a solid. The stress wave will travel the solid several times and distribute the impact evenly. For a liquid drop, being of a short duration of about 1 μs, fracture in the solid occurs at discrete points where the impact pulse interacts. To understand this fully it is necessary to establish the source and nature of the stress. As a liquid drop impacts the surface of a solid, the waves transmitted to the body are mainly longitudinal and transverse. Apart from these, the solid will be subjected to what is known as the Rayleigh surface wave. Except under certain conditions, the longitudinal wave, although the strongest, does not contribute to fracture of the solid directly, being compressional in nature. Both the transverse and the Rayleigh waves, however, are such that tensile forces are associated with them and fracture of surfaces becomes a strong possibility.

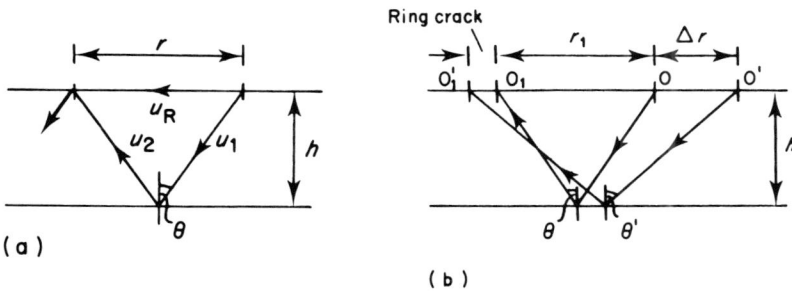

Fig. 5.20 Reflection of the longitudinal component u_1 of an acoustic pulse into a tension component u_2. u_R is the Rayleigh surface velocity. (a) Point contact; (b) reflection from a finite area. (Bowden, F. P. and Field, J. E. (1964). *Proc. Roy. Soc. A* **282**, 331.)

The longitudinal component u_1 of the acoustic pulse contributes to the formation of a ring crack indirectly by being reflected as a tension component u_2 (Fig. 5.20(a)) and $u_1 > u_R$ where u_R is the Rayleigh surface velocity. The three pulses enclose a circle of radius r (Fig. 5.20(a)). For a target of thickness h, from geometry,

$$r = 2h\,[(u_1/u_R)^2 - 1]^{-1/2} \qquad (5.47)$$

Rather than make a point contact, the water droplet will meet the target at a finite area of width Δr as shown in Fig. 5.20(b). The longitudinal wave will be reflected from the whole area of contact so that the reflected tension wave will strike an annular area of width O_1O_1'. Cracks will appear in this annulus if the induced tensile stresses are greater than the critical stress of the material to initiate fracture and it can be shown that

$$O_1O_1' = \Delta r [(u_1/u_R)^2 - 1]^{-1} \qquad (5.48)$$

The values of u_1 and u_R for various materials can be evaluated so that a fair estimate of the width of the ring crack can be obtained from Equation 5.48, although other factors such as duration of the pulse will modify the relationship considerably. If the velocity of impact is high, the longitudinal wave will be reflected many times until the natural attenuation of energy will exhaust the intensity of the reflected waves and no more cracks will form.

The small discrete circumferential cracks are attributed to the interaction between surface flaws and the Rayleigh waves. The surface flaws are in the form of microcracks or inclusions which act as sources for stress concentration.

Interest in impact by liquid drops has increased because of what is termed rain erosion of high speed aircraft [31]. Rain erosion is a high velocity impact process and the wear rate depends on the following factors:

(1) velocity of impact;
(2) angle of impact;
(3) density of the liquid;
(4) size of the drop;
(5) gas content of the rain;
(6) ambient temperature.

Laboratory studies show that there is an incubation period but weight loss of the solid then increases in a linear manner with the number of collisions.

In order of increasing superiority of resistance to rain erosion are glass, plastic, ceramics and metals. There is evidence to suggest that, with metals, a forest of dislocations build up during the incubation period and the residual stress increases to a value when cracks nucleate. Erosion resistance of steels can be improved by heat treatment and that of aluminium alloys, by using a galvanically deposited coating of nickel. Ceramics are inherently brittle, and additionally, their mode of manufacture is by the powder metallurgical technique. This introduces porosity and the resistance to rain erosion is sacrificed. Regarding plastics, polystyrene is brittle and fails readily, but rubber-like poly-urethane is superior to it by three orders of magnitude.

There is a threshold velocity v_c below which there is no erosion. A way of minimizing erosion damage is to break up the natural rain into small droplets. The radomes of aircraft are fitted with spikes which generate shock waves and the rain drops are broken up. This is highly beneficial, particularly during supersonic flights when the impinging velocity is high.

The mean eroded depth h due to attack by rain at a velocity v_i greater than the threshold velocity v_c is given by

$$h = K(v_i \cos \theta - v_c)^n (2r)^m \rho v_i \cos \theta t \qquad (5.49)$$

where, K, n and m are constants depending on the eroding material; r = radius of a rain drop of density, ρ; θ is the angle of impact and t is the time at which h is measured.

Maximum damage due to liquid particles occurs at an impact angle of $90°$; that is, when the drop strikes the target normally. Loss of material from the target has been shown to increase with the velocity of impact, reach a maximum, followed by a fall. Reduced erosion at very high velocities is attributed to the break up of a liquid drop to finer particles.

We have already noted that cavitation bubbles produce depressions and they can be much smaller in diameter than the liquid drop [32]. Another source of surface damage, especially in soft metal surfaces, is the high edge pressure exerted by the spreading liquid.

References

1. Johnson, W. (1972). "Impact Strength of Materials", Edward Arnold, London.
2. Miller, D. R. (1965). *J. Australian Inst. Metals* **10**, 295.
3. Hutchings, I. M., Winter, R. E. and Field, J. E. (1976). *Proc. Roy. Soc. A* **348**, 379.
4. Winter, R. E. and Hutchings, I. M. (1975). *Wear* **34**, 141.
5. Hutchings, I. M. (1975). *Wear* **35**, 371.
6. Winter, R. E. and Hutchings, I. M. (1974). *Wear* **29**, 181.
7. Finnie, I. (1960). *Wear*, **3**, 87.
8. Tilly, G. P. and Sage, W. (1970). *Wear* **16**, 447.
9. Tabor, D. (1951). "The Hardness of Metals", Clarendon Press, Oxford.
10. Tilly, G. P. (1973). *Wear* **23**, 87.
11. Goodwin, J. E., Sage, W. and Tilly, G. P. (1969). *Proc. Inst. Mech. Eng.* **184**, 279.
12. Sheldon, G. L. and Kanhere, A. (1972). *Wear* **21**, 195.

13. Bitter, J. G. A. (1963). *Wear* **6**, 5.
14. Neilson, J. H. and Gilchrist, A. (1968). *Wear* **11**, 111.
15. Tilly, G. P. (1969). *Wear* **14**, 63.
16. Finnie, I., Wolak, J. and Kabil, Y. (1967). *J. Mater.* **2**(3), 682.
17. Siebel, E. and Brockstedt, H. C. (1941). *Verchleissminderung, Maschinen-bau* **20**, 457.
18. Finnie, I. (1972). *Wear* **19**, 81.
19. Gardner, F. W. (1932). *The Engineer* **153**, 147, 174, 202.
20. Worthington, A. M. (1877). *Proc. Roy. Soc.* **25**, 261.
21. Worthington, A. M. (1877). *Proc. Roy. Soc.* **25**, 498.
22. Engel, O. G. (1955). *J. Res. Nat. Bur. Stand.* **54**(5), 281.
23. Engel, O. G. (1958). *J. Res. Nat. Bur. Stand.* **61**(1), 47.
24. Toepler, A. (1866). *Prog. Ann. Phys. Chem.* **127**, 556.
25. Bowden, F. P. and Brunton, J. H. (1961). *Proc. Roy. Soc. A* **263**, 433.
26. Jenkins, D. C. (1955). *Nature* **176**, 303.
27. Honegger, E. (1927). *Brown Boveri Rev.* **14**, 95.
28. de Haller, P. (1933). *Scweizerische Bauzeitung* **101**, 243, 260.
29. Adler, W. F. (1977). *J. Mater. Sci.* **12**(6), 1253.
30. Bowden, F. P. and Field, J. E. (1964). *Proc. Roy. Soc. A* **282**, 331.
31. Proceedings: 4th International Conference on rain erosion and associated phenomena, Neusschloss, Meersburg (1974).
32. Proceedings: 3rd International Conference on rain erosion and associated phenomena, Elvetham Hall, Hampshire, U.K. (1970).

6 Fretting

We shall define fretting as a type of wear arising out of slip by a small amplitude of two surfaces relative to each other. Like all tribological systems, the surfaces, of course, are loaded, but the important differentiation between fretting and general reciprocating motion is that the former undergoes very small amplitude motion. The movement is reciprocating but the amplitude is small, i.e. in the range 20–400 μm. It is necessary to state that as the amplitude is increased the material loss by fretting tends to be similar to that due to reciprocating sliding under comparable conditions of load and environment. The loss of material due to small amplitude reciprocating motion is small. The popular terminology for fretting is small amplitude oscillatory motion, and one serious problem is that fretting results in localized pits which act as centres for initiation of fatigue cracks. Fatigue is a serious disadvantage in such situations as bolted flanges and riveted joints where small amplitude movement is not by design. In certain examples, such as flexible couplings, small amplitude motion is unavoidable and although fatigue may not be of importance here, seizure of the parts may occur. Seizure of two surfaces is believed to be facilitated by the debris which remains trapped at the interface, a peculiarity of fretting. For steels, the debris is red in colour and has been variously termed as cocoa, red-mud, etc., while with aluminium the product is black. The debris shows a large amount of oxides and hence a quite commonly used name is fretting corrosion. However fretting, that is, wear due to small amplitude oscillatory motion, can occur in an inert atmosphere or for couples made of gold or platinum. Loss of material in unidirectional or other forms of sliding also happens on surfaces with oxides and hence we have used the word fretting only to identify the phenomenon. Like most wear processes, there is no neat theory for fretting but a large number of investigators have carried out experimental study.

6.1   Experimental techniques

While designing an apparatus the first question to ask [1] is what is the minimum amplitude at which the motion gives rise to sliding. We shall discuss the effect of amplitude later, but it is not uncommon to design machines capable of providing a range of amplitude from 5 to 200 μm. It is not safe to be dogmatic, but in certain aluminium alloys the fatigue strength is lowered in the amplitude range of 8 to 14 μm. At higher amplitudes, the nuclei for cracks are removed by the work hardened or oxide debris and the propensity to fatigue is reduced. We could say that in this particular case, reciprocating wear takes over beyond a slip amplitude of 14 μm. Whereas the amplitude is very important in demarcating between fretting and reciprocating sliding wear, the apparatus must be able to vary the normal load, amplitude and the frequency of oscillations. Both the atmosphere and the surrounding temperatures are also varied. One apparatus [2, 3] uses an upper specimen which is in the form of an annulus resting on a lower flat specimen. The annulus has a width of 1 mm and the mean diameter of the bottom face can be about 15 mm. A dead weight is applied through holes in the centres of the two specimens and can be varied. The oscillatory motion is applied through a lever attached to the top specimen. The lever is linked to an eccentric drive and the specimen surfaces can be flooded with a lubricant if desired, or the fretting experiments can be carried out in a controlled atmosphere. The surface should show an oscillatory circumferential slip, but because of variable friction during motion at the rubbing interface, the system fails to be perfectly dynamically symmetrical. This results in a small amount of superimposed radial motion. The slip amplitude is maintained at 0·25 μm.

Rather than use bush-type specimens, annular friction surfaces have been produced by recessing ends of solid cylinders [4]. The cylinders are 25 mm in diameter so that the amount of wear can be assessed by the weight loss technique. The specimens are mounted horizontally and are capable of slip amplitudes of up to 200 μm.

Since fretting involves contact between asperities under load, attempts have been made to find some trend in the degree of adhesion

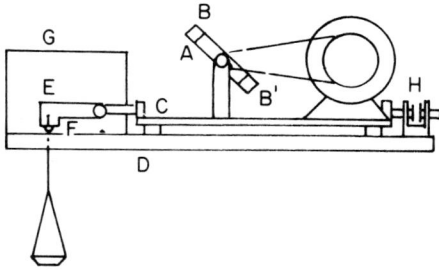

Fig. 6.1 A fretting apparatus which can also measure adhesion. (Bethune, B. and Waterhouse, R. B. (1968). *Wear* **12**, 289.)

between two components in fretting and their propensity to wear. In an apparatus designed [5] to measure the degree of adhesion under a small amplitude oscillatory motion (Fig. 6.1), a hemispherical rider F forms the upper surface which rubs against a plate bolted to D, the bed of the machine. The fretting motion is provided by a rotating arm A, carrying two weights, B and B', which can be in a state of out-of-balance by various amounts. The base plate C, bolted to D, is vibrated if the arm rotates with the out-of-balance weights. The hinged arm E transmits the vibration to the hemispherical rider. The atmosphere around the friction couple can be controlled by using a suitably sealed perspex box G and the amplitude is measured by a transducer, H. The frequency of vibration is the same as the speed of the rotating arm A. The normal load is applied by putting weights over the scale pan which is attached to the hinged arm. The apparatus is run for a predetermined number of cycles and after stopping the machine the adhesion between the rider and the plate is measured by noting the vertical force necessary to separate the couple. This can be done simply by passing a nylon chord, attached to the arm E but directly above the interface, over a pulley vertically above the rider. The other end of the chord is tied to a container which is filled with water from a burette to provide the vertical load of separation.

An interesting situation arises when a steel sample is fretted against a polymer. Ignoring the damage caused to the polymer for the moment, oxides of iron are seen to transfer to the counterface. The wear of steel due to fretting against the polymer is measurable. An apparatus used [6] for these experiments employs a 20 W power amplifier with a built-in low distortion oscillator which drives an electromagnetic vibrator to provide the oscillatory motion to an upper steel test specimen. The lower flat specimen is a polymer and is mounted rigidly, the normal load being applied by dead weights. A maximum frequency of 150 Hz can be obtained and the slip amplitude can be varied in the range 2–30 μm. The amount of metal lost from the upper specimen is expressed in terms of the area of the oxide transferred to the polymeric counterface.

A good way of producing vibration is to employ the piezoelectric method [7] and this has been used to study frictional forces under fretting. A piezoelectric barium titanate crystal with a cross-sectional area 1 mm × 1 mm is glued to a metal base. The material whose frictional force is to be measured is produced in the form of a thin flat specimen and is glued to the top surface of the crystal. A wire is drawn over the exposed top surface of this sample and the frictional force is transmitted to the crystal. A voltage is generated in the transducer which is displayed on the oscilloscope. The normal load is constant so by measuring the tangential force to cause slipping, the coefficient of friction can be recorded continuously. The damage produced to the sample is estimated visually, but the amount of wear can also be expressed quantitatively. A machine has been designed [8] which uses a strain gauge device to measure friction.

A quantitative investigation of fretting and measurement of friction have been possible by fretting a 12·7 mm long cylinder with a diameter of about 25 mm oscillating against two pads (Fig. 6.2) [9–11]. The pads

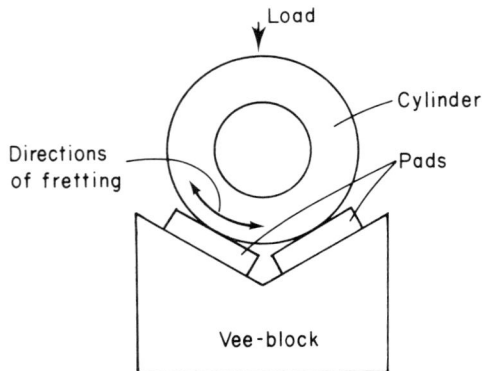

Fig. 6.2   A fretting apparatus using a vee-block. (Halliday, J. S. (1957). Proceedings: Conference on Lubrication and Wear, Institution of Mechanical Engineers, London, p. 640.)

are mounted against the sloping sides of a vee-block. A vibration generator via a rigid lever, not shown in Fig. 6.2, applies the oscillation to the cylinder with varying amplitudes between 2 and 420 μm and frequencies of up to 500 Hz. A transducer measures the amplitude and is a second small vibration generator. As friction is overcome, the cylinder slips over the pads and the critical r.m.s. current i_K to do this is noted from the driving generator. If the output voltage of the measuring transducer is plotted against the input current of the driving generator, a sudden change is noted at i_K as shown in Fig. 6.3.

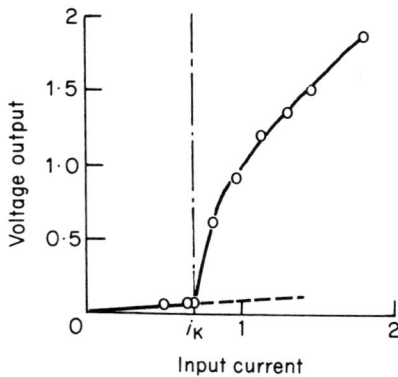

Fig. 6.3   Variation of r.m.s. output volts with r.m.s. input current for mild steel specimens. (Halliday, J. S. and Hirst, W. (1956). *Proc. Roy. Soc. A* **236**, 411.)

If α is the force exerted by the driving vibrator, say, in kg/ampere, r.m.s. at the operating frequency, it has been shown [9] that the mean frictional force F is,

$$F = 9\alpha i_K \qquad (6.1)$$

The frictional force acting at the two line contacts is

$$F = \mu W \sec \theta \qquad (6.2)$$

where μ = coefficient of friction, W = normal load, and θ = slope of the sides of the vee-block, e.g., 30° in Fig. 6.2. Combining Equations 6.1 and 6.2,

$$\mu = 7 \cdot 8 \alpha i_K / W \qquad (6.3)$$

6.2 Experimental studies

An important parameter in fretting experiments is the amount of wear, and this has been expressed in a number of ways. The most simple way is to obtain the weight loss and plot this against the cumulative number of cycles. Unfortunately, loss of material is usually very small, and to get a meaningful curve the experiment has to be conducted for a long period of time. One approach has been [12] to take repeated Talysurf traces of the interface over a large number of parallel tracks. This should register the wear pits as valleys and also any lumps in the form of metal transfer. If this happens, apart from a quantitative evaluation of wear, some basic information regarding the nature of the interfacial interaction may emerge.

The use of Talysurf for this type of work is extremely tedious and it is not always easy to establish the loss of material from the profilometric traces. The areas of the pits can also be measured optically. The depth of the deepest pit is obtained by lapping the surface of the sample. If weight loss is also noted due to lapping, a change of mass of 0·2 mg corresponds to the removal of a layer of 100 μm for an apparent surface area of 230 mm^2. A method such as this is subject to error if simply because it needs great care to keep on lapping the surface until the maximum depth just disappears. Hence a very large number of measurements is necessary to take a statistical mean of the values obtained. If A is the area of the largest pit and d is the dimension of the deepest valley, experiments show that $Ad =$ constant and using the product Ad, the volume loss of the largest pit has been examined in terms of normal load and other variables. The radioactive tracer technique appears to be a very satisfactory method and it has been used to study fretting of steels [1].

The effect of load in the range 10–60 kg has been studied at a constant number of cycles (50 000) and a slip amplitude of 25 μm. If W is the applied load, empirically,

$$A \propto W^{0.8}$$

$$d \propto W^{0.2}$$

so that

$$Ad \propto W$$

This shows that wear is a function of load, which influences the area of the pit rather than the depth, d.

Laboratory studies have shown [13] that a plot of weight loss against contact pressure is linear for steel samples fretting in a bath of lubricant. However, loss of material is generally high and one wonders if the lubricant is in fact corrosive. Weight loss or depth of pitting would increase with normal load but the characteristic of the individual apparatus must be taken into account [4] when interpreting fretting results. One problem is that the interfacial motion can be lost elastically in various members of the machine. This means that the actual slip will be smaller than what is believed to be the case due to the externally applied force. However, if the slip amplitude is maintained, the wear increases linearly with displacement, other things being equal. This is also confirmed [11] for steel on steel and steel on duralumin couples using the cylinder and pad apparatus (Fig. 6.2).

At a load of 19 kg and for a total running time of 10^6 cycles for each experiment Fig. 6.4 shows that there is hardly any wear at amplitudes below 100 μm. This is rather a high value and the explanation may not lie in the elastic absorption of the driving power. Beyond this amplitude, however, the amount of wear increases continuously. The suggestion for low wear at the amplitude range 100–120 μm is that a very large proportion of product of wear rolls at that degree of oscillation, and hence wear and surface damage are low. The high wear with duralumin probably occurs as a result of direct abrasion of the interface by hard particles of aluminium oxides. Similar wear amplitude plots have been observed for fretting steels in a bath of lubricant [8].

As expected, the amount of wear increases with the cumulative number of cycles, that is with the total sliding distance. A pattern for weight loss with the number of cycles has been presented to suggest four different stages of wear (Fig. 6.5). The first stage, OA shows a rapid rise in wear and is possibly initial metal transfer. It is suggested that as wear particles detach and remain trapped in the interface some abrasion occurs, resulting in a rapid rise of wear, AB. It is then assumed that the abrasives become ineffective after a period resulting in a declining rate of wear, BC, followed by the final steady state CD. Examination of the authors' [14] experimental points shows, however, that we could easily draw a simple curvilinear running-in period followed by a steady state. The experimental set-up utilized two annular surfaces under a normal load of about 435 kg with a frequency of about 1·3 Hz and slip amplitude of 45 μm. Using a similar apparatus a near linear relationship between the product Ad and the number of cycles N is shown

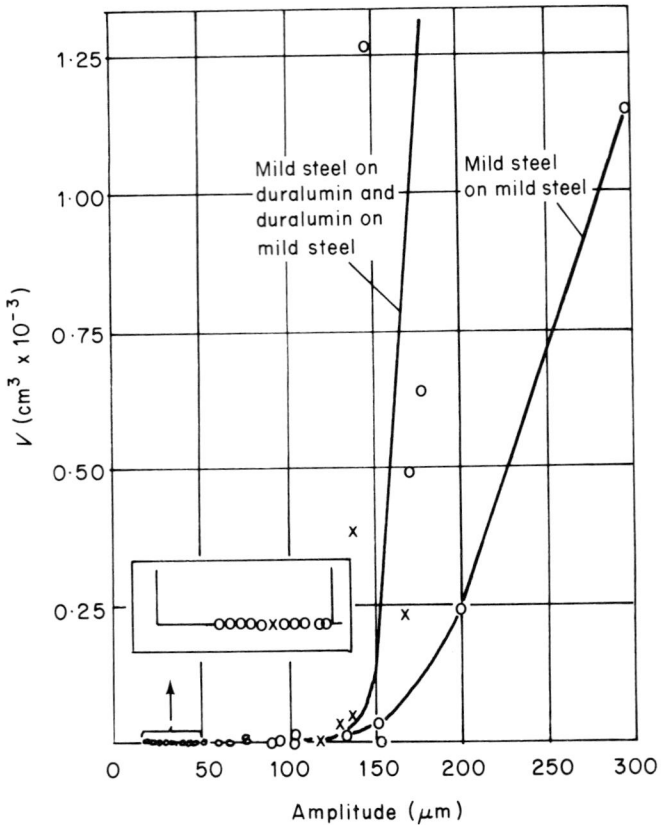

Fig. 6.4   Variation of fretting, V, with slip amplitude. Load 19 kg; 10^6 cycles. (Halliday, J. S. (1957). Proceedings: Conference on Lubrication and Wear, Institution of Mechanical Engineers, London, p. 640.)

elsewhere [2]. Thus,

$$Ad \propto N^{0.9} \quad \text{and} \quad A \propto N^{0.4}, \qquad d \propto N^{0.5}$$

We see that the effect of prolonging the fretting operation is to increase both the area and the depth of damage. Unfortunately, direct proportionality with load is not observed. A phenolic resin fretting against a nickel silver surface shows [7] an increase in the amount of wear with the number of cycles, but wear is more likely proportional to $N^{1/3}$. On the other hand samples produced by sintering aluminium powder show a rapidly increasing rate of wear when interacting with themselves or

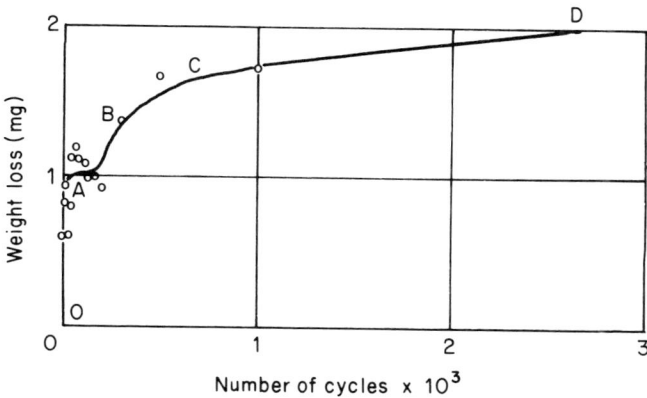

Fig. 6.5 Amount of fretting of steel with running time. (Feng, I. M. and Rightmire, B. G. (1956). *Proc. Inst. Mech. Eng.* **170**, 1055.)

zirconium or steel [15]. It at once suggests that the wear process in materials due to small amplitude oscillatory motion is complex and there is no general law which can be applied to any material combination.

Although all sliding involves an oxide layer at the interface, much is made of the influential role of the oxide in fretting. Indeed, as the metallic asperities are sheared, they oxidize and provide a protective blanket. If the interface moves at low frequency, the time of a complete cycle is prolonged, facilitating the formation of an oxide film. Effective oxidation is the reason put forward for the greater amount of wear of steel at low frequency [16]. We must note that oxidation in this case is not beneficial, unlike in sliding wear. Fretting, which becomes reciprocating sliding beyond a critical amplitude, must be a special process and our understanding of the role of the amplitude is inadequate.

The existence of oxidation products has been a ready means of identifying a fretting process. An example of substantial practical hazard is reported [17] where a black fretting product was found in large amounts in ships' holds. Aluminium boxes packed with explosives were stacked one upon the other in the holds. The structural work of the holds was of steel, so that fretting occurred both as aluminium rubbing on itself and aluminium on steel. The wear product was identified as a mixture of pure aluminium, its oxide and about 1% elemental iron. The particles had an average diameter of 0·15 μm and could easily be airborne. In the confined atmosphere of a ship's hold, an external source of fire such as an electric spark has been known to ignite aluminium dust. This itself is a hazard which can prove catastrophic if explosives are present in the vicinity.

The nature of oxidation of steel surfaces under small amplitude oscillatory motion has been studied in some detail. A hard metal such as steel produces very fine debris [18] comparable in thickness to that of its oxide formed by frictional heating. It seems that as vibration starts, the oxide film is initially disrupted. Thereafter, the exposed asperities oxidize so that rubbing is between oxide layers, which are removed by a process of abrasion. Fretting of mild steel on itself in air produces debris which have two different colours, viz., black and red. The former has been identified as iron and the red product is α-Fe_2O_3.

Naturally, the role of environment has been investigated. Fretting phosphor bronze [19] on steel shows that the surface area of damage is larger in a non-oxygen bearing atmosphere, but the depth of damage is low. The conclusion is that whereas metal transfer will occur by the adhesive mode, fretting is low. Metal transfer in this manner should produce rough surfaces and this is seen in experiments with a protective atmosphere such as nitrogen, although mild steel on itself produces a much lower wear rate when compared with experiments in air. A very important environmental influence in fretting is the role of humidity. Fretting decreases with an increase in the humidity, the most likely reason being the lubricating action of water. The beneficial role of humidity is not observed in an atmosphere of nitrogen only. The reason is not clear, but the presence of oxygen is essential.

Oxidation is facilitated at high temperatures, and therefore this aspect is important. Before examining the effect of high temperature we should note that oxidation may not be essential for fretting. Firstly experiments [20] on steels show that fretting can be high at a temperature of $-150\,°C$, falling continuously as the temperature rises to $0\,°C$. Secondly, non-metals such as glass, quartz, ruby and mica show fretting damage [21]. These observations need further investigation, but we can make use of the comfortable evidence [22] submitted for steel regarding the beneficial role of oxidation at high temperature. As Fig. 6.6 shows, at least at temperatures greater than $200\,°C$, fretting accelerates in an atmosphere of argon and reduces to near negligible proportion at about $500\,°C$ when experiments are carried out in air. It is suggested [23] that above $200\,°C$, it is the properties and the nature of the oxide films which determine the magnitude of fretting for like combinations of mild steel.

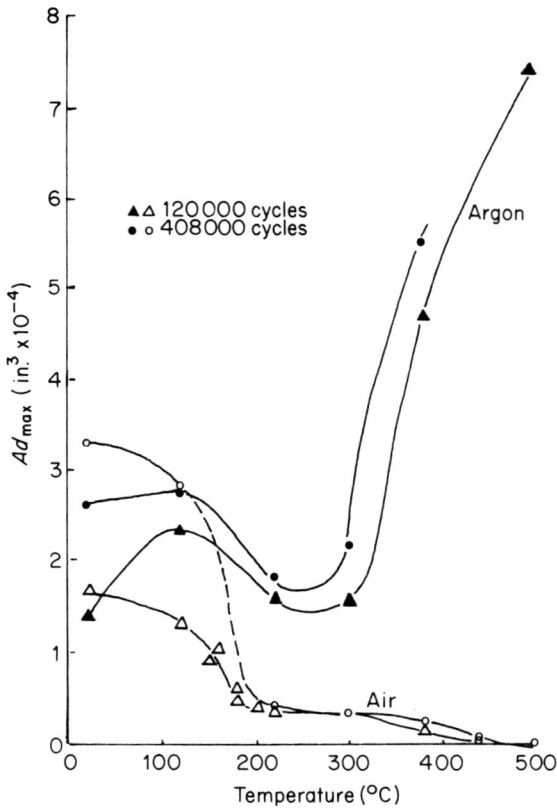

Fig. 6.6 Variation of fretting with temperature. (Hurricks, P. L. and Ashford, K. S. (1969). *Proc. Inst. Mech. Eng.* **184**(3L), 165.)

Oxidation is facilitated by mechanical working of the asperities. Thus as the machine starts for the first time, a limited amount of metal to metal adhesion does occur, but with time oxidation of the interface accelerates when interfacial interaction is essentially between oxides. Assuming that the adhesion zones are converted to oxides, for a given temperature, the rate of oxidation and that of wear can be correlated. Denoting the thickness of an oxide by z and taking d as the deepest pit due to fretting, for a given temperature, we could say that,

$$\text{Fretting wear} \propto \frac{1}{z} \qquad (6.4)$$

For mild steel over the temperature range 200–500 °C, $z^2 = Kt$ where t is the time for the oxide film to attain a thickness z. K is a rate constant and is given by

$$K = A \exp(-Q/RT) \tag{6.5}$$

where, $A =$ Arrhenius constant, $T =$ absolute temperature, $Q =$ activation energy for the oxidation process, and $R =$ gas constant. Since the amount of fretting is inversely proportional to z, we could use a term $1/q$ for resistance to fretting and

$$\frac{1}{q} = K_1 z$$

since

$$z = (Kt)^{1/2},$$

$$(1/q) = K_1 K^{1/2} t^{1/2}$$

That is,

$$(1/q) \propto K^{1/2} \propto [A \exp(-Q/RT)]^{1/2} \quad \text{or} \quad (1/q) = B \exp(-Q/2RT) \tag{6.6}$$

Calculations show that Q remains constant up to a temperature of 380 °C but increases above this temperature. The suggestion is that the value of $(1/q)$ will be governed by the rate with which the exposed metal interface due to mechanical movement will oxidize up to 380 °C. Beyond this temperature, the oxidation rate, as obtained from experiments without the aid of mechanical disruption of surfaces, will decide on the wear resistance.

6.3   Adhesion

At large amplitudes, the situation is that of reciprocating sliding and this can be confirmed by experiments [24]. Since fretting also entails disruption of surfaces, we shall examine if the intrinsic propensity to adhesion between a friction couple can offer an explanation of their fretting behaviour. Surface damage by metal transfer is readily observed in many friction couples under unidirectional and small amplitude oscillatory motion. Fretting experiments with copper on itself, glass and steel respectively show [25] that the immediate effect is metal transfer with a concomitant rise in the coefficient of friction, μ, which is about unity, being the same as under unidirectional sliding. It then reduces to

0·5 as the amount of debris in the interface increases. The debris is shown to comprise cuprous and cupric oxides and fretting of the substrate continues. It is known, however, that surface oxide films are disrupted by oscillatory motions and extensive cold welding results which produce a strong adhesive bond between a pair [26].

One way of obtaining indirect evidence regarding the nature of interaction during fretting is to measure the frictional resistance with time [11]. A plot of the coefficient of friction μ against the cumulative number of cycles, N, shows that for a mild steel couple the effect of vibration is to break down the oxide film. This exposes metallic asperities which weld under normal load and μ increases by a factor of 3. If the amplitude of vibration is not high, the oxide appears to remain intact so that rubbing is between oxide films and the friction is low. We must recall at once that a low amplitude as registered by a transducer of a machine may simply be the elastic accommodation of driving parts. That is, there is no slip at all at the interface. This is a problem with fretting experiments. The other question which urgently needs an answer is what is the amplitude of slip, at and above which the phenomenon ceases to come under fretting motion? The pattern of $\mu - N$ plots for mild steel is repeated by many metal pairs but there are important differences. With like combinations of duralumin and aluminium respectively, there are sudden increases in μ when large metallic junctions form after a number of cycles. Copper on copper behaves in this erratic manner but nickel vibrating on itself gives a consistently low coefficient of friction. Conclusions of the authors [10, 11] from experiments such as these are that there are two stages in fretting. In the first stage, metallic debris are produced. The second stage then follows and the debris comprise largely particles of oxides which are believed to be produced by one or both of the following processes:

(1) frictional heat oxidizes the metallic wear debris very quickly;
(2) the interface itself oxidizes so that the wear process is now that of removal of oxide films.

Oxidation of metallic debris is not just peculiar to fretting and it is always a difficulty to be certain as to which of the above two processes results in oxidized debris. It is hard to imagine that at the moment of failure of the oxide film, the exposed metallic asperities do not interact. We may conclude that adhesion between metallic junctions will occur, always provided that oxide films will dilute the effect.

There is evidence [27] that adhesion is facilitated by lateral movement of the interface because inhibiting contaminants are removed [28] from the surface mechanically. The adhesive force develops because of plastic deformation [29] and the degree of adhesion is high for metals capable of work hardening.

Using the apparatus shown in Fig. 6.1, the coefficient of adhesion, λ for interfaces subjected to oscillatory motion has been defined [5] as

$$\lambda = \frac{\text{Load to separate}}{\text{Normal load}} \tag{6.7}$$

λ increases with the number of cycles and, not unexpectedly, experiments in air show an overall decrease in adhesion between two surfaces.

Experiments with copper, duralumin and steel show clearly that the adhesion coefficient does not correlate with the hardness of the metal as measured on the work hardened layer, or if the hardness is that of the age hardened condition. On the other hand, the coefficient of adhesion shows a decided correlation with the hardness values of annealed and overaged alloys. As far as is known, sliding provides a hardness gradient from the surface to the centre of a longitudinal section of, say, a pin, the maximum hardness being in the surface layer. This does not appear to be so when samples subjected to small amplitude oscillations are examined (see Fig. 6.7). For an age hardening alloy, the surface overages and an initially cold drawn steel work softens so that the wear track assumes low hardness values. Fretted surfaces of mild steel show recrystallization so that an interfacial temperature of about 480 °C is expected [30]. This seems a high estimate but we can accept that an

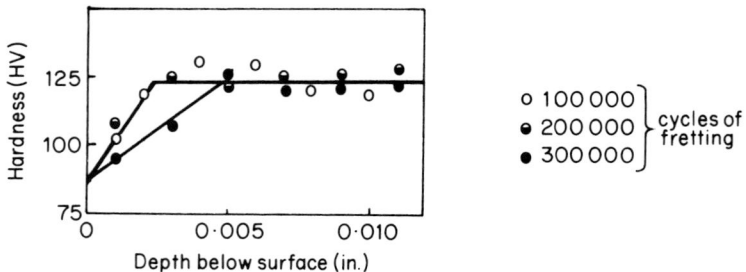

Fig. 6.7   Microhardness variation from the surface to the centre of a fretted specimen (fully aged Al–4% Cu alloy). (Bethune, B. and Waterhouse, R. B. (1968). *Wear* **12**, 289.)

initially annealed component will work harden due to normal and tangential stresses. Mechanical work will supplement the frictional heat resulting in recrystallization and softening of a very thin surface layer. For small amplitude displacement, there will be a fatigue action which the components will undergo under high strain loading. Figure 6.7 shows the soft layer becoming deeper with increasing cycles of fretting.

Experiments [5] with mild steel show that the coefficient of adhesion increases with the amplitude of slip and a maximum is reached followed by a fall. The suggestion is that the shearing action at an interface exposes fresh metallic asperities for cold welds to form and thus provide strong adhesion between a pair. Oscillatory motion also means that the welded junctions are subjected to an alternating stress which is expected to be proportional to the amplitude of motion. The presence of an alternating stress supplies the right condition for a fatigue crack to initiate at appropriate junctions. This will reduce the strength due to adhesion. Since a steel has a fatigue limit, there will be a critical amplitude of slip depending on the load below which there will be no fatigue cracks in the asperities. In that case adhesion will continue to improve with the amplitude as more and more of the surface areas will be available for interaction. The ultimate strength of the junctions, however, will depend on these two opposing effects and adhesion will decrease to zero when the amplitude is high enough for the fatigue cracks to be induced readily. We should add that in the authors' speculation [5] of a fatigue crack, a junction does not remain intact. That is, since an interface slips, a junction will form momentarily, separate and then re-form when the statistical chance allows this. Continuous interaction of a junction with the counterface will induce a fatigue crack so that at the next encounter or so, the junction will fracture and thus will not be available for adhesion. If the intrinsic hardness of a series of alloys is plotted against their coefficient of adhesion a correlation is obtained [31] with the softer metals. The suggestion is that it will be the softer metal which will yield. This will lead to the disruption of its oxide and adhesion with the steel surface will be relatively high. The oxide films on the harder alloys will remain intact and adhesion will be low .

Further work on the idea of adhesion coefficient and the rate of fretting should be carried out. At the moment we can only say that both λ and the volume loss of material tend to show similar trends regarding load and amplitude of slip. There is some indication [32] that fretting decreases with increasing hardness as does adhesion. We must emphasise, however, that it is unwise to think that there are some generalized laws. Fretting, like any other form of wear, should be examined for individual friction couples and, hopefully, the next decade or two will produce some reliable correlations between theory and practice.

6.4 Mechanism

There is no universal theory of fretting at the moment but we note, for example, from Fig. 6.5 that there is a running-in period of wear followed by a steady state. This, at a first glance, suggests that there is little difference between the mechanisms of fretting and sliding wear. Unlike sliding, however, the wear debris remains trapped at the interface for a much longer period and the nature of damage is localized. The localized pits are characteristic of fretting although they are also noticed in gear teeth undergoing sliding and rolling. In the latter, the pits are believed to be there because of the influence of Hertzian stress resulting in an intense subsurface shear stress. Various hypotheses regarding the mechanism of wear due to small amplitude oscillatory motion have been reviewed, for example, by Waterhouse [1] and Hurricks [33], but we shall examine the deliberations of Uhlig [20] and reproduce his derivation of a law for fretting if only to highlight the limitations of his model.

Uhlig [20] points out that fretting is low in a vacuum or in an inert atmosphere and is aggravated in air or oxygen provided the humidity is low. This suggests that there is a chemical process involved in fretting. However, metallic particles are also noted in the debris and relatively noble metals and many non-metals also undergo fretting. Therefore, there must be a mechanical action as well. We can thus represent the loss of material by fretting as Q such that

$$Q = Q_c + Q_m \tag{6.8}$$

where Q_c = wear due to the removal of the oxide film, and, Q_m = wear due to shearing of junctions formed by adhesion.

Consider a bottom surface (Fig. 6.8) covered with a layer of sorbed gases or oxide. Consider a top surface with asperities making circular contact areas of diameter c. Let the asperities be apart and equidistant by an amount s. Suppose the top surface moves with a linear velocity v_0. Each asperity is assumed to remove an amount of material from the counterface and behind it the exposed surface of the metal adsorbs gases

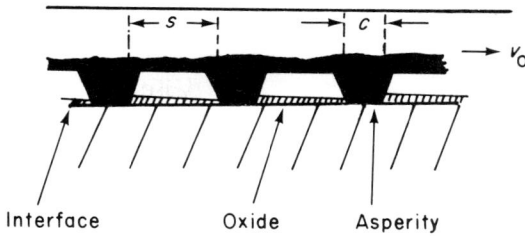

Fig. 6.8 An idealized model of a metallic surface undergoing fretting. (Uhlig, H. H. (1954). *J. Appl. Mech.* **21**, 401.)

or forms a layer of oxides. The next asperity will remove this gas or oxide layer and the process of oxidation and removal of oxide will occur as motion of the interface continues. Removal of material in this manner constitutes the chemical component and involves scraping an area given by the width of the asperity c and the length traversed.

Ignoring the adsorption of gases and considering oxidation only, the time t during which an oxide layer forms is

$$t = s/v_0 \qquad (6.9)$$

The above law is equally applicable to the case of gas adsorption as is the relationship which will now be derived for Q_c. If the length removed is l, the area involved is cl. Assuming a steel surface, Uhlig [20] uses the empirical law of oxidation as follows:

$$Q_c = clK \ln\left(\frac{t}{\tau} + 1\right) \qquad (6.10)$$

where τ and K are constants.

Taking the value of t from Equation 6.9,

$$Q_c = clK \ln\left(\frac{s}{v_0\tau} + 1\right) \qquad (6.11)$$

The fretting motion is sinusoidal so that the total length traversed per cycle is $2l$. The linear displacement x from the midpoint of traverse at a time t, for an angular displacement of θ, is represented as

$$x = \frac{l}{2}\cos\theta$$

Therefore,

$$\frac{dx}{dt} = v_0 = -\frac{l}{2}\sin\theta\,\frac{d\theta}{dt} \qquad (6.12)$$

If the frequency of vibration is f, the angular velocity $d\theta/dt$ is,

$$\frac{d\theta}{dt} = 2\pi f \qquad (6.13)$$

Using Equations 6.12 and 6.13, the average velocity v is expressed as

$$v = -\frac{\pi l f \int_0^\pi \sin\theta\,d\theta}{\pi} = 2lf \qquad (6.14)$$

If the number of contacts per unit area of the interface is n, from Equation 6.11, Q_c per single cycle is

$$Q_c = 2nlcK\,\ln\left(\frac{s}{v_0\tau}+1\right)$$

Substituting in the above for $v_0 = v$ from Equation 6.14,

$$Q_c = 2nlcK\,\ln\left(\frac{s}{2lf\tau}+1\right) \qquad (6.15)$$

To obtain an expression for mechanical wear Q_m, suppose each junction produces wear throughout its interface. The area of each junction is $\pi(c/2)^2$ and $2l$ is the length per cycle. Therefore, the amount of wear per cycle of movement is

$$Q_m = 2K'nl\pi(c/2)^2 \qquad (6.16)$$

The area $n\pi(c/2)^2 = W/H$, where W is the applied load, H is the hardness of the metal and K' is a constant. Therefore,

$$Q_m = (2K'/H)lW \quad \text{or} \quad Q_m = K_2lW \qquad (6.17)$$

where $K_2 = 2K'/H = \text{constant}$.

Expanding the log term in Equation 6.15 in the form

$$\ln(x+1) = x - (x^2/2) + (x^3/3) - \ldots$$

where

$$x \equiv (s/2lf\tau)$$

the term $(s/2lf\tau)$ is much smaller than unity so that its higher powers can be neglected. Therefore,

$$Q_c = \frac{ncKs}{f\tau} \tag{6.18}$$

Considering an unit area, the number of asperities along one edge is $\sqrt{n}$ so that

$$s + c \simeq \frac{1}{\sqrt{n}} \quad \text{or} \quad s = \frac{1}{\sqrt{n}} - c \tag{6.19}$$

Now,

$$\frac{n\pi}{4} c^2 = \frac{W}{H}$$

Therefore,

$$c = \frac{2W^{1/2}}{\sqrt{(\pi H)}} \cdot \frac{1}{\sqrt{n}} \tag{6.20}$$

Substituting for s and c from Equations 6.19 and 6.20 in Equation 6.18,

$$Q_c = n \frac{2W^{1/2}}{\sqrt{(\pi H)}} \cdot \frac{1}{\sqrt{n}} \cdot \frac{K}{\tau} \cdot \frac{1}{f} \left(\frac{1}{\sqrt{n}} - c \right)$$

Substituting for c in the above expression from Equation 6.20,

$$Q_c = \frac{K_0 W^{1/2}}{f} - \frac{K_1 W}{f} \tag{6.21}$$

Where K_0 and K_1 are constants given by

$$K_0 = \frac{2}{\sqrt{(\pi H)}} \cdot \frac{K}{\tau} \quad \text{and} \quad K_1 = \frac{4}{\pi H} \cdot \frac{K}{\tau}$$

Therefore, for a total number of cycles N, the total wear from Equations 6.8, 6.17 and 6.21 is

$$Q = (K_0 W^{1/2} - K_1 W) \frac{N}{f} + K_2 l W N \tag{6.22}$$

Uhlig [20] points out that Equation 6.22 is in reasonable accord with experiments. Thus Q varies linearly with N, the total number of cycles. This is true with most fretting experiments. Wear is hyperbolic with the frequency and parabolic with load. The former is generally true but the load effect is not general for all materials. In fact, there is a report [34] which states that the wear rate actually decreases with increasing load, but this may be because slip is hindered as discussed already.

6.4.1 Metal–polymer system

Hemispherical steel riders have been fretted against various types of polymers [35, 36]. With some of the polymers, an annular ring forms with time which separates a central apparently undamaged area from the rest of the polymer sample. The steel specimen wears and the debris is transferred into this annulus as small platelets, 35–50 nm in diameter. The debris is predominantly α-Fe_2O_3 and the amount of wear is expressed by measuring the area covered by the oxidized debris. Plotting this against the number of cycles, produces three regimes, viz.,

(1) a period of incubation during which stage any transfer of metal oxide to the polymer counterface does not occur;

(2) a period of running-in showing a decreasing loss of metal oxide with the number of cycles;

(3) a steady state wear when the amount of material loss from the steel rider increases at a constant rate.

If the wear scar on the polymer is followed, the annulus created is continuously covered with oxide debris and the steady state begins once the annulus is full of wear debris. Wear continues by increasing the width of the annulus. The amount of wear of the steel varies according to the type of polymer and some of the polymers also undergo significant surface deterioration. The steel surface shows characteristic pits, but unlike metal–metal couples, moist gases, in particular oxygen, accelerate the wear rate of the steel. The role of moisture is not clear but the suggestion [6] is that metal wears by oxidation in one half cycle followed by its deposition on the polymer in the next half cycle. The moisture probably facilitates the creation of loosely held α-Fe_2O_3 on the steel substrate. There is a strong probability that the wear of steel is controlled by the adhesion capability of a particular polymer to the oxide.

6.5 Prevention

To continue our theme of polymer–metal system, PTFE does not cause wear of the steel, at least for 10^6 cycles or so. A PTFE coating or a sheet of PTFE between two potential surfaces appears to be a solution to the problem of fretting. However, it is not known how long the PTFE will last and all designers can hope is to minimize fretting. Prevention is only possible by designing so that there is no relative slip between the two surfaces. For our purposes we are only concerned with fretting in so far as, if not minimized, it will give rise to excessive wear and possibly seizure. A common approach to minimize fretting is to allow movement but reduce the friction. Polymeric inserts are used, but we have already seen that they are quite capable of inducing fretting. The alternative, and quite an ancient approach is to use lubricants in the form of liquid, grease or a coating of, say, graphite. It should be mentioned that the surface speed in fretting is low, being of the order of 20 mm s^{-1}, and therefore a liquid lubricant will work only in the boundary regime and may be ineffective. Metallic coatings are used and the principle is twofold:

(1) a soft coating such as lead is applied to both surfaces so that the friction is reduced since shearing is confined to the lead layers. This is a proposition for surfaces which are designed to undergo relative movement;

(2) there are interfaces which are designed to remain rigid, a typical example being the underside of a bolt head against the flange upon which it rests. In that case both surfaces can be suitably plated to increase the interfacial friction and prevent slip.

Fretting is very common in engineering components such as splined connections and ball races. Loss of fit due to surface damage gives economic disadvantage, but a big problem is that the fretted regions are those where potential fatigue cracks originate [37]. Attempts are therefore made to minimize fretting, and a great deal of attention has been given to coatings. In contradiction to the delamination theory of wear, fretting is known to increase with decreasing thickness of the coating. Of the metal coatings, in decreasing order of effectiveness on steel, are silver, copper, lead, tin, zinc, nickel and chromium [38], the latter two failing because of their tendency to crack. Apart from polymers, various non-metallic coating processes such as phosphating, sulphidizing, bonding an elastic material to a surface, etc. are available for use to minimize fretting.

References

1. Waterhouse, R. B. (1972). "Fretting Corrosion", Pergamon Press, Oxford.
2. Wright, K. H. R. (1952). *Proc. Inst. Mech. Eng.* (1B), 556.
3. Tomlinson, G. A., Thorpe, P. L. and Gough, H. J. (1939). *Proc. Inst. Mech. Eng.* **141**, 223.

4. Uhlig, H. H., Tierney, W. D. and McClellan, A. (1953). ASTM Special Technical Publication No. 144, 71.
5. Bethune, B. and Waterhouse, R. B. (1968). *Wear* **12**, 289.
6. Stott, F. H., Bethune, B. and Higham, P. A. (1977). *Tribology Int.* **10**, 211.
7. Mason, W. P. and White, S. D. (1952). *The Bell System Tech. J.* **31**(3), 469.
8. McMath, R. R. (1961). *Trans. ASLE* **4**, 197.
9. Halliday, J. S. (1956). *J. Sci. Instrum.* **33**, 213.
10. Halliday, J. S. and Hirst, W. (1956). *Proc. Roy. Soc. A* **236**, 411.
11. Halliday, J. S. (1957). Proceedings: Conference on Lubrication and Wear, Institute of Mechanical Engineers, London, p. 640.
12. Wayson, A. R. (1964). *Wear* **7**, 435.
13. Reed, F. E. and Batter, J. F. (1960). *Trans. ASLE* **2**, 159.
14. Feng, I. M. and Rightmire, B. G. (1956). *Proc. Inst. Mech. Eng.* **170**, 1055.
15. Commissaris, C. P. L. and deGee, A. W. J. (1966). *Proc. Inst. Mech. Eng.* **181**(30), 41.
16. Feng, I. M. and Uhlig, H. H. (1954). *J. Appl. Mech.* **21**, 395.
17. Andrew, J. F., Donovan, P. D. and Stringer, J. (1968). *Brit. Corros. J.* **3**, 85.
18. Wright, K. H. R. (1954). *Corros. Prevention and Control* **1**, 465.
19. Sakmann, B. W. and Rightmire, B. G. (1948). National Advisory Committee for Aeronautics, Technical Note 1492.
20. Uhlig, H. H. (1954). *J. Appl. Mech.* **21**, 401.
21. Godfrey, D. (1951). National Advisory Committee for Aeronautics, Report 1009.
22. Hurricks, P. L. and Ashford, K. S. (1969). *Proc. Inst. Mech. Eng.* **184**(3L), 165.
23. Hurricks, P. L. (1974). *Wear* **30**, 189.
24. Ohmae, N. and Tsukizoe, T. (1974). *Wear* **27**, 281.
25. Godfrey, D. and Briley, J. M. (1953). National Advisory Committee for Aeronautics, Technical Note 3011.
26. Bethune, B. and Waterhouse, R. B. (1965). *Wear* **8**, 22.
27. Nicholas, M. G. (1963). *Trans. Met. Soc. AIME* **227**, 250.
28. Anderson, O. L. (1960). *Wear* **3**, 253.
29. Sikorski, M. E. (1964). *Wear* **7**, 144.
30. Waterhouse, R. B. (1961). *J. Iron and Steel Inst.* **197**, 301.
31. Bethune, B. and Waterhouse R. B. (1968). *Wear* **12**, 369.
32. Wright, K. H. R. (1957). Proceedings: Conference on Lubrication and Wear, Institution of Mechanical Engineers, London, p. 628.
33. Hurricks, P. L. (1970). *Wear* **15**, 389.
34. Weatherford, W. D., Valtierra, M. L. and Ku, P. M. (1966). *Trans. ASLE* **9**, 171.
35. Higham, P. A., Bethune, B. and Stott, F. H. (1978). *Wear* **46**, 335.
36. Higham, P. A., Bethune, B. and Stott, F. H. (1977). *J. Mater. Sci.* **12**, 2503.
37. Fenner, A. J., Wright, K. H. R. and Mann, J. Y. (1956). Proceedings: International Conference on Fatigue of Metals, Institution of Mechanical Engineers, London, pp. 386–93.
38. Waterhouse, R. B., Brook, P. A. and Lee, G. M. C. (1962). *Wear* **5**, 235.

7   Rolling Contact

Examples of rolling contact proliferate in human society in day to day living much more than those involving sliding. The rural life in the Indian subcontinent would become very difficult without the use of bullock carts, whose design has probably remained unchanged, with the possible exception of the steel rim on the wheels, through millenia. All sectors of the globe now enjoy the dubious achievement of individual ownership of automobiles; the rolling wheel has great technological importance. Of specific value to the economy of nations is rolling stock, whether used for passenger or for goods transport. There are rollers on household articles such as beds, sofas and lawn mowers, although the first two articles remain static most of the time. Surprisingly, friction and wear studies regarding rolling do not seem to constitute a routine topic in tribological establishments and are pursued by specialist investigators only. One reason may be that the optimum loads and speeds are known for the common rolling elements and if these are not exceeded, the amount of friction and wear is small. Typically, a correctly run ball bearing may easily give a service life of 30 000 h before it wears sufficiently to lose operational efficiency. Even assuming continuous running, this means a life of $3\frac{1}{2}$ years, which can be regarded as very satisfactory for small ball bearings. These have often been termed frictionless bearings and are known to have low wear rates. A study of rolling motion in terms of friction and wear is necessary, however, because there are many instances such as with automobile tyres, railway wheels and tracks, or heavy rolling element bearings, where failure does occur as a result of relative motion between opposing surfaces. As with most cases of wear, loss of material depends on load, speed, temperature and cleanliness of the environment apart from the intrinsic properties of the materials constituting the friction couple. In this chapter, we shall examine as many of these aspects as possible from available published literature. An important aspect of rolling processes is slip which involves a small amount of sliding. This will be examined in some detail.

7.1   Instantaneous centre

Rolling motion has been broadly divided [1] into pure or free rolling, and rolling with imposed surface traction forces. The former is the case when a cylinder or a ball rolls in a straight line along a plane without any hindrance. A typical example of the latter is an automobile tyre where normal and tangential forces are applied at the interface between the track and the wheel.

Although bearings involving rolling motion have been termed frictionless, an understanding of the factors which may be responsible for interfacial resistance to motion is important because in reality there is always friction. If a sphere rests on a plane surface the effect of a tangential pull T will merely be skidding of the ball if there is no frictional resistance at the interface. If the sphere is given a twist about its vertical axis, it will spin, and the spinning and skidding will go on indefinitely in the absence of friction. In practice, there is always a finite but small amount of frictional force F at the interface acting in a direction opposite to T. This introduces a couple, and provided $T \not> F$, the ball will roll in the direction of the externally applied tangential force. Rolling occurs about an instantaneous centre O which is the point of common contact between the sphere and the plane. An instantaneous centre is a point about which the relative velocity between the sphere and the plane is zero. If, however, both rolling and sliding take place, the instantaneous centre is not at the mutual point of contact and the instantaneous axis of rotation is not in the plane of contact.

7.2   Grooved tracks

A ball bearing is a good example of a rolling element moving in a grooved track. The grooved tracks are provided by the two races. The inner race is attached to the shaft so that it rotates with the latter, and the outer race is stationary. As the shaft rotates, the balls roll along the grooves of both races. Consider a sphere under load in conformal contact with a groove (Fig. 7.1). The contact area will be elliptical and

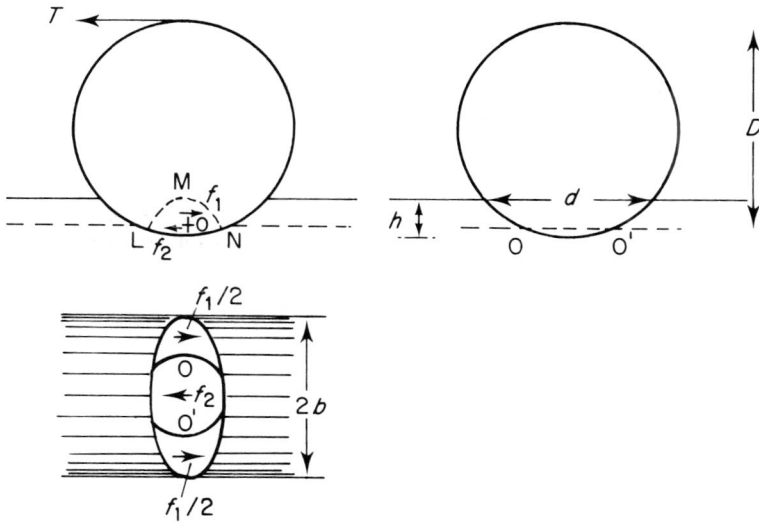

Fig. 7.1 A ball resting in a groove, making an elliptical contact area. (Bowden, F. P. and Tabor, D. (1964). "The Friction and Lubrication of Solids", Part 2, Oxford University Press, Oxford.)

an appropriate tangential stress will roll [2] the sphere about an instantaneous axis OO' passing through O. The regions remote from this axis of rotation may slip but can be prevented due to accommodation by elastic deformation of the contact zone. A cylinder and a sphere on a plane surface will give rise to a line and a point contact respectively. However, because of deformation, there will be an appreciable spread of material at the interface and the line will expand to a rectangle and the point to a circle. We have seen in Fig. 7.1 that a sphere in a groove produces an elliptical contact. As the sphere rolls the material ahead of it will be compressed, but will relax as the rolling body moves on. Certain parts of the contact zone will slip, the magnitude being decided by the product of the angular velocity and the distance of the region in the elliptical area from the axis OO' (Fig. 7.1) in the x–z plane. The external force T will result in a couple, G, which will resist rotation of the sphere. The couple is established because of the two force components f_1 and f_2 above and below the axis of instantaneous rotation respectively. Thus, at equilibrium, $T = f_1 - f_2$. Since both f_1 and f_2 are very large compared with T, the latter can be ignored and $f_1 \simeq f_2$. For a sphere of diameter D, rolling begins when $TD = G$.

Portions of the elliptical contact area remain locked with the sphere and are referred to as the stick area, while the remainder is available for interfacial skidding. That is the components undergo differential slip at the contact interface and the coefficient of friction μ as a result of this, is given by

$$\mu = 2f_1/W = 2f_2/W$$

where W is the applied normal load. If h is the depth of the deformed track, the forces f_1 and f_2 act a distance of about $h/2$ apart. Therefore,

$$G = f_1(h/2) = (\mu W/4)h$$

Rolling ensues when

$$T = G/D = (\mu Wh/4D)$$

It is easier to measure the major diameter $2b$ of the ellipse than the depth of the indentation, and ignoring the deformation of the sphere, $h = b^2/D$ since h is small.

Therefore,

$$T = \frac{\mu W}{4}\left(\frac{b}{D}\right)^2 \tag{7.1}$$

From a knowledge of the geometry of contact the normal load and the force required just to initiate sliding, the coefficient of friction due to differential slip can be calculated. This differential slip is known as Heathcote slip, and the full extent of this phenomenon can be realized with elastic materials such as rubber provided the grooved track is deep. Slip occurs over the trailing edge of the ellipse [3] and not, of course, over the entire area.

The tangential pull T can maintain rolling successfully only if it is strong enough to overcome the friction due to Heathcote slip. Frictional resistance has also been attributed to hysteresis loss and this is important while examining rolling of rubber. Rubber is discussed in a later chapter, but ball bearings are usually made of steel, as are many engineering components undergoing rolling motion. In such cases, the width of the contact area and the height of the instantaneous centre from the bottommost point of the sphere are several orders of magnitude lower than the ball radius. Under operational conditions, the contact time for a point on the sphere with the counterface is very small and the sliding speed over the slip areas is low, being of the order of 10^{-2} mm s^{-1} for steel on steel [1].

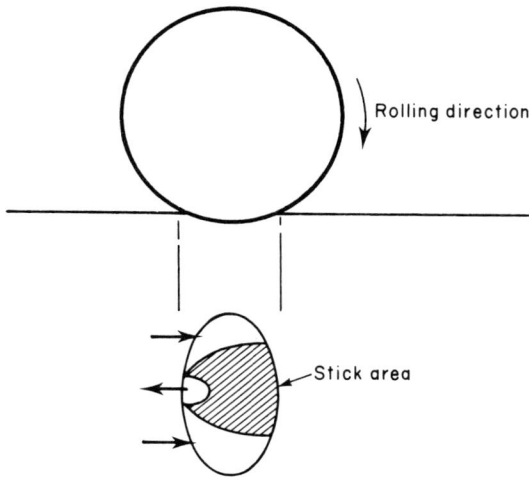

Fig. 7.2 Stick and slip regions of an interfacial area due to conformal contact.

The stick–slip phenomenon in a contact zone, or micro-slip as it is often called, is complicated and has been studied. A typical stick and slip region for conforming contact is shown in Fig. 7.2. The shaded portion of the elliptical contact is known as the stick area where the sphere and the groove move together in the direction of rolling. One way of describing the slip area is to say that the part of the groove at the trailing edge moves backward, but the fractions at the sides move forward.

7.3 Flat counterface

Although a sphere in a groove is quite a common situation in roller bearings, micro-slip is also a possibility if a sphere rolls on a plane elastic surface. This is Reynolds' slip [4], and occurs as a result of a differential strain pattern at the interface. The source of the slip process can be understood if we examine Fig. 7.3 where a hard sphere A causes a conformal depression in a block of soft rubber B under a normal load. It is evident that the block B extends more in position 1 than at either of the points 2 and 3. If the sphere attempts to roll, the interface will slip because of this differential extension.

Reynolds' slip is very small, and like Heathcote slip, does not contribute significantly to the loss of energy during rolling. It has been

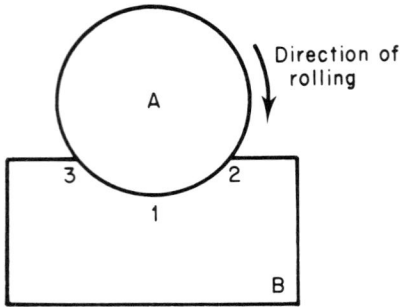

Fig. 7.3   A cylinder, A, rolling in an elastic block, B.

shown [2] that the frictional resistance during rolling can be attributed mostly to the hysteresis loss due to deformation. The phenomenon of hysteresis can be appreciated if we examine the elastic input energy during rolling. Thus consider a cylinder of radius R resting on the flat surface of a highly elastic block of material. Let the load per unit length of the cylinder be W producing a contact area of width $2a$ (Fig. 7.4). If the interaction is elastic, the pressure σ at a distance x from the centre of the contact region is $\sigma = \sigma_m[1-(x^2/a^2)]^{1/2}$ where σ_m is the maximum

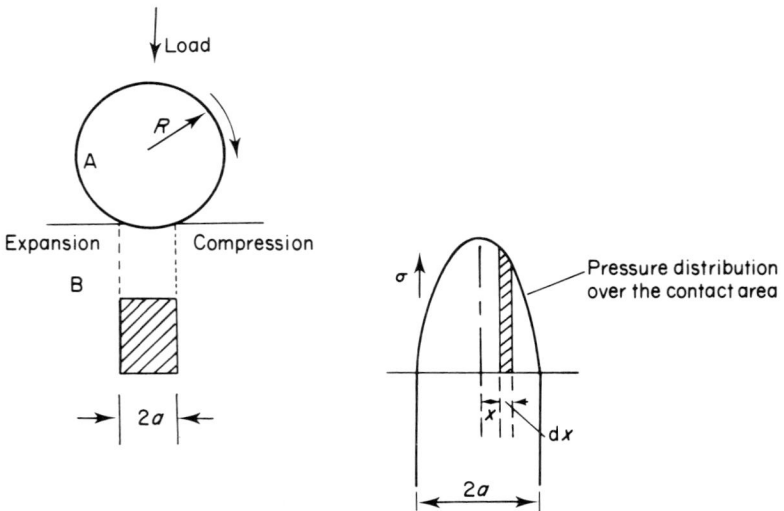

Fig. 7.4   A cylinder of radius R resting on a flat elastic block. $2a$ is the width of the contact area and the pressure σ is parabolic.

stress at the centre of the rectangular area of contact and $\sigma_m = 2W/\pi a$. Considering an elemental segment of width dx at a distance x from the centre, the normal force on this is σ dx if the strip is of unit length. During the forward compression of the interface, there will be a resisting couple dG about the centre of the contact zone and

$$dG = \sigma x\, dx$$

Therefore, the total couple exerted by the region in front of the centre line is

$$G = \int_0^a \sigma x\, dx = 2Wa/3\pi \tag{7.2}$$

As the cylinder rolls forward a distance x, the front region is compressed and the elastic input energy due to this is

$$Gx/R = 2Wax/3\pi R$$

Therefore, the elastic input energy ϕ per unit distance of rolling is

$$\phi = \frac{2Wa}{3\pi R} \tag{7.3}$$

By suitable experiments, the rolling friction can be measured at various loads and if the respective contact widths are measured the values of ϕ can be calculated from Equation 7.3. A plot of ϕ and rolling friction against load shows that rolling friction constitutes about 30% of the elastic input energy. Experiments carried out by loading and unloading a sphere under static conditions show that the hysteresis loss is about 9% of the elastic input energy. This leads to the conclusion [2] that rolling friction is largely due to hysteresis losses.

Considering Equation 7.3 again, the elastic input energy can be expressed in terms of intrinsic properties of a material. Thus assuming the cylinder to remain rigid under load, from Hertzian analysis,

$$a = \frac{2}{\sqrt{\pi}}\left[WR\left(\frac{1-\nu^2}{E}\right)\right]^{1/2}$$

Where ν and E are the Poisson's ratio and the Young's modulus respectively. Substituting this value of a in Equation 7.3,

$$\phi = \frac{4}{3\pi^{3/2}R^{1/2}}(W^{3/2})\left(\frac{1-\nu^2}{E}\right)^{1/2} \tag{7.4}$$

For the general case where the cylinder undergoes deformation as well, using suffixes 1 and 2 for the roller and the flat respectively,

$$\phi = \frac{4}{3\pi^{3/2}R^{1/2}}(W^{3/2})\left[\frac{1-\nu_1^2}{E_1}+\frac{1-\nu_2^2}{E_2}\right]^{1/2} \tag{7.5}$$

Although the above relationship provides us with the means of evaluating rolling resistance, the situation is complex. For example, the preceding analysis does not take into account the relaxation effect of the rear of a moving roller and the assumption that hysteresis loss is independent of time and temperature is not correct. The other factors which add to the complexity are the possibility of energy being dissipated due to plastic deformation in the region of maximum shear stress below the surface, although the other parts of the two solids are largely elastic. We should also consider that the hysteresis loss rate is dependent on the rate of strain, and that about three times as much energy is lost by hysteresis in rolling as in a simple tension test.

7.4   Plastic interaction

Although rolling situations are designed to avoid plasticity of the interface, it is instructive to consider plastic interaction. This will happen if a hard sphere of diameter D is rolled on the flat surface of a soft metal. The first effect is to form a groove by ploughing [5]. If $2a$ is the width of the track, the area supporting the load is $\pi a^2/2$ since the load W is supported by the front half of the ball only. If H is the hardness of the soft metal, the interface will yield when

$$H = 2W/\pi a^2 \tag{7.6}$$

From geometry, the cross-sectional area AOB (Fig. 7.5) of the groove is

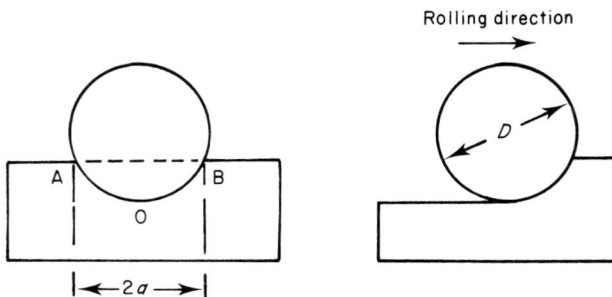

Fig. 7.5   A hard sphere rolling on a soft material.

$4a^3/3D$. If p is the resistance offered to the sphere by the wall of metal ahead of it that is, if p is the ploughing stress,

$$p = F/(4a^3/3D) = 3FD/4a^3,$$

where F is the net frictional resistance offered to the motion of the ball. There is no hysteresis term since the interaction is entirely plastic. Thus the frictional resistance is

$$F = \frac{4a^3 p}{3D}$$

Substituting for a^3 from Equation 7.6 in the above expression,

$$F = \frac{4p}{3D}\left(\frac{2W}{\pi H}\right)^{3/2} \quad \text{or} \quad F = K\frac{W^{3/2}}{D} \tag{7.7}$$

where K is a constant and $K = (4p/3)(2/\pi H)^{3/2}$.

The first traversal will produce a groove and the interaction is plastic, so that according to Equation 7.7, the frictional resistance is proportional to $W^{3/2}$. For sliding contact, the proportionality is directly to the applied load, and another important and notable difference is that the ball geometry is pertinent for rolling contact. Thus the larger the sphere, the smaller is the frictional resistance. We should note that the effect of interfacial adhesion is not apparent in Equation 7.7.

A limitation of expressing the frictional resistance in terms of the load and the diameter of the ball only is that we ignore the fact that F is governed by the track width as well. Experiments show that as rolling continues the width of the track increases, reaching a constant equilibrium value after a certain number of traversals. The probable reason for the increase in the frictional resistance, occurring initially, lies in the changing track width and in the fact that the counterface takes the form of a groove. The equilibrium radius of curvature of the track, however, is found to be some 10% larger than the radius of the rolling sphere.

7.5 Work hardening

Consider again a very hard sphere making a conformal contact with an equally hard counterface so that the interaction is elastic and an area below the interface is subjected to the maximum shear stress. Let the sphere make a conformal contact with the counterface creating a contact width $2a$ (Fig. 7.6). If the load is high the subsurface region will yield at a distance $0.705a$ from the centre of the contact. The maximum

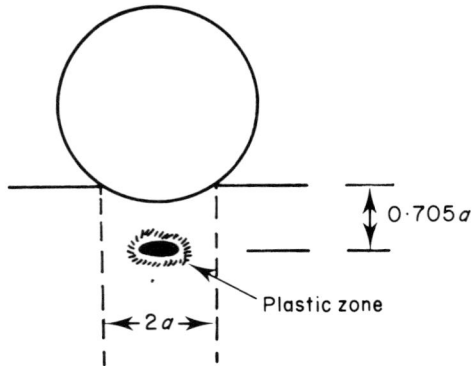

Fig. 7.6   A hard sphere contacting an equally hard counterface.

Hertzian stress necessary to cause plastic yielding is $3 \cdot 1\tau$, where τ is the shear stress of the material. Since the yielding is below the surface, a plastic zone will be held within an elastic hinterland. The effect of a tangential pull will, not unexpectedly, lead to the elastic outer layer being displaced in the forward direction of the rolling sphere, and this has been observed experimentally [6–8]. A mathematical analysis has been given [9] for the case of a rigid cylinder rolling on a flat surface, assuming that plane strain deformation takes place and the material under observation is isotropic. A final constraint in this mathematical treatment is that the material is elastic-perfectly-plastic that is it does not work harden.

When the roller is above a point such as is shown in Fig. 7.6, the region below will yield if the normal stress is high enough. As the roller moves on, the area at the rear is unloaded, but since there was plastic yielding, some residual stress will remain. This loading and unloading may continue for a while until the material reaches what has been termed its shake-down limit. This is the limiting yield stress attained by the material as a result of the normal compressive loading and the counterface becomes non-elastic again if the load necessary for this shake-down limit is exceeded. Plastic yielding will result in an appreciable shear in the direction parallel to the surface upon which the cylinder is rolling. Once the rolling element leaves the contact zone, the stress on it reduces to zero, but some permanent strain is left in the component in the forward direction.

7.6 Bearing life

A common type of failure in rolling elements is pitting followed by loss of material by spalling. In some quarters, this is referred to as fatigue failure, but we shall include this under wear for the following reason. A fatigue failure in engineering components manifests itself by complete fracture of the component. In a rolling element, complete failure in this manner may occur in which case we shall call it a failure by fatigue. If the effect of the repetitive stresses is spalling of surface layers, we are justified in calling it fatigue wear. The phenomena of micro-slip and spinning should give rise to sliding wear, but the amount will be small. We can say, therefore, that spalling or fatigue wear is the major source of loss of bearing life, which however will also depend on complete fracture of a rolling element, if that happens. Fatigue wear, that is the removal of surface layers by spalling, is a function of the number of revolutions. We do not visualize a progressive loss of material with number of revolutions as in sliding wear, but spalling, from the practical point of view, occurs suddenly and the rolling element loses surface layers. It is not known as to whether spalling is progressive once it starts or whether it goes on in an intermittent manner, but a bearing may become inefficient once spalling has begun, even if only on a small part of its surface.

The normal life N of a bearing is the number of revolutions it will sustain until spalling of the surface occurs. The bearing manufacturers plot N_n/N against % n as shown in Fig. 7.7, where N_n is the life of n% of bearings which have been tested. The testing conditions have to be kept constant to be able to plot a curve as shown in Fig. 7.7 and a large number of bearings are tested. A $B10$ life is defined as where 90% of the bearings used will give a life of N revolutions or more. Thus in Fig. 7.7 for a $B10$ life, $n = 90$% and the ratio N_n/N is unity. An average life N_m is defined as where 50% of the bearings have failed and N_m is five times the $B10$ life for an identical set of bearings. When one considers that a bearing may last 20 years or so, a considerable amount of time and effort is required using some form of accelerated laboratory testing to obtain a life curve.

Experiments show that for rolling, ball bearings give a life–load relationship as follows:

$$NW^3 = \text{constant} \tag{7.8}$$

Evidently, the load at the interface is critical.

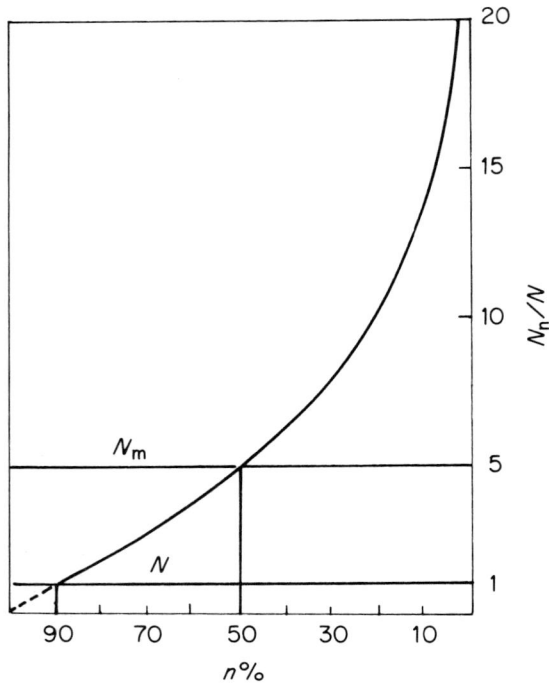

Fig. 7.7   Life curve for a rolling element. n = number of bearings still operational with life N_n; N = nominal life; N_m = mean life.

Spalling or fatigue wear is experienced even under fully developed elastohydrodynamic lubrication so that wear occurs as a result of induced normal and tangential stresses and direct physical contact between the opposing surfaces is irrelevant. The stress, which is Hertzian, can be increased locally if there is misalignment during assembly of a bearing and the failure mode is aggravated by the presence of metallurgical inhomogeneities in the material. Roller bearings are used under lubricated conditions, and at low speeds fatigue wear occurs relatively easily if there is high cage friction or if the lubricant contains water. When conditions at the interface are mostly rolling, as they should be, the surfaces should be finished smooth. A rough topography may mean that a selected number of asperities will carry the load, resulting in high stress locally. This will increase the probability of wear. High speeds increase the frictional resistance [10] of roller bearings and the possibility of high interfacial temperatures will need to be considered. Pitting and spalling occur also in cams and gears where the problem is aggravated because of high applied tangential pull, and the onset of failure can be observed visually in many cases. Among the aspects which indicate wear of bearing are:

(1) appearance of a large amount of wear debris;
(2) increased power consumption to maintain the same load and speed;
(3) increase in vibration of the system.

Of these, the wear debris can be detected if fluid lubrication is employed, but energy loss is not easy to ascertain in low powered systems. There is the possibility that excess power consumption due to energy loss will result in heat generation and a temperature measuring device for the contact area will indicate wear when efficiency is impaired. Monitoring vibration is a matter of experience, but whatever effect is utilized, the measuring device must be amenable to automation. That is as soon as a predetermined amount of wear has been reached the machine should be stopped automatically by, say, operating a trip mechanism for the driving motor.

7.7 Wear mechanism

Basic studies regarding the mechanism of wear become difficult because of the presence of flaws in the experimental component which act as sources of stress concentration. This has motivated studies of single crystals of MgO after being subjected to rolling traction. Single crystals of MgO have high elastic moduli and hence they can sustain high contact stresses. For a curved component, 1, of radius R indenting a flat surface, 2, the maximum shear stress region in the latter under a normal load W is

$$\left[\frac{3WR}{32}\left(\frac{1-\nu_1^2}{E_1}+\frac{1-\nu_2^2}{E_2}\right)\right]^{1/2}$$

where the E and the ν terms are the elastic moduli and Poisson's ratios respectively. Rolling steel balls, 6·25 mm in diameter, over single crystals of MgO under a normal load of 244 g, the depth of maximum shear stress in one experiment [11] was found to be 0·02 mm. The Young's modulus of the MgO produced by the authors [11] was 1·26 times that of the steel. Taking the Poisson's ratio of MgO as 0·25 and that of the steel as 0·30, the region of maximum shear stress is, by calculation, 0·017 mm below the surface. Agreement with experimental results is very good and evidence of plastic deformation is found by noting the dislocation density in those regions. Dislocations are generated in the subsurface region of maximum shear stress and crystallographic slip occurs on the $\{1\ 0\ 0\}$ planes of easy glide oriented at near 45° to the rolling interface [12]. The depth of slip cannot be expressed by a single formula because it depends on the rolling velocity as well as the normal load (Fig. 7.8). The effect of load is to increase the depth where dislocations generate, but the slip depth becomes shallower as the ball rolls with higher velocities. Although the rolling track retains its equilibrium depth, at low velocity, a transverse section of the rolling track of the MgO single crystal shows a deep and wide slip region (Fig. 7.9). The results in Figs. 7.8 and 7.9 are from single pass experiments and are explained [12] by making references to the velocity dependence of dislocation on the applied shear stress τ. If v is the velocity of dislocations, $v \propto \tau^{23}$. The subsurface of the solid will have a

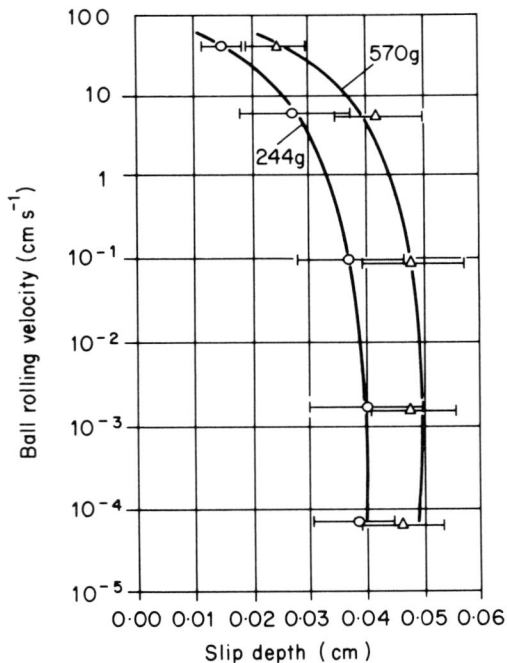

Fig. 7.8   Variation of slip depth with rolling velocity at loads of 244 and 570 g respectively. (Dufrane, K. F. and Glaeser, W. A. (1976). *Wear* **37**, 21.)

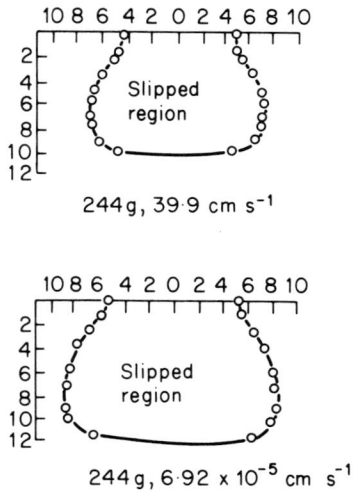

Fig. 7.9   Boundaries of dislocation loops below the surface. (Dufrane, K. F. and Glaeser, W. A. (1976). *Wear* **37**, 21.)

well-defined region where the shear stress effect is maximum, so that the generated dislocations will have a high velocity initially. The dislocations, however, will slow down and a great deal of time will be necessary for deep penetration. If the ball indents the surface and rolls quickly, the time available for the dislocations is inadequate and penetration is shallow.

It follows that there will be some increase in the slip depth with repeated rolling of the same track (Fig. 7.10). If rolling is continued for, say, 10^6 cycles or so, it is noted that at about 10^3 cycles, the slip depth increases a little more steeply than before. The track also widens with repeated cycles but there again appears to be a second regime where the increase in width is relatively high. It is in this second regime of change of track width with rolling contact cycles that spalling occurs. The micro-hardness of the track also increases with the running time.

Increased micro-hardness of the track suggests severe lattice strain due to plastic deformation. Typical of these rolling experiments is the fact that very soon after rolling begins, the generated dislocations interact extensively and their density increases in the same manner as the hardness. That is, after an initial increase, the density of dislocations levels off to a constant value and then increases again rapidly at about 10^5 rolling cycles, depending on the applied load. Microscopic

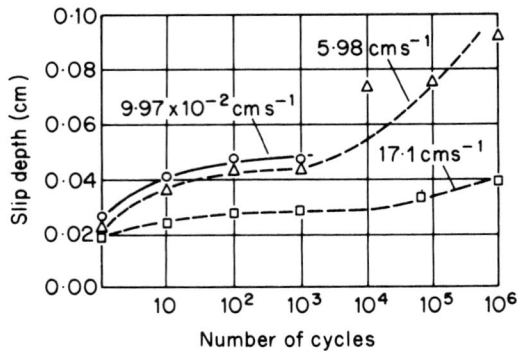

Fig. 7.10   Variation of slip depth with the number of rolling cycles at three rolling velocities. (Dufrane, K. F. and Glaeser, W. A. (1976). *Wear* **37**, 21.)

examination using polarized light shows that although the subsurface reveals the presence of residual compressive stress, tensile stress patterns appear after a number of cycles, when spalling occurs. The mechanism thus seems to be that rolling contact produces a network of high density dislocation in a localized area. As the dislocations interact further, a residual tensile stress is left which is responsible for nucleating cleavage-type cracks resulting in spalling A spalled layer may be produced by the propagation of a single crack or by joining together of a number of cracks.

Unfortunately, elements in bearings or other components such as gears are produced from steels which are heat treated to produce complex microstructures, and microscopic examination to produce evidence for dislocations may not be an easy exercise. Furthermore, the presence of inhomogeneities creates additional nucleation sites for crack initiation. Bearing steels have, however, been studied by metallurgical techniques and useful information has been provided to elucidate mechanism of fatigue wear [13].

Experiments with disc-type machines show two typical patterns of residual stresses measured both in axial and circumferential directions. In one case, there is some tensile residual stress just below the surface while the other example shows the subsurface to possess a wholly compressive residual stress. The nature of the stress pattern is important because conditions which give rise to a tensile peak stress promote failure by spalling. Formation of a peak tensile stress is obviously a result of strain hardening when dislocations interact, but experiments suggest that a tensile stress can be obviated by introducing compressive stress in the specimen by suitable heat treatment. The introduction of compressive stress in the surface prior to rolling may complicate the situation, but it is interesting to note that tensile stress is absent when the interaction is between unlubricated rough surfaces. The situation thus is that as the roller contacts an area on the counter-face, the region is subjected to a compression load. As the ball moves on, the region undergoes a tensile pull. This alternate compression and tension in a part of the component should initiate a fatigue crack which will soon propagate, forming a wear product. The fatigue crack probably initiates very near the surface and the nucleation centre could well be an inclusion or a microcrack, both left in the body of the component during manufacture.

A very worthwhile contribution to the understanding of fatigue wear due to rolling contact has been made by Hirano *et al.* [14]. The experiments using discs operating under elastohydrodynamic lubrication show four types of subsurface deformation:

    (1) the subsurface shows severe plastic flow in the direction of rolling while a thin layer of the surface is unaffected;

    (2) this type shows forward flow of the surface layer and subsurface deformation up to a substantial depth;

    (3) only a thin surface layer is deformed;

    (4) plastic deformation is not apparent.

Subsurface deformation such as those shown by types 1 and 2 is favoured at high contact loads, whereas under comparable experimental conditions, smooth surfaces show little deformation during rolling contact under elastohydrodynamic lubrication. If hardness is plotted from the surface to the centre, a peak hardness for the types 1, 2 and 4 is below the wear track, possibly in the region of the maximum shear stress. For an initially rough surface such as type 3, the surface is the hardest and the material becomes softer in the interior. The initially rough surfaces facilitate work hardening and a work hardened surface prolongs the fatigue life of rolling discs [15]. Work hardening is encouraged when the surfaces are rough because the ratio r increases, where

$$r = \frac{\text{Roughness of the mating discs}}{\text{Elastohydrodynamic oil film thickness}} \qquad (7.9)$$

When $r \simeq 0.4$, the discs' surfaces are completely separated by the oil film. In that event, the surfaces do not strain harden because they do not make contact although subsurface flow occurs if the contact stress is high enough. The suggestion is that fatigue wear will still be experienced but the life of the discs will be prolonged if the surfaces are rough, as finished, so that they work harden during running-in. We must note, however, that work hardening *per se* is not the criterion for improved fatigue life. If the result of deformation is a build-up of residual compressive stress only fatigue life may improve.

7.7.1 Slide–roll

If two discs of unequal diameters rotate with respective surface speeds v_1 and v_2, the velocities of rolling or sweep may be different for the two discs and this causes an amount of sliding between them. Gear teeth undergo slide–roll motion, and pitting, which may be the forerunner of spalling, is quite common. The causal factor for pitting must be the establishment of alternate residual tensile and compressive stresses. That is to say that dislocations emanate from the subsurface zone of maximum shear stress. These interact and eventually nucleate cleavage-type cracks. Application of an external tangential traction aggravates the condition in two ways. Firstly, the region of maximum shear stress moves towards the surface. Secondly, sliding friction causes heat generation at the interface [16]. This results in the breakdown of the elastohydrodynamic film. Possibly localized softening facilitates pitting if it is true that hard surfaces prolong the life of the components.

7.8 Slip

Pure rolling seldom occurs and some sliding takes place, which as we have already seen is known as interfacial slip. Apart from spalling due to residual stresses during rolling, some sliding wear is expected due to slip, although the magnitude of this is small.

To restate the phenomenon of slip, when a rolling element interacts with another surface under radial loads, there may be a portion of the contact area where the two elements adhere, while in the remainder they slide. The relative motion of the two surfaces due to the phenomenon of slip has also been termed creep, and the width d of the locked region of two cylinders rolling under a load W can be related [17] to the width a, of the interfacial area as,

$$d = a\left(1 - \frac{T}{\mu W}\right)^{1/2}$$

(7.10)

where μ is the static coefficient of friction and T is the applied tangential force. μW is the frictional resistance F, so that if $T = \mu W$, $d = 0$ and slip occurs over the whole contact width a. This is skidding on a large scale.

A static load on a rubber ball placed on a piece of inked linen stuck on a flat surface will show up the circular area of contact. If a tangential force is now applied [18], a crescent-shaped slipped region is seen at the trailing edge. The direction of slip and the amount of creep depend on the magnitude and the direction of the surface forces transmitted by T in Equation 7.10. As shown in Figs. 7.1 and 7.2, the experiments [18] confirm that, for a ball in a conforming cup, the contact area is elliptical and slip occurs in the central and the outer regions. It is also confirmed that the slip direction is opposite in the central area to that in the outer regions. That is, there is a negative amount of creep, which is slipping in the backward direction of the rolling element. Plotting the rate of wear of the ball and creep against the ratio (ball radius)/(groove radius), there is no slip at a characteristic ratio. This ratio is 0.5, and since creep is zero, there should be no wear. This is not so, although the wear rate is a minimum at the conformity where creep is zero. Ignoring experimental error, we could suggest that, at zero creep, there is a small amount of wear of the ball because some material from it is transferred to the counterface due to the normal load only. It has been shown that the rate of wear q is not linear with the applied load W and the experimental results fit the following relationship:

$$q = KW^2 \tag{7.11}$$

where K is a constant.

Generally, an incremental wear rate dq for elemental strips of the contact area is shown [19] to be dependent on the contact stress σ and the slip velocity v, so that

$$dq = K\sigma^n v^m \tag{7.12}$$

where, K, m and n are constants and the wear rate for the total contact area is obtained by integrating Equation 7.12.

We may note that it is unusual to see a velocity term in a wear equation as shown above. For sliding, wear rates decrease with speed presumably because of the protective effect of the oxide film covering the metal surface, formed as a result of a rise in the interfacial temperature. However, these velocities are very high and the probable slip velocity for a ball in a groove is about 0.2 mm s^{-1}. Lightly loaded ball and roller bearings show significant cage slip, being influenced by the surface speed of the shaft [20]. At a certain speed range, the cage slips by a maximum amount and the balls give wear rates of several hundred µg per hour. For example, at light loads, at a speed of about 16 000 r min^{-1}, a ball may lose some 400 µg h^{-1} for a running period of 25 h. This is a large amount of wear, but the balls do not lose material to any substantial degree upon further running. Wear due to slip is regarded by some as a running-in phenomenon, but slip encourages pitting and spalling, the latter being the most destructive. The races also wear albeit by small amounts, and the wear of balls is also influenced by the cage material. For example, when brass cages are replaced by steel, the amount of running-in wear of the balls is increased by a factor of five. Apart from the materials of the friction couple, other factors to consider are trapped debris and the presence of water in the environment. Debris, if trapped, will add an abrasive component in the wear process. The effect of water in the system will be discussed later, as even small amounts of water can cause an almost instantaneous increase in the wear rates by two orders of magnitude for balls, the cage and the races, including the cases where brass cages are employed.

7.8.1 *Sidecutting*

Sidecutting is the terminology used to describe wear between the wheel flanges and the side of the rails, resulting in a substantial amount of financial loss. Sidecutting is aggravated at sharp corners and studies [21] show that wear can be related to both the contact stresses and to creep. Creep, in this context, is defined as the fractional difference in the rolling speeds between two surfaces, and there will be two types of creep in a wheel-rail contact, viz.,

(1) longitudinal creep;
(2) lateral creep.

Representing the contact between a wheel and a rail as in Fig. 7.11, the longitudinal creep, expressed as a percentage is

$$\% \text{ longitudinal creep} = \left(\frac{2\pi n R_1 \cos \theta - 2\pi W R_2}{2\pi n R_2} \right) \times 100$$

or

$$\% \text{ longitudinal creep} = \frac{n R_1 \cos \theta - W R_2}{n R_2} = \left(\frac{R_1}{R_2} \right) \cos \theta - \left(\frac{W}{n} \right)$$

$$(7.13)$$

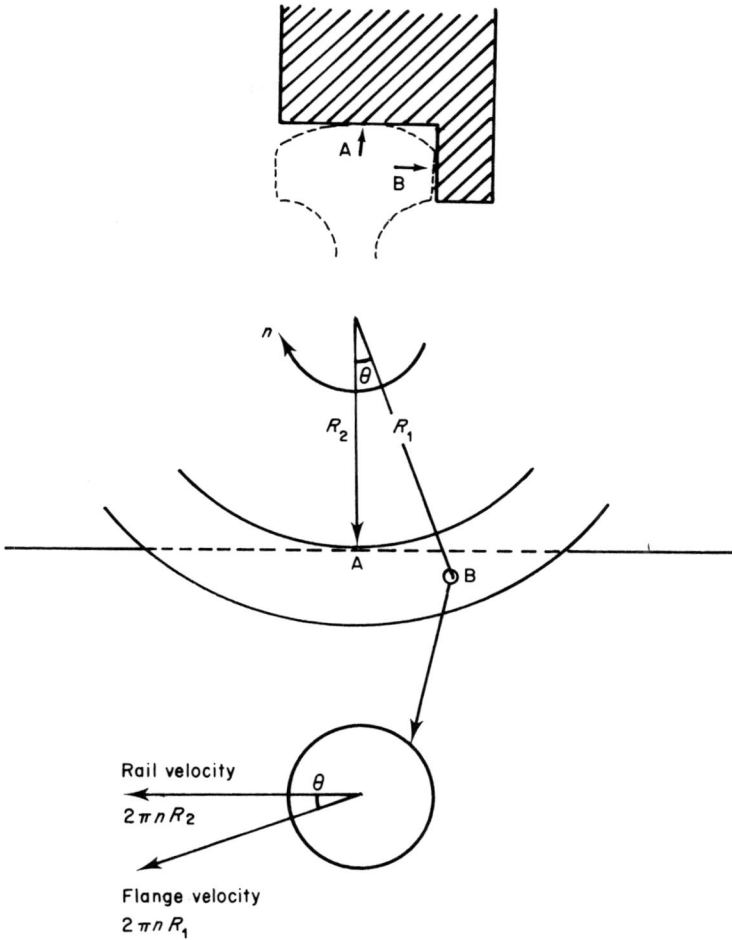

Fig. 7.11 Relative movement between flange and rail. (Beagley, T. M. (1976). *Wear* **36**, 317.)

where R_1 is the distance of the contact from the centre of the wheel of radius R_2 rotating at n r min^{-1}. A and B are the contact points between the wheel and the rail, and θ is the angle as shown. W is the axle load and θ is usually small. The lateral creep is small and is defined as

$$\% \text{ lateral creep} = \frac{2\pi n R_1 \sin \theta}{2\pi n R_2} \qquad (7.14)$$

If the coefficient of friction μ defined as the applied tangential force divided by the normal force, is plotted against the longitudinal creep δ, μ increases initially but becomes constant at $\delta \simeq 3\%$.

Laboratory disc tests allow wear rates to be measured accurately by the weighing technique. Converting the load into Hertzian contact stress, wear is seen to increase with load, and considerably with the amount of slip. This is shown in Fig. 7.12 where the wear particles were continuously removed during the test by brushing. This simulates the actual situation where the contact areas are subjected to twists and spins which help to shear off adherent debris. At low contact stresses and creep, the roller surfaces remain polished in appearance and the debris constitutes about 80% oxide and 20% metal. If the degree of creep or the contact stress increases, the roller surfaces develop a matt grey appearance and the wear debris is largely metallic, indicating severe wear. If the interface is lubricated, higher contact stresses can be sustained although wear by the fatigue mode is experienced if a critical normal load is exceeded. An interesting conclusion from laboratory studies [22] is that a low friction of $0 \cdot 05$ can be obtained between rolling surfaces in the presence of water which apparently forms a low viscosity paste with the wear debris at the rail–tyre interface.

The wear of the sides of the rails can be as much as three times that of the heights of the rails, and increases rapidly as the curvature of the rail falls below 1 km. Wear increases during climbing a gradient and during braking and is aggravated by sand at the interface. The track also loses material due to corrosion if the interface temperature rises above 200 °C. A very decisive effect is seen by atmospheric pollution. For example, due to the relatively clean environment, the rails in an underground train system suffer the least amount of wear. In tunnels full of steam and smoke such as are produced by steam trains, the height loss of rails can go up by a factor of about 10^2.

7.9 Materials

Taking the rolling elements as examples, a high carbon through hardening steel with 1% C and $1 \cdot 5\%$ Cr, heat treated to 900 HV, is the most popular. The hardness thus achieved is lost at elevated temperature and in these circumstances high speed tool steels containing tungsten and molybdenum are used. For a maximum rolling contact fatigue life, heat treatment is very important to obtain the desired hardness.

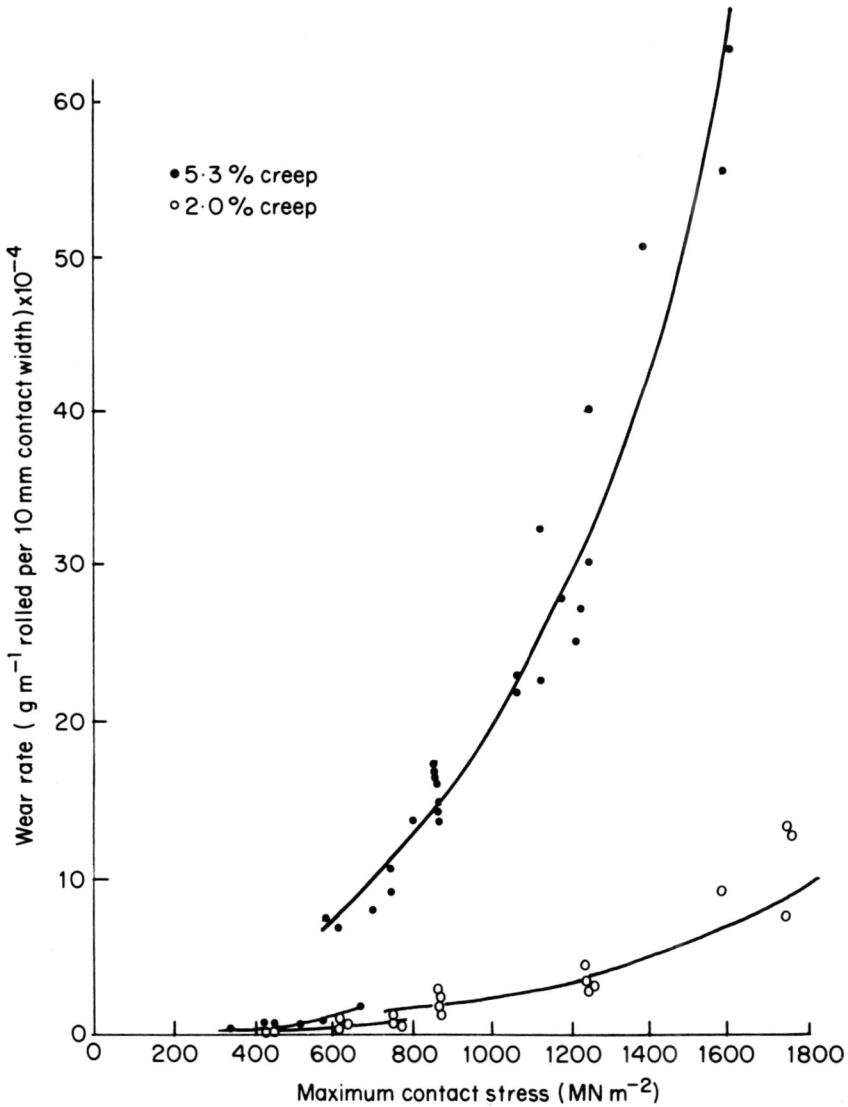

Fig. 7.12   Wear rates of rail steel as a function of contact stress. (Beagley, T. M. (1976). *Wear* **36**, 317.)

Whereas hardness is a useful criterion, the combination of materials constituting the friction couple such as a ball and a race is shown to be of fundamental importance [23]. In experiments with a four ball machine, the width of the rolling track shows a relationship with pitting failure. A narrow track width is produced by tungsten carbide rolling on itself and pitting occurs quickly. The material has a high Young's modulus so that the track width, that is the interfacial contact area, is small. It follows that a tungsten carbide ball running against a race produced from the same material will exert high contact stresses, which is unfavourable to long life. Contact stresses can only be reduced by a degree of deformation of one of the elements so that some plastic yielding is sought. Excessive plastic flow may result in undesirable subsurface tensile residual stresses so that a compromise is the only alternative. The recommended practice for ball bearings is to have a ball with a hardness about 10% higher than the race.

The harmful effect of inclusions has been fully appreciated and much attention has been devoted to producing clean steels by, for example, vacuum melting and casting [24–26]. Careful melting and casting techniques also ensure that the ball bearings contain a minimum amount of harmful gases. Presence of water at the interface has been noted to be deleterious [27–29] because of the hydrogen which is liberated at the interface and then dissolved in rollers and races.

One of the harmful gases which reduces the life of rolling elements is nitrogen. Nitrogen, like hydrogen, can diffuse into the body of a component at elevated temperatures and it has been shown [30] that plastic deformation produces defects in α-iron which trap nitrogen. Loss of life with the nitrogen content of the metal has been demonstrated [31]. Results of experimental investigation to date have led to the speculation that rather than the total nitrogen content of the steel, the interstitial gas should pin down the generated dislocations. This delays any recovery and recrystallization process and fatigue cracks are readily nucleated. An interesting report [32] has appeared recently which has evaluated high speed steel rolling elements manufactured by the powder metallurgy technique. These show a poor performance at ambient temperature, but the mean life is considerably improved if tests are carried out at 200 °C. Examination of samples from rolling experiments reveals that a finely dispersed phase is precipitated. This is probably silicon nitride and is less harmful than fine nitrides or interstitial nitrogen.

The loss of rolling elements due to contamination by water from the lubricant is undoubtedly a serious issue. Running of rolling elements in a water contaminated environment results in pick-up of hydrogen by the steel and the propensity to fatigue increases. Hydrogen appears to penetrate *in situ* in atomic form in the rolling element since plastic deformation aids diffusion through vacancies. The practical solution is to incorporate additives in the lubricant which impede access of gas in the body of the metal.

Of the inclusions, mainly in the form of oxides and silicates, the shape and size seem the dominating factors in effect on the fatigue life of rolling elements. Brittle angular silicates are harmful and they appear as small particles, being products peculiar to air-melted steels. From the fatigue point of view of steels, particles less than 10 μm in diameter are less harmful for 1·5% Ni–Cr–Mo type steels. Since the mechanism is similar, the conclusions should apply to fatigue wear as well. As a general rule, highly stressed steels will not tolerate large inclusions. Metallographic examination [33] of used rolling elements shows that cracks can appear at both the surface region and below the surface, and the sites always appear to be those of inclusions. Surface cracks appear at the edge of the contact area and are influenced by the environment. They spread into the body of the component at an acute angle to the rolling direction. The cracks then change direction as they encounter the region of maximum shear stress and propagate parallel to the surface producing a thin layer of wear debris. The presence of hydrogen may cause the crack to spread deeply into the component.

Some other tribological components used under mainly rolling contact are gears, cams and tappets. In the former, the adhesive mode of interaction due to lubrication failure gives rise to scuffing. Apart from this, pitting is common in gears but may become arrested. Arrested pits are harmless from the fatigue point of view, but progressive pitting may result in spalling. In low duty applications, a wide range of materials from polymer to cast iron are used for gears and a carefully selected steel is the usual material for arduous service conditions. Unless both members of a gear pair are surface hardened, it is usual to have two different types of steel from the composition point of view. This reduces scuffing, and for long life the member of the pair which has the most heavy duty to perform at a high speed, e.g., the pinion, is made harder than the other mating gear.

Cams and tappets can fail by scuffing, which happens because of intimate metallic contact at regions of lubricant starvation. The possibility of scuffing during the running-in stage is real if the surfaces are rough, allowing highly stressed contacts at the asperities. It is noted, however, that we have already examined the view that running-in of rolling elements with rough surface finish will result in work hardening and hence improved fatigue life. Nevertheless, pitting is common if the contact stress is high. The spalled surface layers are helped to separate from the substrate apparently by lubricants which can act as levers. A form of wear called polish wear is quite common in these components and is believed to fall intermediate between pitting and scuffing. An amount of chemical wear due to the lubricant is suspected in polish wear.

A component such as cam undergoes rolling and sliding and the maximum Hertzian stress σ_m is calculated [34] from a knowledge of the load W between a cam and tappet as,

$$\sigma_m = K\left(\frac{W}{Rb}\right)^{1/2} \tag{7.15}$$

where $R =$ cam radius of curvature at the contact point; $b =$ cam width, and K is a constant depending on the material combination.

The safe stress depends on the material property and includes such surface treatments as phosphating. Note that in rolling combined with sliding, surface traction forces also determine the propensity to failure, e.g., by reducing the friction coefficient at the interface. The sliding friction, of course, will also be governed by the lubricant, in particular, by its viscosity at the operating temperature. Naturally, the permissible contact stress is low if there is excessive interfacial slip. Very good surface finish commensurate with a stable elastohydrodynamic oil film separating the surfaces appears to be the practice although quantitative information is sparse. From the materials point of view, some typical combinations are a phosphated chill-cast iron tappet with a carburized steel cam for high contact stress and large relative sliding, or grey cast iron on tool steel for low contact stresses.

References

1. Halling, J. (Ed.) (1975). "Principles of Tribology", Macmillan, London, p. 174 ff.
2. Bowden, F. P. and Tabor, D. (1964). "The Friction and Lubrication of Solids", Part 2, Oxford University Press, Oxford.

3. Johnson, K. L. (1959). *Proc. Inst. Mech. Eng.* **173**, 809.
4. Reynolds, O. (1876). *Phil. Trans.* **166**, 155.
5. Eldredge, K. R. and Tabor, D. (1955). *Proc. Roy. Soc. A* **229**, 181.
6. Crook, A. W. (1957). *Proc. Inst. Mech. Eng.* **171**, 187.
7. Welsh, N. C. (1957). Proceedings: Conference on Lubrication and Wear, Institution of Mechanical Engineers, London, p. 701.
8. Hamilton, G. M. (1963). *Proc. Inst. Mech. Eng.* **177**, 1.
9. Merwin, J. E. and Johnson, K. L. (1963). *Proc. Inst. Mech. Eng.* **177**, 676.
10. Palmgreen, A. (1945). "Ball and Roller Bearing Engineering", S.K.F., Philadelphia, PA.
11. Amateau, M. F. and Spretnak, J. W. (1963). *J. Appl. Phys.* **34**, 2340.
12. Dufrane, K. F. and Glaeser, W. A. (1976). *Wear* **37**, 21.
13. Muro, H., Tsushima, T. and Nagafuchi, M. (1975). *Wear* **35**, 261.
14. Hirano, F., Yamashita, N. and Kamitani, T. (1971). Tribology Convention, Institution of Mechanical Engineers, London, p. 151.
15. Hirano, F., Kuwano, N. and Ichimaru, K. (1966). *Proc. Inst. Mech. Eng.* **181**(30), 85.
16. Hirano, F., Sakai, T. and Kamitani, T. (1971). Tribology Convention, Institution of Mechanical Engineers, London, p. 86.
17. Johnson, K. L. (1958). *J. Appl. Mech.* **25**, 339.
18. Brothers, B. G. and Halling, J. (1965). Lubrication and Wear Third Convention, Institution of Mechanical Engineers, London, p. 134.
19. Giolmas, S. N. and Halling, J. (1965). Lubrication and Wear Third Convention, Institution of Mechanical Engineers, London, p. 145.
20. Herkert, B. (1972). Tribology Convention, Institution of Mechanical Engineers, London, p. 33.
21. Beagley, T. M. (1976). *Wear* **36**, 317.
22. Beagley, T. M. and Pritchard, C. (1975). *Wear* **35**, 299.
23. Scott, D. and Blackwell, J. (1966). Lubrication and Wear Fourth Convention, Institution of Mechanical Engineers, London, p. 36.
24. Uhrus, L. O. (1963). "Clean Steel", Iron and Steel Institute Special Report 77.
25. Scott, D. and Blackwell, J. (1963). *Proc. Inst. Mech. Eng.* **178**(3N), 81.
26. Scott, D. (1969). *Vacuum* **19**, 167.
27. Grunberg, L. and Scott, D. (1958). *J. Inst. Petroleum* **44**, 406.
28. Grunberg, L., Jamieson, D. T. and Scott, D. (1963). *Phil. Mag.* **8**(93), 1553.
29. Ciruna, J. A. and Szieleit, H. J. (1973). *Wear* **24**, 107.
30. Wriedt, H. A. and Darken, L. S. (1965). *Trans. Met. Soc. AIME* **233**, 111.
31. Scott, D. and McCullagh, P. J. (1975). *Wear* **34**, 227.
32. Scott, D. and Blackwell, J. (1974). *Wear* **34**, 149.
33. Scott, D., Loy, B. and Mills, G. H. (1966). *Proc. Inst. Mech. Eng.* **181**(30), 94.
34. Neale, M. J. (Ed.) (1973). "Tribology Handbook", Butterworths, London.

8 Metals and Alloys

The evolution of metallic materials from the point of view of wear resistance has undoubtedly been governed by the supposition that the hardness is the criterion. This has led to the use of white iron or Hadfield manganese steel for abrasion resistance. Although many tribological components are hard, for example, gears, cams, tappets etc., it does not always follow that a hard surface will automatically resist wear. We therefore wish to know what parameters will give a satisfactory answer while designing for minimum wear. The answer probably is that such parameters are not known at the moment and each friction couple should be assessed individually. It follows that both members of a metallic friction pair will influence the friction and wear behaviour. That is to say that it is the nature of the metallurgical interaction at the interface which probably decides on the tribological properties of a metallic friction couple. One effect of sliding a metal under load is to deform the interface. Deformation causes work hardening and we have already seen that the hardness of this deformed layer may influence the wear rate of the component. The ability for a metal to work harden depends on the crystal structure and there is some information on its role in friction and wear of metals. Another property which has been investigated is the mutual solubility of the metals constituting the friction couple. The literature pertaining to the role of crystal structure and solid solubility is not prolific but certain general conclusions are possible. These will be discussed in the sections of this chapter, followed by a summary of laboratory studies of some tribologically interesting alloys such as brass, cast iron and aluminium–silicon alloys. The usual experimental approach is to vary certain engineering parameters such as load and speed to see their effect on rates of wear. Normally one or both members are examined metallurgically to establish the nature of the subsurface deformation or of the wear debris to help elucidate mechanisms.

8.1   Crystal structure and solubility

We have already seen that the plasticity of a metal is greatly aided by a favourable crystal structure. Easy glide of slip planes should introduce a low friction material but quantitative wear rates have not been a pressing area of study as far as the role of crystal structure is concerned. This is because the work is considered to be too mechanistic in nature and the experiments are difficult to conduct. One of the difficulties is that with certain combinations of metals, severe stick–slip even in room atmosphere produces excessive vibrations in wear machines so that the pin may be out of contact with the counterface most of the time. In that case plotting the amount of wear against sliding distance becomes misleading. Seizure of the interface is very common, particularly if experiments are attempted in high vacuum.

Although quantitative wear rates have not been attempted, there is the possibility that hexagonal structures give low wear [1]. Graphite is a typical material with hexagonal structure and the tribology of it is discussed in a later chapter. The possibility of low friction and wear has been demonstrated for certain rare earth metals and friction experiments have been conducted for cobalt [2]. Cobalt is a useful tribological material and is an alloying constituent for many hard facing materials.

Experiments with mutually solid soluble metals are also difficult because of the propensity to seizure but it is known that the best friction couple is that which comprises two metals showing no solid solubility. Qualitative studies [3, 4] have been carried out to demonstrate the concept of solubility by using resistance to surface damage as the yardstick. The authors use the term *score resistance* for experiments on steel discs supporting 156 mm² square specimens at variable normal loads of up to 540 kg and at a fixed surface speed of $23 \cdot 3 \text{ m s}^{-1}$. The ability of a friction pair to survive was classified by stating the magnitude of the score resistance as follows:

(1) very poor—seizure at a load of 135 kg;
(2) poor—a load of 135 kg could not be sustained without seizure for a sliding time of one minute;

(3) fair—a load of 225 kg could not be maintained for more than one minute;

(4) good—sliding could be sustained up to one minute at a normal load of 225 kg.

The authors [4, 5] examined some 38 different metals sliding against the steel discs and came to the general conclusion that materials which had limited solubility in iron or which formed intermetallic compounds show good resistance to surface damage. However, the results show certain inconsistencies. For example, calcium and barium are immiscible in iron but they show very poor and poor score resistance, respectively, while cadmium and copper failed to provide reproducible results. All this suggests that wear is a complex process and caution is necessary when interpreting results from a particular experiment. A curious aspect of these experiments is the very high load which the friction couples were able to sustain. A much smaller load of 0·27 kg and a surface speed of only 5 mm s^{-1} have been used [6] to carry out similar experiments but the apparatus was kept at a low pressure. It is again shown that mutually soluble metals form coherent junctions and wear readily.

8.2 Brass

Brasses have been popular tribological materials, particularly the 60/40 variety with up to 2% lead. This has also been used as a material to study mechanisms in laboratory machines. Plotting the weight loss against the sliding distance for dry sliding produces a linear wear pattern (Fig. 8.1) and the rate increases with load. This agrees with the wear law (Equation 3.7). Experiments with radioactive pins have shown that as the machine starts some brass is deposited on to the steel counterface. The deposit is then progressively removed, possibly by a process of abrasion, by the work hardened pin. Thus, the implication is that the brass wears by both the adhesive and the abrasive mode. Both these laws state that the wear rate is directly proportional to the load. This is broadly true for brass and a plot of rates of wear against load shows three regimes of wear (see Fig. 8.2), which have been explained as follows.

Up to a load of about 7 kg, the brass loses material by adhesion and abrasion and the rate of wear is decided by the applied normal load. The applied stress, of course, will always give rise to plastic and elastic yielding of the true areas of contact. In the case of brass, there is evidence to show that marked plastic yielding of the surface layers

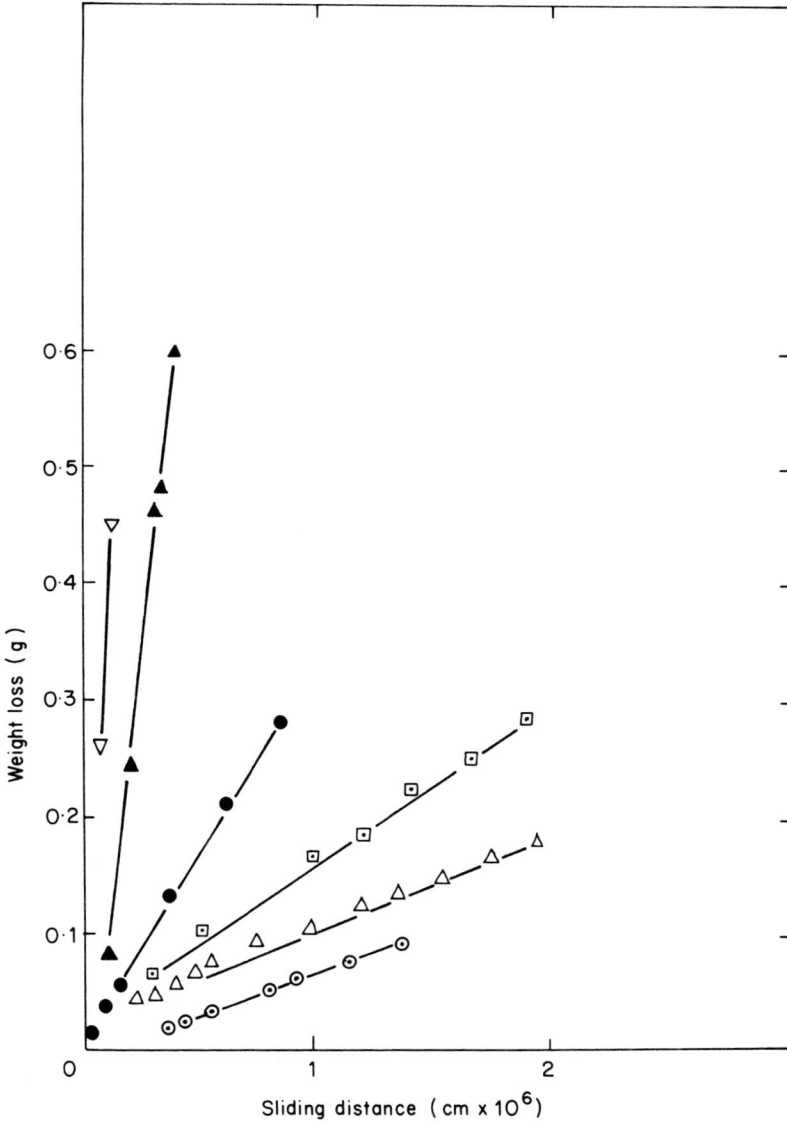

Fig. 8.1 Weight loss of 60/40 brass with 0·5% lead, sliding against hard steel. Loads in kg: ○ 1; △ 3; □ 5; ● 7; ▲ 9; ▽ 11.

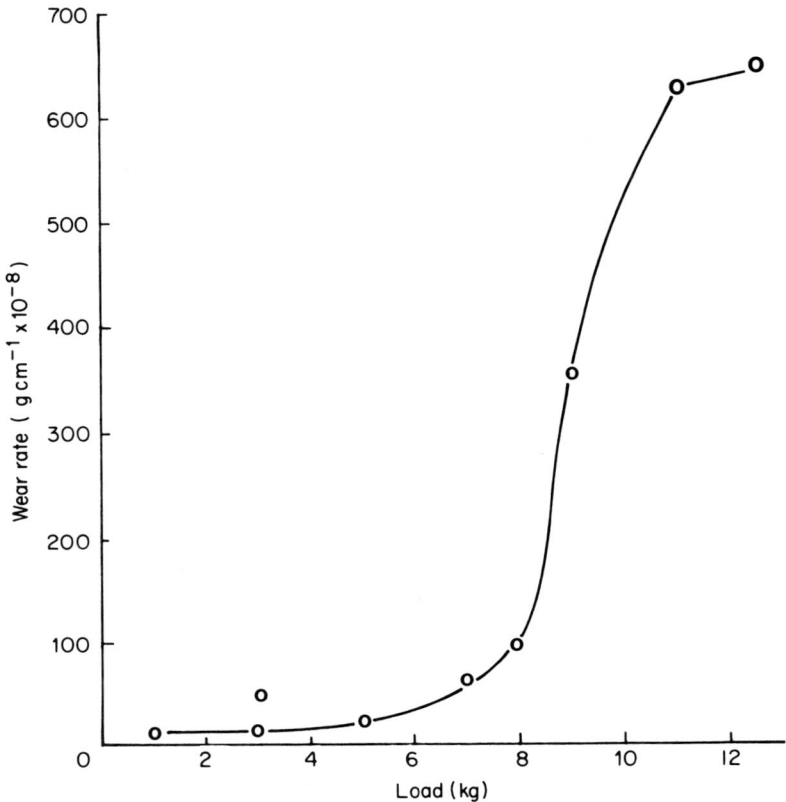

Fig. 8.2   Wear rate of 60/40 brass with 0·5% lead, sliding against hard steel as a function of applied normal load. Pin diameter 6·25 mm.

begins at a characteristic load. This should accelerate the rate of wear and in Fig. 8.2 a different but accelerated wear rate continues beyond a load of 8 kg. The third regime shows a lower wear rate than the previous two regimes and it has been speculated that a hard equilibrium layer forms on the brass pin which resists wear. In other words, the suggestion is that there are three different modes of wear for brass depending on the applied normal load.

Taking the hardness of the brass as 112 kg mm^{-2} and density $\rho = 9$ g cm^{-3}, the wear rate q can be expressed as $q = \beta(W/3H)\rho$. This gives a value of β of the order of 10^{-4} up to a load of about 8 kg, assuming a linear wear rate. As Fig. 8.2 shows, however, this is not so and we can

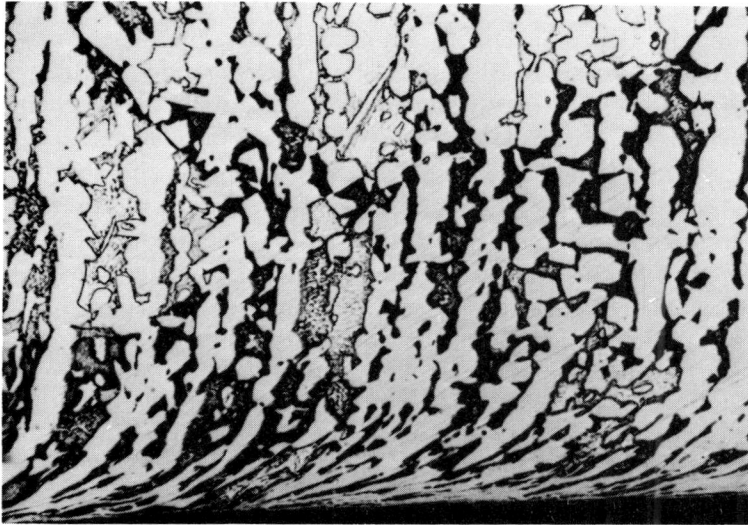

<----Direction of sliding

Fig. 8.3   A longitudinal section through a worn 0·5% leaded 60/40 brass pin. Hard alloy steel bush; load 3 kg on a pin 6·25 mm in diameter; surface speed, 141 cm s^{-1}.

say that the wear rate of brass is only approximately linear with the applied normal load.

Apart from examination of wear rates, basic information is also provided by metallographic studies of both the wear scar and of sections through members of friction couples. For example, a longitudinal section through a 60/40 brass pin in the very early stage of sliding shows that at a load of 3 kg marked bending in the direction of motion occurs (Fig. 8.3). This is below the transition load but a thick work hardened layer in the third regime of wear is readily seen in a longitudinal section of a pin slid at a very high load, say, of 16 kg on a 6 mm diameter sample. The depth of this deformation varies in the range 100–200 μm and the hardness increases to about 165 HV from a prior value of 112 HV. The hard deformed layer fractures and provides the major source of wear debris.

8.3 Steel
Some early work [7] on steel sliding on steel has led to the conclusion that there are two types of wear, viz. adhesive and abrasive wear. The

adhesive mode is experienced when mild steel runs on itself and an effect of it is to cause severe vibrations of the wear machine, particularly if the applied normal load is high. The adhesive mode produces severe surface damage which can be reduced if the mechanism changes to abrasive wear. Abrasive wear is attributed to the presence of oxide layers in the surface, the oxide particles acting as the abradants. A typical counterface for which mild steel undergoes a low rate of abrasive wear in this manner comprises a hard chromium steel disc. Abrasion of both members by oxidized particles has also been suggested [8] from experiments on hard steels sliding on a similar counterface. For a mild steel pin sliding on a hard steel disc, in all probability metallic lumps are transferred from the former on to the latter [9]. If the applied load is not high so that the rate of wear is low, the transferred metal oxidizes and then detaches as wear debris. As fresh surfaces of the counterface are thus exposed, further deposit from the pin occurs and the sequence of events is repeated in this manner. The wear debris has been identified as black α-Fe_2O_3 and the rate of metal transfer and of wear are controlled by the rate of oxidation of the interface, which is markedly influenced by the surrounding atmosphere and temperature.

Atmospheric nitrogen has been shown [10] to reduce the wear rates of mild steel, presumably by forming nitrides. Experiments with 0·12% carbon steel show that the wear rate is high consistent with the time of running when the applied load is low. At high loads, a hard surface layer is formed and the initially high wear rate diminishes to a lower value with time of running. Plastic deformation of a steel facilitates solution of nitrogen and it is likely that iron nitrides form which are wear resistant.

A great deal of valuable information has been obtained by employing the technique known as ferrography [11]. Essentially, a lubricant from a machine which holds the wear particles produced in service is pumped out at a low rate of 0·25 ml min^{-1}. The lubricant, thus extracted, is diluted and the particles are subjected to a magnetic field so that the magnetized debris adheres to a transparent substrate. The wear particles attached to the slide are washed clean of lubricants etc. and the quantity and size of the wear debris can be ascertained by optical density measurements. The particles thus separated may also be examined by microscopical methods.

Samples collected from actual machines can be identified as being due to various modes of wear [12]. Gears may produce platelets (Fig. 8.4) which are typical products of adhesive wear. Abrasion of the steel

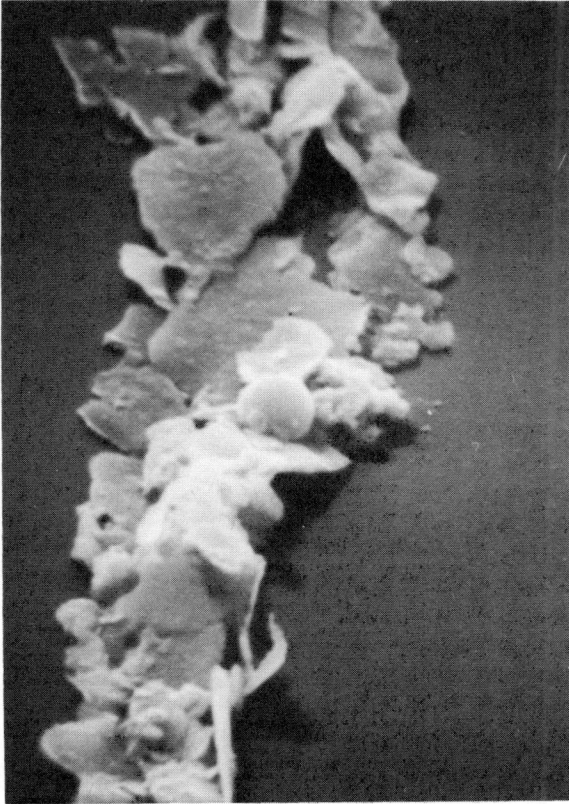

Fig. 8.4   Strings of platelet-type wear particles from the lubricant of a full scale gear test (×6000). (Scott, D. (1975). *Wear* **34**, 15.)

surface shows micro-chipping as expected due to the ploughing action. A ground surface or a well run-in surface exhibits a Beilby layer, but if running is continued at low loads the surface may produce wear particles in the form of platelets as shown by ferrography.

Spherical particles of wear debris of worn roller elements demonstrate fatigue wear during rolling contact (Fig. 8.5). The spheres are believed to form from tongues of metal as deformation continues. The rolling or the balling up process produces a layer-like structure of spheres. The balling up process happens in the surface cracks which first form due to fatigue action. A platelet may form, say, by spalling,

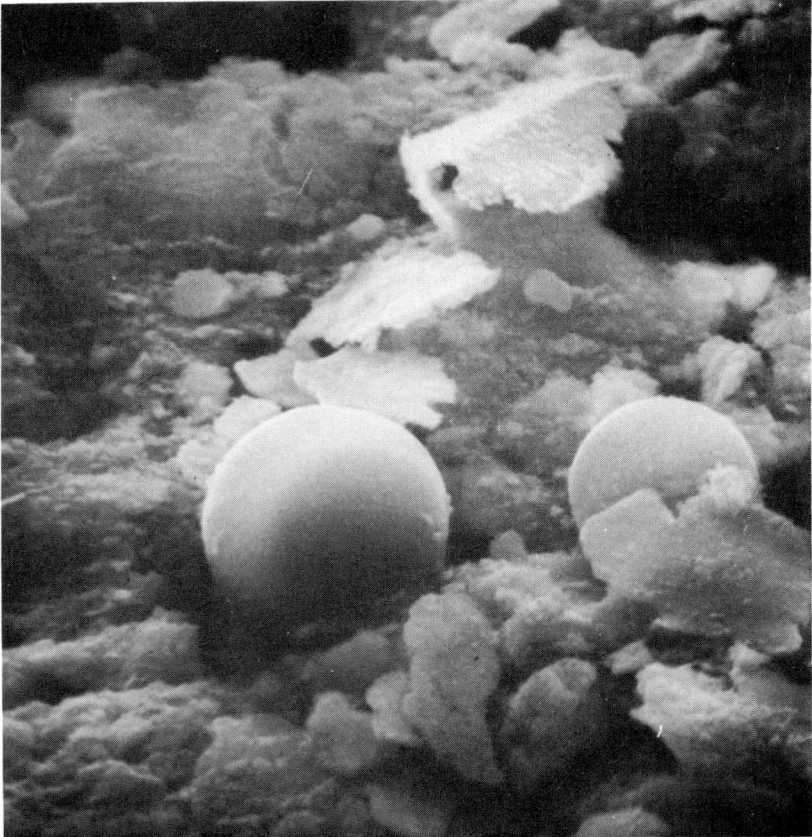

Fig. 8.5   Spherical and platelet debris of worn roller elements. The spherical particles demonstrate fatigue wear (×3300). (Scott, D. (1975). *Wear* **34**, 15.)

and this is then carried into the crack by the lubricant. As the crack undergoes deformation the platelet becomes spherical and escapes into the lubricant before the crack opens up further. It is assumed, of course, that the platelet is malleable enough to be capable of deformation.

8.4   Cast iron
Some typical steels and cast irons have been discussed from the point of view of abrasion resistance in Chapter 4. In this section we shall review the available literature to see if wear rates can be correlated with bulk

hardness and composition. Prior to this, we may consider the wear rates of cast irons with flake and nodular graphite respectively. Experiments with a pin–bush machine, where the bush is a hard steel, show the weight loss to vary linearly with sliding distance as expected for both these irons. Plotting the wear rates against load, however, tends to show a similar overall pattern as that of the brass in Fig. 8.2. Laboratory experiments suggest that the load carrying capacity of spheroidal graphite iron is slightly superior to that of an iron with flake graphite, but direct proportionality with load is not shown by these materials. The marginal superiority of SG irons is believed to be due to the shape of the graphite particles.

The general role of graphite must be as a lubricant since cast iron plain bearings are successfully used without the aid of an external lubricant. We shall see in a later chapter that the bearing interface must be contaminated with, say, moisture for graphite particles to be effective in lubricating the interface. Dry graphite crumbles readily under the action of normal and tangential forces but even a moist interface will give rise to wear of the graphite phase. This is beneficial under lubricated sliding since the voids left behind by the graphite flakes or nodules will hold oil and act as reservoirs for the lubricant and be helpful during boundary lubrication.

Like most industrial alloys, cast iron has a number of micro-constituents with varying degrees of hardness. As we have seen in Chapter 4, a high bulk hardness appears to improve the wear resistance of cast iron in various abrasive situations. The recommended hardness for both plain carbon and alloyed white irons is in the range 400–900 HV, depending on the service requirement. Ni-hard, a cast iron with 5% Ni and 2% Cr achieves its wear resistance due to its martensitic structure but irons with up to 30% Cr are used in brick making dies where corrosion problems are common. The role of chromium is to form carbides in a matrix of ferrite, and irons incorporating chromium only are not as hard as, e.g., Ni-hard. Plain carbon and alloyed cast irons with flake and nodular graphite structures respectively have been evaluated as ploughshare materials by large scale field trials coupled with accelerated laboratory experiments [13]. Various compositions were used with carbon 2·9–3·7%, phosphorus up to 0·12% with very low amounts of nickel, chromium and molybdenum additions. If the irons were cast in sand moulds, a correlation between the bulk hardness and wear rate was obtainable under laboratory tests. If the iron was chilled so that the carbides were distributed in a haphazard manner in a pearlitic matrix, the laboratory tests showed poor wear resistance. An improperly chilled iron will have free graphite but if the casting conditions are controlled to produce a completely white cast iron, field trials confirmed its superior abrasive wear resistance.

There is a suggestion [14] that the wear resistance of cast iron should correlate with the algebraic sum of the hardnesses of all the micro-constituents dispersed in the matrix. That is, if R is the abrasive resistance of the component and R_i represents the wear resistance of the ith constituent and $i = 1, 2, 3 \ldots n$,

$$R = \sum_{i=1}^{i=n} V_i R_i \qquad (8.1)$$

where V_i = fractional volume occupied by the ith constituent; n = the number of micro-constituents in the iron.

The validity of Equation 8.1 has been examined [14] by carrying out abrasive wear studies of iron under reciprocating sliding conditions. The abrasives comprised 30% of the total lubricant introduced at the interface and consisted of equal parts of silica sand, corundum, iron filings and mill scale. The abrasives were chosen with the hope of simulating the interface of machine tool slideways. Calling the cumulative hardness of all the micro-constituents as H_c, the conditional hardness, and taking the hardness of the graphite phase as zero, it was shown that

$$H_c = \sum_{i=1}^{i=n} A_i H_i \qquad (8.2)$$

where A_i and H_i are the area occupied by and the hardness of the ith micro-constituent respectively. Equation 8.2 shows that the conditional hardness increases with the amount of hard phase in the matrix. Finally, if the friction materials do not contain any residual stresses or cracks, the wear resistance R of the component could be expressed as

$$R = B \frac{H_c}{\log m} \qquad (8.3)$$

where B is a constant and m is a measure of the number of graphite particles in the wearing interface. Note that Equation 8.3 shows that the higher the conditional hardness of the iron, the greater is its wear resistance. An increased amount of graphite means a lower resistance to wear suggesting that graphite is ineffective under lubricated abrasive conditions.

The parameter H_c has not caught on, but the bulk hardness has been shown to correlate well with wear resistance even in a corrosive environment [15]. It is important, however, that care is taken to avoid residual stresses in the component. This is demonstrated in poor tribological performance of cast iron components used in the flame hardened condition.

8.4.1 *Lubricated sliding*

A component invariably manufactured from grey cast iron is the machine tool slide. A correlation of the wear rates of these components in service has been attempted with the composition of the cast iron [16]. Some 30 components were observed and the general conclusion was that microstructure was of secondary importance provided the prior surface finish of the slide was fine and the interface was kept free from grit and vibration during service. From the point of view of composition, a low phosphorus iron wears rapidly. Although microstructure is believed to be unimportant, the amount of ferrite should not exceed 10%. The deleterious effect of low phosphorus content and of ferrite in excess of 10% is also evident for cast iron cylinders. Measurements [17] on marine engine liners and experiments with similar materials on a pin–disc machine have provided useful results. Wear was unidirectional and the laboratory machines simulated wear due to contaminated lubricant, the contaminant being actual debris collected from engines. The results show that undercooled interdendritic graphite was undesirable and wear resistance was associated with finely distributed graphite particles and grain boundary carbides. The morphology of the graphite flakes is not difficult to control if consistent foundry techniques are employed. Grain boundary carbides are produced if small amounts of vanadium, titanium or chromium are incorporated in the metal. Unfortunately, some of these conclusions have been contradicted and in particular there is a report [18] which suggests that undercooled graphite particles provide a wear resistant iron. However, the beneficial role of finely dispersed graphite has often been confirmed and such conclusions are the most accepted. One possible reason for contradictory conclusions from lubricated sliding studies may be because of the variations in the nature of the lubricant between laboratories. The environment of the experimental rig may influence the results. For example, it has been shown [19] that an oxidized oil accelerates the wear rates of cast irons.

The superiority of nodular iron from the point of view of wear resistance has been confirmed [20, 21], so that the shape of the graphite phase seems important. Rolling wear tests [22] on both flake and nodular irons show that as in sliding [23] a nodular iron in a pearlitic matrix is undoubtedly superior to an iron with flake graphite as the micro-constituent.

Typical examples of dry sliding contact involving cast iron are found in automobiles and commercial vehicles. The problem in these components is heat checking which results in surface cracks and broadly these occur because the surface layers are heated to a high temperature due to friction. Brake application gives rise to heating, and subsequent cooling of the surface layer produces martensite. The differential volume change between the surface layer and the bulk of the component results in cracking. Heat checking, although a tribological problem, is not wear but surface damage and progressive loss of material also occur in these components. Heat checking is aggravated if the phosphorus content is increased because of a reduction in the thermal shock resistance of the iron. Wear resistance on the other hand doubles for railway brake shoes when the phosphorus content is increased from 0·3 to 0·8% and the wear rate is diminished by a factor of six if the phosphorus is increased to 1·5% [24]. Increasing the phosphorus content in an iron excessively may create casting difficulty in heavy sections. However, an iron containing 0·7% phosphorus with copper and chromium is known [25] to provide a good wear resistant material for brake shoe application. Definitive conclusions regarding the effect of phosphorus on the wear rates of cast irons will be of value.

8.5 Aluminium–silicon alloys

Replacement of cast iron by aluminium alloys for pistons in automobiles was first considered in the early twentieth century. The main attraction for these alloys is their lightness and the British practice is to use an aluminium–silicon alloy with a silicon content in the range 11–13%. The alloy can be modified with sodium and the pistons are cast in permanent or semi-permanent moulds. Other advantages of the alloy are a low coefficient of thermal expansion and good corrosion resistance. Further weight saving is now possible by casting the automotive cylinder heads in an aluminium alloy with a silicon content in the range

17–25%. The high silicon alloys are modified with phosphorus and the forecast is that these alloys will find increasing applications in the form of cylinder heads and blocks. Some other commercial alloys used for tribological application contain various proportions of nickel and copper. Improved mechanical properties obtain when the alloys are used in the age hardened condition and a magnesium content between 0·8 and 1·5% of the alloy is necessary.

Since the silicon content can be varied, its effect on the wear resistance of sliding couples has been observed. A laboratory investigation [26] finds that the effect of silicon is not significant and the important aspect is the distribution of the silicon particles in the matrix, a fine dispersion improving the wear resistance. On the other hand a detailed laboratory study of a few commercial alloys with silicon contents up to 21·60% shows that the distribution of the silicon phase is immaterial and it is the silicon content which influences the rate of wear [27–29], a high silicon content imparting a high resistance to wear. There is confirmation of the beneficial role of a high silicon alloy in such applications as automotive pistons, cylinder liners and clutches.

It is quite likely that forged or cast aluminium–silicon alloys will find increasing applications in tribological components but results of experiments quoted so far are contradictory. It is apposite, therefore, to summarize the results from a recent programme of study undertaken on both age hardening and cast aluminium–silicon alloys [31–34].

8.5.1 *Phenomenology*
Observations on the nature of interfaces at low and high magnifications during a wear run provides valuable information. The aluminium alloy first deposits some metal on cast iron and steel counterfaces but both of these wear later. The wear on these counterfaces is never uniform over the apparent contact area if loss of material is expressed in terms of loss of height from a datum line. If a steel pin is loaded against an age hardened hypoeutectic aluminium–silicon alloy bush and the latter observed as sliding progresses, the change in surface topography provides interesting information. The material ahead of the pin flows plastically producing a hump which then work hardens. Many such high spots are produced in this way. These high spots are removed as sliding proceeds, but the track becomes wavy elsewhere. Wear of the aluminium alloy continues in this manner and the track continues to spread laterally, reaching a final steady value. At heavy loads, the

Fig. 8.6   An aluminium alloy pin showing plastic flow (×4). (Clarke, J. and Sarkar, A. D. (1979). *Wear* **54**, 7.)

waviness is accentuated and the amount of wear is high. It is certain that the first effect of sliding under load is plastic flow of the aluminium (Fig. 8.6) alloy whether it is in an age hardened or in as-cast condition. Debris is generated at the onset of sliding, but a significant quantity is deposited on the steel or cast iron counterface (Fig. 8.7). Plastic flow occurs because of the low yield strength of the aluminium alloys and can be conclusively demonstrated by observing the nature of the friction surface under a scanning electron microscope. The surface asperities can be seen clearly to deform and fold in the direction of sliding. Similar flow patterns are observed with alloys of aluminium. Although a harder counterface receives some transferred material, this is soon removed and wear of the counterface is progressive with time (Fig. 8.8). From observations on sections through wear tracks it would appear that wear occurs in a number of ways.

As expected, the subsurface shows a well-defined deformed layer (see Fig. 8.9) and at any load, the work hardened layer may fracture and detach in the form of a wear particle (Fig. 8.10). Detachment of a wear particle occurs in this manner irrespective of the silicon content or

Fig. 8.7   Aluminium alloy deposit on a hard steel bush (×4). (Clarke, J. and
Sarkar, A. D. (1979). *Wear* **54**, 7.)

whether the alloys are as-cast or heat treated. However, as the silicon
content increases, the fracture of the silicon particles is evident (Fig.
8.11). The question which needs answering is, Where does the crack
originate? We have already seen the separation of the deformed layer
from the bulk for both brass and aluminium alloys. We could explain
this in terms of the delamination theory but a model proposed by Sarkar
especially for aluminium–silicon alloys is plausible [31]. In this, the
model surface is represented by asperities which are covered with
micro-microasperities. The effect of a normal load is to cause the
asperities to flow. The normal load will be transmitted via the asperities

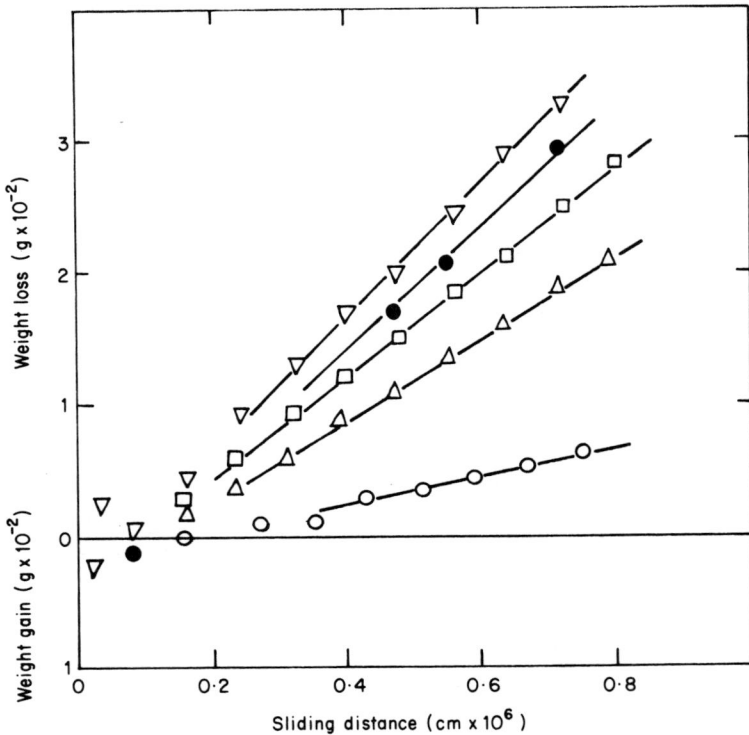

Fig. 8.8 Weight variation with sliding distance of a grey iron bush sliding against an aluminium alloy. Loads in kg: ○ 6; △ 1·6; □ 2·6; ● 3·6; ▽ 4·1.

of the counterface so that we can assume that contact is Hertzian as the aluminium alloy asperities are flattened and work hardened. Since the aluminium alloy work hardens, the interaction is elastic, so that a crack below a hemispherical steel or cast iron asperity will be expected to nucleate in the region of the maximum shear stress. The crack can propagate in isolation or several of them may join up to give rise to a long platelet of wear debris. Although much work remains to be done to confirm some of the observations summarized in the foregoing, the mechanism of the interfacial interaction can be described as follows.

With the onset of sliding aluminium–silicon alloy on steel or cast iron counterfaces, the first event is deposition of the former due to its plasticity at the prevailing stresses on to the latter. As sliding continues, the aluminium pin work hardens and a layer or a particle of debris is

Fig. 8.9   Section through a worn surface of an aluminium–16% silicon alloy, showing a deep work hardened layer. The dark horizontal patches within this layer are stratified primary silicon (×300). (Sarkar, A. D. and Clarke, J.)

produced because of nucleation and propagation of subsurface cracks. Debris is also produced immediately the machine starts by ploughing of the soft metal by the asperities on the steel surface. The soft deposit also work hardens and is probably removed by two-body abrasion and by a fatigue mechanism. As the work hardened layers are removed from the aluminium alloy, the exposed subsurfaces deposit fresh metal on the opposing surface. This metal transfer is small and the counterface wears progressively as a steady state wear rate is established.

8.5.2   Wear debris
For steel–aluminium alloy couples, the wear debris will vary in shape and size and will comprise particles of steel apart from products of the aluminium alloy. This can be confirmed by X-ray diffraction analysis. Another constituent of such debris is γ-Al_2O_3.

In an experiment to evaluate the role of silicon, a comprehensive particle size analysis was carried out using a Quantimet [33]. Some

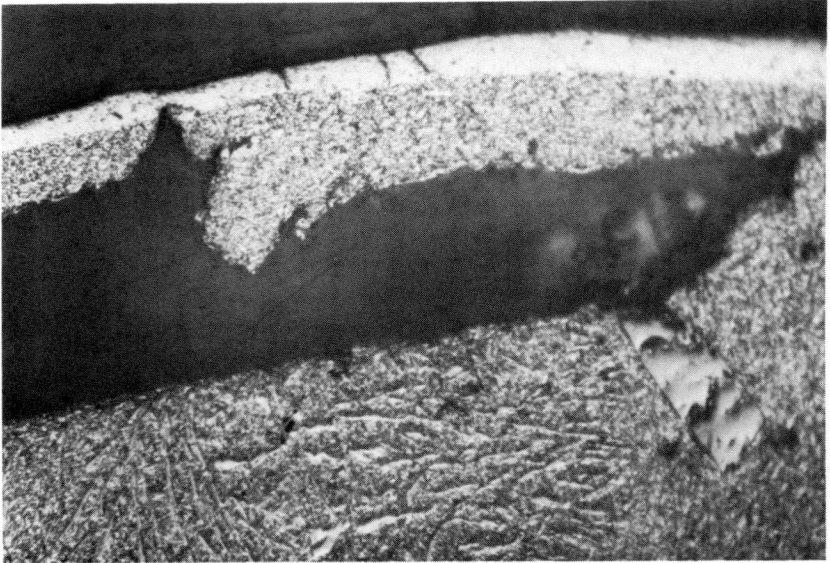

Fig. 8.10   The same alloy as in Fig. 8.9, showing detachment of wear debris
(×325). (Sarkar, A. D. and Clarke, J. [34])

results are shown in Table 8.1 for various silicon contents and a range of
normal loads. The method of expressing the results is to work out a
mean particle size and obtain a percentage of this in terms of all the
debris collected. An isolated case from Table 8.1 has been selected for a
21% Si as-cast alloy where the sliding distances for the various loads are
roughly similar. Plotting the actual number of particles (Fig. 8.12), we
see that the smallest size is the most numerous. The higher the load, the
larger is the number of particles, particularly the ones with small
average diameters. This load dependence, however, is not generally
borne out as Table 8.1 shows. Particle size should be dependent on load
and there does not seem to be any relationship here. Empirically,
however, for a complete wear run, N, the percentage number of
particles in a given size, can be expressed [34] as

$$N = A \exp(-Kd) \tag{8.4}$$

where d = mean diameter of all the particles in a size range and A and K
are constants. Further work is necessary along these lines.

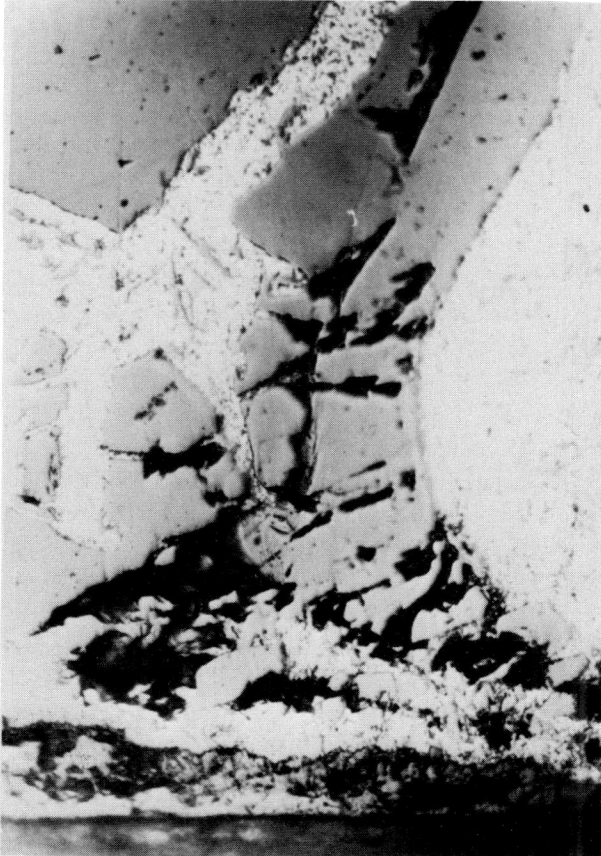

Direction of sliding ⟶

Fig. 8.11 Worn subsurface of an aluminium–21% silicon alloy pin, showing distorted and fractured primary silicon crystals in a coarse eutectic matrix (×590). (Sarkar, A. D. and Clarke, J. [34])

8.5.3 Effect of silicon

This has been studied for both age hardening [31] and as cast [33] aluminium–silicon alloys. If aluminium pins are run on binary aluminium alloy bushes with varying silicon content, the wear rate increases to a maximum with load, followed by a fall and a rise. The speculation is that the initial rise in the wear rate is due to the load effect, but a fall results because of work hardening. A further increase in load results in thermal softening when the wear rate rises again. The overall

Table 8.1 Particle size distribution of wear debris at various loads for aluminium–silicon alloys with varying silicon content (pin–bush machine with surface speed 196 cm s^{-1})

Materials	Load (kg)	Per cent of total number of particles of a given mean particle size (mm $\times 10^{-2}$) for a material at a given load								Total sliding distance (cm $\times 10^4$)
		1·13	3·4	6·8	13·6	27·2	54·3	111·2	225·0	
Pure aluminium	0·5	31·0	22·4	19·8	16·7	8·2	1·65	0·12	0·00	96·191
	2·0	32·8	24·2	22·8	13·2	5·2	1·35	0·25	0·03	47·581
Al+2% Si	0·5	33·3	24·5	20·4	15·1	5·6	0·80	0·13	0·00	75·471
	1·0	29·8	27·2	23·7	14·1	4·3	0·70	0·03	0·00	88·638
	1·5	31·0	27·5	21·9	13·1	4·8	1·20	0·10	0·02	90·074
	2·0	28·8	26·3	24·3	14·2	5·7	0·30	0·20	0·02	56·582
Al+6·5% Si	0·5	32·8	25·1	20·9	14·3	5·4	1·20	0·20	0·00	89·811
	1·0	27·6	23·8	22·1	17·9	6·5	1·40	0·40	0·10	83·850
	1·5	31·8	27·3	23·5	12·2	3·8	1·00	0·25	0·03	103·564
	2·0	29·0	25·3	23·4	16·5	5·1	0·50	0·10	0·00	64·662
	2·5	30·3	24·8	21·7	13·6	6·3	2·30	0·90	0·00	7·876
Al+12% Si	0·5	37·5	23·3	18·0	13·6	6·9	0·58	0·00	0·00	71·880
	1·0	36·2	28·2	20·6	10·6	3·6	0·62	0·00	0·00	87·381
	1·5	26·0	24·0	22·7	18·1	8·1	0·64	0·30	0·04	101·745
	2·0	26·7	24·2	23·2	18·2	6·5	1·00	0·06	0·00	71·856
	2·5	29·3	25·2	23·0	15·7	5·7	0·90	0·05	0·00	97·017
	3·0	27·7	24·5	21·8	15·9	7·6	2·30	0·13	0·00	83·826
Al+21% Si	1·0	27·5	24·5	22·7	16·5	7·1	1·40	0·14	0·00	97·591
	2·0	28·7	25·1	23·3	15·5	6·0	1·10	0·14	0·02	101·769
	2·5	32·0	25·0	22·4	13·8	5·0	1·20	0·26	0·07	101·505

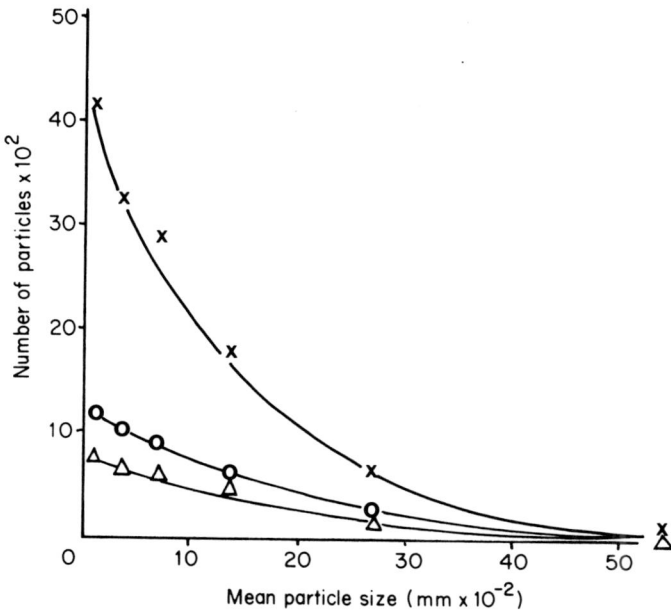

Fig. 8.12   Mean particle size distribution of wear debris from an aluminium–
silicon alloy pin, 6·25 mm in diameter. Loads in kg: △ 1·0; ○ 2·0;
× 2·5. (Clarke, J. and Sarkar, A. D. (1979). *Wear* **54**, 7.)

wear rate is high when the aluminium pin slides on aluminium. This is
expected by the solid solubility concept and can be verified indirectly by
running pure aluminium pins on aluminium–silicon alloy bushes. A
progressively lower rate of wear of the aluminium pin is observed as the
silicon content of the counterface is increased.

Experiments [31] with age hardening alloys show that the wear rate q
is not linear with load W and has been expressed as

$$q = KW^\alpha \qquad (8.5)$$

where K and α are constants. α has probably a value of 0·5 and K
appears to depend on the heat treatment and hence, possibly, on some
mechanical property of the alloy.

A non-linear pattern is also shown by as-cast binary alloys. The pure
aluminium and the low silicon alloys show an increasing rate of wear,
and a diminishing rate results as the silicon content is increased. The
interesting aspect of increasing the silicon content is to decrease the

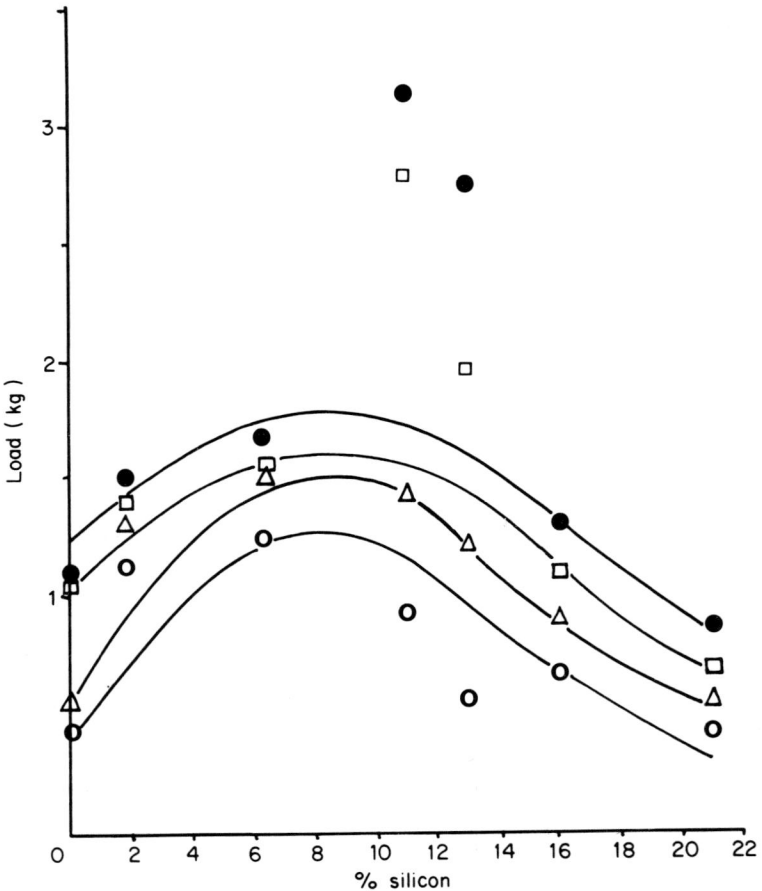

Fig. 8.13   Iso-wear lines of binary aluminium silicon alloys. Wear rates in $(g\ cm^{-1}) \times 10^{-9}$: $\bigcirc$ 15; $\triangle$ 20; $\square$ 25; $\bullet$ 30. (Clarke, J. and Sarkar, A. D. (1979). *Wear* **54**, 7.)

wear rate to a minimum value at a characteristic silicon content, depending on the applied load. The silicon content at which minimum wear occurs varies from about 7% at a load of 1 kg to 11% at a load of 2·5 kg on 6·25 mm diameter pins. There is much controversy about the effect of silicon on the wear rates of the alloys, but we can state that silicon does affect wear rates, although by only a factor or two. One way of assessing the role of silicon content is to plot a series of iso-wear rates with the normal load as the ordinates and per cent silicon as the abscissa.

This is represented in Fig. 8.13 which shows a series of curves, each curve representing one rate of wear for various combinations of load and silicon content. The useful conclusion is that the load bearing capacity of the alloys is the highest for a near-eutectic alloy. A low wear rate of course imposes the condition of light service load. This is fortunate since an eutectic alloy has the most favourable castability. The high silicon alloys have poor foundry characteristics and they cause a high amount of wear of counterfaces made from steel or cast iron.

References

1. Buckley, D. H. and Johnson, R. L. (1965). *Trans. ASLE* **8**, 123.
2. Buckley, D. H. (1968). *Cobalt* **38**, 20.
3. Roach, A. E., Goodzeit, C. L. and Totta, P. A. (1953). *Nature* **172**, 301.
4. Roach, A. E., Goodzeit, C. L. and Hunnicutt, R. P. (1956). *Trans. ASME* **78**, 1659.
5. Goodzeit, C. L., Hunnicutt, R. P. and Roach, A. E. (1956). *Trans. ASME* **78**, 1669.
6. Coffin, L. F. (1956). *Lubrication Engineering* **12**, 50.
7. Mailander, R. and Dies, K. (1943). *Arch. Eisenhüttenwesen* **10**, 385.
8. Archard, J. F. and Hirst, W. (1957). *Proc. Roy. Soc. A* **238**, 515.
9. Kerridge, M. (1955). *Proc. Phys. Soc. B* **68**, 400.
10. Welsh, N. C. (1957). *J. Appl. Phys.* **28**, 960.
11. Scott, D., Seifert, W. W. and Westcott, V. C. (1974). *Scientific American* **230**(5), 88.
12. Scott, D. (1975). *Wear* **34**, 15.
13. Mohsenin, N., Womochel, H. L., Harvey, D. J. and Carleton, W. M. (1956). *Agricultural Eng.* **37**, 816.
14. Talanov, P. I. and Chelushkin, A. S. (1964). *Russian Castings Production*, 127.
15. Deardon, J. and Swindale, J. D. (1957). *J. Iron and Steel Inst.* **185**, 227.
16. Angus, H. T. (1957). *Wear* **1**, 40.
17. Toresson, S. and Olsson, B. (1956). Conference on Engineering Properties and Uses of Iron Castings, BCIRA, London.
18. Dumitrescu, T. (1961). *Revue Roumaine de Metallurgie* **6**, 47.
19. Takeuchi, E. (1970). *Wear* **15**, 201.
20. Lyadskii, V. B. (1963). *Metal Sci. and Heat-Treatment* Nos. 11–12, Nov–Dec, p. 656.
21. Serpik, N. M. and Kantor, M. M. (1964). *Metal Sci. and Heat-Treatment* Nos. 7–8, Jul–Aug, p. 451.
22. Vashchenko, K. I. and Zhuk, V. Ya. (1961). *Russian Castings Production*, 496.
23. Stahli, G. (1965). *Giesserei* **52**, 406.

24. Sulea, P. (1964). Metallurgical Researches, Conference at Bucharest, Paper No. 8, pp. 79–93.
25. Nakai, M., Saito, S. and Okabayashi, K. (1961). *Imono* **33**, 24.
26. Vandelli, G. (1968). *Alluminio and Nuova Metallurgia* **37**, 121.
27. Okabayashi, K., Nakatani, Y., Notani, H. and Kawamoto, M. (1964). *Keikinzoku* (Light Metals Tokyo) **14**, 57.
28. Okabayashi, K., Nakatani, Y., Notani, H. and Kawamoto, M. (1964). *Keikinzoku* (Light Metals Tokyo) **14**, 71.
29. Okabayashi, K., Kawamoto, M. and Notani, H. (1966). *Keikinzoku* (Light Metals Tokyo) **16**, 38.
30. Stonebrook, E. E. (1960). *Modern Castings* **38**, 111.
31. Sarkar, A. D. (1975). *Wear* **31**, 331.
32. Sarkar, A. D. (1977). *Tribology International*, **10**, 235.
33. Clarke, J. and Sarkar, A. D. (1979). *Wear* **54**, 7.
34. Sarkar, A. D. and Clarke, J. (1980). *Wear* **61** (1), June.

9 Heat Transfer

The sources of friction in tribological components are the true areas of contact which adhere and separate. As the junctions separate, heat is generated due to frictional work but we should note that contact life of two opposing asperities can be as small as 10^{-4} s. We can state that heat is generated at these true areas of contact and it then diffuses to the surroundings to establish a consistent bulk interfacial temperature once a steady state of heat transfer is achieved. The bulk temperature of the moving bodies can be measured but a knowledge of the temperatures attained by the junctions will provide much sought after basic information. Since the junctions attain a peak temperature quickly and since their lives are short, the true contact areas are said to attain flash temperatures. Although attempts have been made, experimental evaluations of flash temperatures always leave a nagging doubt as to the accuracy of the values obtained. Various heat transfer models have also been proposed and some of them predict values which do not appear unreasonable. Mathematical treatment of heat transfer is the realm of specialists and readers interested in very rigorous, full-scale analysis will find the volumes by Ingerson, Zobel and Ingersoll [1] and by Carslaw and Jaeger [2] useful. For our purpose we shall only provide an outline of a few selected models and see if they can predict realistic values of bulk and flash temperatures.

9.1 A cylinder on a plane
Bowden and Tabor [3] have considered the case of a simple cylinder (Fig. 9.1) moving unidirectionally under a normal load W and a surface speed v. It is assumed that all the frictional work is dissipated as heat

217

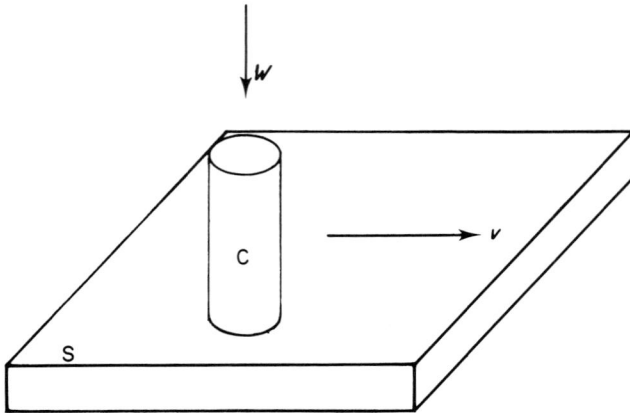

Fig. 9.1   A simple cylinder under a load W moving horizontally over a flat
surface at a velocity v (after Bowden and Tabor [3]).

and the interface of the cylinder is receiving a fraction f of the total heat
generated, Q. That is, the heat entering the cylinder is fQ and that
diffusing into the plate is $(1 - f)Q$. If the thermal conductivity of both
the cylinder and the plate is the same, a reasonable assumption is that
$f = 0.5$.

The cylinder receives heat by conduction from the sliding interface
which should be lost by radiation and convection. Heat loss by the latter
mode is ignored by the authors who consider that heat is lost by
radiation only and arrive at an expression as follows:

$$\theta - \theta_0 = \frac{f \mu W g v}{J \pi a} \sqrt{\left(\frac{1}{2 \sigma k a}\right)} \qquad (9.1)$$

where θ is the temperature of the flat end of the cylinder in contact with
the plate, θ_0 is the ambient temperature, $f = 0.5$ for similar metals, μ is
the static coefficient of friction, g is the acceleration due to gravity, a is
the radius of the cylinder, and J is the mechanical equivalent of heat. k
is the thermal conductivity of the cylinder which is a mean value in a
temperature range and σ is the cooling coefficient, assuming Newton's
law of cooling. The authors [3] consider a 1 mm diameter constantan
cylinder rubbing against steel at a velocity $v = 200$ cm s^{-1} and a normal
load $W = 0.1$ kg. Taking $\mu = 0.3$, $k = 0.05$ cal cm^{-1} s^{-1} °C^{-1}, $\sigma =$
0.001 cal cm^{-2} °C^{-1}, $J = 4.8 \times 10^7$ and $g = 981$ cgs units, from Equation

9.1, $(\theta - \theta_0) = 180\,°C$. According to this the interface can attain quite a high temperature even at a small contact pressure of about 31 g mm^{-2}.

9.2 Flash temperature

Whereas a knowledge of the bulk temperature is a necessity in industrial sliding components, the flash temperatures probably influence the magnitude of friction and wear. The heat flow phenomena at asperity level have been analysed rigorously by Blok [4] and by Jaeger [5]. The analyses, however, are difficult to comprehend but Archard's treatment [6] of the phenomena is manageable and we shall reproduce selected areas of his work. The starting point of Archard's model is that a moving heat source is transferring an amount of energy to a stationary surface at the sliding interface so that two extreme limits of speed need considering.

9.2.1 Slow speed

Let an asperity B indent a circular contact area A of diameter $2a$ (Fig. 9.2). Assume that the area of contact is a plane source of heat which is generated as a result of frictional work. The heat evolved is of course conducted away into the asperity and into the component providing the flat surface.

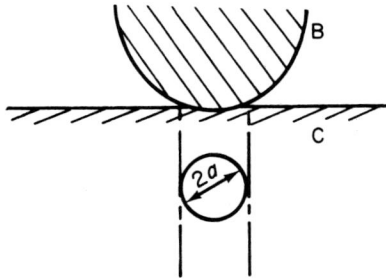

Fig. 9.2   An asperity, B, indenting a circular area of radius a in a solid, C (after Archard [6]).

At very slow speeds, sufficient time is available for equilibrium conditions to be established and we can say that the body C receives energy from a stationary contact of area A for all practical purposes. Using an electrical analogy where current flows through a resistance element,

$$\theta_m = \frac{Q_C}{4akc}$$ (9.2)

where θ_m = mean temperature of the contact area, Q_C = quantity of heat supplied to the body C, and k_C = thermal conductivity of the body C.

9.2.2 High speed

As the sliding velocity v is increased, an equilibrium condition at the asperity level is not feasible because of the very short duration of a junction life. To appreciate the boundary line which demarcates a slow speed from a high speed, a dimensionless parameter L is defined such that $L = va/2\alpha$, where $\alpha = (k/\rho c)$ is the thermal diffusivity of a material and k is its thermal conductivity, ρ, the density and c is its specific heat.

The mean temperature θ_m can be calculated from Equation 9.2 when $L < 0 \cdot 1$. For values much higher than this at $L > 5$, the speed is high and the mean temperature of the contact area must be calculated from a different relationship.

For a very fast moving source, we have to visualize the generation of instantaneous heat at a time $t = 0$. Heat will be conducted below the surface of the body C and the maximum effect will be realized at a depth z after a time $t = t_1$. It is stated [6] that

$$t_1 = z^2/2\alpha$$ (9.3)

If $z = a$,

$$t_1 = a^2/2\alpha$$ (9.4)

For the contact spot to move a distance a, the time t_2 is

$$t_2 = a/v$$ (9.5)

Since $L = va/2\alpha$, from Equations 9.4 and 9.5,

$$L = t_1/t_2$$ (9.6)

Note that as the speed increases, t_2 decreases and the time t_1 taken for the heat to diffuse below the surface becomes large compared with the time of contact. At high speeds, the heat flow can be regarded as linear, that is, there is no radial flow and the heat travels unidirectionally towards the interior of the body.

It can be shown that if the rate of heat supply q per unit area is constant, the temperature θ for the interface of the body C is

$$\theta = 2qt^{1/2}/(\pi k_C \rho_C c_C)^{1/2} \tag{9.7}$$

where the suffix C is used to denote the respective parameters for the body C in Fig. 9.2 and t is the time during which period the heat is supplied to C. The average temperature θ_m over the area A can be obtained from a knowledge of the mean effective time during which the heat is supplied. Such an effective value of the time is considered, because on a microscale the area A can be subdivided into many smaller areas, and each such area will have its own time during which an amount of heat Q is supplied. This mean effective time is shown to be equal to $\pi a/4v$ and, since $q = Q_C/\pi a^2$, from Equation 9.7, the average temperature θ_m over a circular area of diameter $2a$ is

$$\theta_m = \frac{2Q_C \pi^{1/2} a^{1/2}}{\pi a^2 2v^{1/2}(\pi k_C \rho_C c_C)^{1/2}}$$

$$= \frac{0 \cdot 32 Q_C}{a} \cdot \frac{1}{(va)^{1/2}(k_C^2/\alpha)^{1/2}}$$

or

$$\theta_m = \frac{0 \cdot 32 Q_C}{a k_C}\left(\frac{\alpha}{va}\right)^{1/2} \tag{9.8}$$

9.2.3 Calculation

It is suggested [6] that to obtain the true temperature θ_m, we should calculate θ_B for body B under low speed condition using Equation 9.2. The next step is to obtain θ_C for body C from Equation 9.8. Since the assumption is that the generated heat is divided equally between the two bodies,

$$\frac{1}{\theta_m} = \frac{1}{\theta_B} + \frac{1}{\theta_C} \tag{9.9}$$

Factors such as α, v and, hopefully, the true area of contact are known. The unknown quantity is the heat generated due to frictional work, but this can be calculated using the argument outlined in Section 9.1. As a general case, assume that a fraction f of the total quantity of heat generated enters the asperity, B (Fig. 9.2) so that the remainder, $(1-f)$, diffuses into the body C. The total quantity of heat is

$$Q = \frac{\mu W v g}{J}$$

Therefore,

$$\theta_B = f\frac{\mu W v g}{4 a k_B J} \qquad (9.10)$$

and

$$\theta_C = (1-f)\left[\frac{0\cdot 32 \mu W v g}{a k_C J}\left(\frac{\alpha}{va}\right)^{1/2}\right] \qquad (9.11)$$

The fraction f is still uncertain but a broad assumption is

$$\frac{f}{1-f} = \frac{\alpha_B}{\alpha_C}$$

where the suffixes B and C are used for bodies B and C respectively.

The relationships such as those given by Equations 9.7 and 9.11 are not complete in the sense that they cannot be readily used for carrying out calculations. However, Archard has produced expressions which can be used for both low and high speed sliding during either elastic or plastic interaction. Considering plastic interaction only and for high speeds when $L > 100$, not much error is involved if it is assumed that all the heat generated is supplied to the surface below the asperity, so that $Q_C = Q$.

If H is the hardness of surface C, for plastic contact, $a = (W/\pi H)^{1/2}$. Substituting this value of a in Equation 9.11 and since $f = 0$,

$$\theta_C = \frac{0\cdot 32 \mu g}{J} \cdot \frac{(\pi H)^{3/4}}{(\kappa_C \rho_C c_C)^{1/2}} \cdot W^{1/4} v^{1/2} \qquad (9.12)$$

Taking the values of μ etc. for a constantan surface from the end of Section 9.1 and assuming $H = 10 \times 10^6$ g cm^{-2}, $\theta_C \simeq 150\,^{\circ}$C. Note that this value is close to that given by Bowden and Tabor at the end of Section 9.1 if we assume an ambient temperature of $20\,^{\circ}$C, $\rho = 8$ g cm^{-3} and $c = 0\cdot 153$ cal g$^{-1}\,^{\circ}$C^{-1}.

9.3   Dimensional analysis
One advantage of tribological mechanisms is that the interface normally assumes a steady thermal state. It is fair then to ignore the usually

difficult treatments of non-steady state heat transfer and concentrate on the steady state situations. In the absence of a meaningful expression based on theoretical reasoning, quasi-empirical laws provide a basis for calculating temperature rise. The technique of dimensional analysis has been used [7] to obtain an approximate relationship assuming a steady state heat transfer during frictional work.

Let a body 1 approach another denoted by 2. Suppose only two opposing asperities make contact in any one instant at the ith location in the spatial distribution of the protuberances. At the ith contact, let the load at the junction be W_i and the corresponding frictional force F_i so that the coefficient of friction $\mu = F_i/W_i$. Let k be the average thermal conductivity, that is, using the appropriate suffixes, $k = (k_1 + k_2)/2$. Denoting the interfacial and the ambient temperatures by θ and θ_0 assume that $(\theta - \theta_0)$ is a function of a set of independent variables μ, W_i, k, v and A_t where v is the interfacial velocity and A_t is the true area of contact. Thus,

$$(\theta - \theta_0) = \phi(\mu, W_i, k, v, A_t) \tag{9.13}$$

The Buckingham Pi theorem gives two dimensionless quantities π_1 and π_2 such that $\phi(\pi_1, \pi_2) = 0$. It is then assumed that π_1 is equivalent to μ and then shown that

$$\pi_2 = \frac{(\theta - \theta_0)k\sqrt{A_t}J}{W_i vg}$$

Since the dimension of $\sqrt{A_t}$ is length, a generalized length l can be used so that

$$\pi_2 = \frac{(\theta - \theta_0)klJ}{W_i vg} \tag{9.14}$$

Since $\phi(\pi_1, \pi_2) = 0$,

$$\phi\left[\mu, \frac{(\theta - \theta_0)klJ}{W_i vg}\right] = 0$$

This can be written as

$$(\theta - \theta_0) = M\left(\frac{\mu W_i vg}{klJ}\right) \tag{9.15}$$

where M is a constant.

Alternatively, let the heat generated be Q and that being conducted away to a body be Q_C and

$$Q = F_i v = (\mu W_i vg)/J \tag{9.16}$$

$$Q_C = k \frac{(\theta - \theta_0)}{l} A_t \tag{9.17}$$

(A_t/l) has the dimension of length so that Equation 9.17 can be written as

$$Q_C = k(\theta - \theta_0)l \tag{9.18}$$

By the law of dimensional similarity, for a given load, surface speed etc., $(Q/Q_C) = 1/M$ where M is a constant. Therefore, from Equations 9.16 and 9.18,

$$\frac{\mu W_i vg}{Jk(\theta - \theta_0)l} = \frac{1}{M} \quad \text{or} \quad (\theta - \theta_0) = M\left(\frac{\mu W_i vg}{klJ}\right) \tag{9.19}$$

We note that Equation 9.19 is the same as Equation 9.15.

We can take appropriate values of the parameters μ, W_i, k etc. for constantan as before. For a very low sliding speed, taking $M = 0\cdot5$ and $v = 2 \text{ cm s}^{-1}$, $(\theta - \theta_0) = 0\cdot24\,°\text{C}$. This suggests that the interface remains at room temperature, but taking $v = 200 \text{ cm s}^{-1}$ as for constantan in the previous examples, the temperature difference is $24\,°\text{C}$. This is lower than the previous values.

9.3.1 *Interpretation*

Equation 9.19 shows that the temperature rise depends directly on the velocity and inversely to the contact radius, represented by the length dimension. Only the load W_i at the ith contact is shown but the ratio (W_i/l) should be the same as the ratio of the applied load and the sum of the radii of all the true contact areas. In that case, the temperature rise, according to this model, is directly proportional to the applied load.

Archard's model is different from this. It does not incorporate the contact area in any of the expressions, and the load and velocity dependence vary according to the surface speed and the nature of the interaction. The dependences are given as follows:

(1) Low speeds: for plastic deformation, $\theta_m = \phi(W^{1/2}, v)$; for elastic deformation, $\theta_m = \phi(W^{2/3}, v)$.

(2) High speeds: for plastic deformation, $\theta_m = \phi(W^{1/4}, v^{1/2})$; for elastic deformation, $\theta_m = \phi(W^{1/2}, v^{1/2})$.

It is fair to emphasise that Equation 9.1 gives the bulk temperature of the interface. The expressions proposed by Archard and that by dimensional analysis give the flash temperatures, that is, the transient temperatures of the true contact areas. A final comment which should be made is that all the contact spots achieve the same temperature. This is consistent with the absence of any area term in Equation 9.12, but there is a length dimension in Equation 9.15 or 9.19. Incidentally, while calculating the flash temperature using Equation 9.19, a value of $l = 0.05$ cm was chosen arbitrarily. We should note that a value of l smaller than this will predict a much higher $(\theta - \theta_0)$ than 24 °C at a surface speed of 200 cm s^{-1}.

9.4 Temperature measurement

A rise in temperature of sliding interfaces is facilitated by an increase in both load and surface speed. The effect of this temperature rise would in certain cases be catastrophic on machines were it not for the fact that liquid lubricants extract much of the heat generated at the interface. The influence of oxide films on metals is well known and they owe their life to frictional heat. Heat transfer phenomena in non-metals are discussed in a later chapter, but we may wonder if overheating would have destroyed our teeth early in life due to mastication were there not free-flowing saliva in the mouth. Undoubtedly, the synovial joints are protected from temperature effects by the fluid which is continuously replenished at the articulating surfaces.

Measurement of surface temperatures of components is not very difficult and we could probably obtain an estimate of the temperature by observing the thermal degradation of the lubricant. Attempts have been made to measure both surface and flash temperatures and the techniques can be classified under three types.

(1) Measurement of thermoelectric e.m.f. such as is generated in a thermocouple.
(2) Measurement of transient hot spots by exploiting infra-red radiation.
(3) Measurement of hardness in metals.

9.4.1 Thermoelectric e.m.f.

The technique [8] measures the change in e.m.f. when dissimilar metals are heated. The source of heat is frictional work during sliding and the hot junction is provided by a rotating ring and a rod of another metal

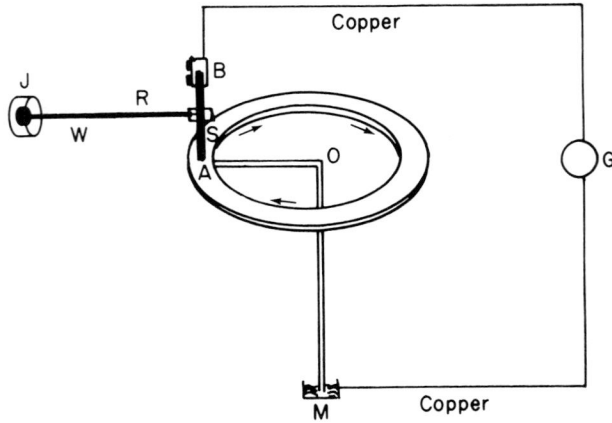

Fig. 9.3   An apparatus utilizing the thermo-electric method to measure bulk temperature. A = annular disc made from one of the metals; B = cylinder made from the second metal (from Bowden and Tabor [3]).

(Fig. 9.3). One of the metals, A, forms the annular disc which can be rotated at a range of surface speeds. A wire of the same metal, A, is in contact with the ring at one end, the other being dipped in a cup, M, containing mercury. A copper lead then connects this to one of the terminals of a galvanometer, G. The disc is one element of this thermocouple system and the second element, B, is in the form of a cylinder resting on the ring and connected to the other terminal of the galvanometer by a copper lead. The hot junction is thus created at S upon rotation and the rest of the metal elements, being at room temperature, form the cold junction. A rigid arm R carried on a gimbal J holds the cylinder, the normal load upon which can be varied by placing weights on the arm at W. The maximum e.m.f. generated at a given load and surface speed is recorded. The temperature can easily be obtained by reading a calibration curve carried out with thermocouples prepared from the same metal pair as the disc and the rod. If the interface work hardens, the e.m.f. will be affected. A noteworthy feature of measuring e.m.f. in this manner is that heating due to frictional resistance is confined to a very thin layer of the interface. Heating facilitates metal transfer and no e.m.f. is then generated because sliding is between like metals. This is the reason why it is necessary to move the cylinder radially so that it forms a spiral track and a fresh surface of the counterface is exposed to A during sliding. One objection to this is, of course, that the surface speed is continuously modified as the rider moves in. A better arrangement would be a rider traversing horizontally as it rubs on a rotating bush.

As Bowden and Ridler [8] point out, the method provides the bulk temperature of the interface. The overall e.m.f. is the integrated temperatures of the many junctions formed and broken over the apparent area. We have already seen that flash temperatures are dependent on normal load and velocity. The actual loads on individual asperities remain an imponderable quantity.

9.4.2 External thermocouples

Rather than use the friction couple to provide the e.m.f., separate thermocouples can be inserted in a component and the variation of the temperature with time can be followed. One possibility is to bury a thermocouple in a component a known distance away from the interface. The temperature of the sliding surface can then be calculated.

A variation of this exercise is to bury a thermocouple just behind the interface of a pin and rub it on a disc until wear is sufficient for the thermocouple hot junction to be poised against the counterface. The hope is that the thermocouple, once exposed, will measure the bulk temperature of the counterface right away. What happens, as expected, is that the hot junction breaks. An approximate indication, however, of the bulk surface temperature of the pin can be taken from the reading immediately before the breakage.

Temperature variation at various distances from the interface has been obtained by burying thermocouples at various positions in the component. The idea is that a material such as steel will undergo phase change when cooled from a high temperature. This causes a variation in hardness so that use of many thermocouples followed by hardness measurements at the relevant distances from the interface will provide a calibration chart. Only the hardness need be measured on components which will provide a value of the bulk temperature during sliding. However, the layer of importance in tribological components is very thin and it is likely that the hot junction will be larger than the longitudinal dimension of this deformed layer. Lack of room alone will make measurement of a temperature gradient from the interface mean-ingless. Furthermore, hardness of steels due to sliding will be influenced by other factors, e.g., by work hardening. The technique, therefore, is of doubtful value and reliability.

9.4.3 Infra-red radiation

Infra-red sensitive photocells such as the lead sulphide cell have been used [9] to measure flash temperatures. There are two conditions. The first is that the cell must have a small time constant, e.g., 10^{-4} s, so that it can register the transient hot spot quickly. The second requirement is that there must not be any opaque obstruction between the cell and the heat source. In one experiment [9] a pin is slid against a 2 mm thick glass disc and the cell is mounted below this disc where the wear track will be. The cell is calibrated against a platinum filament. For a particular metal, if the speed of the glass disc is fixed but there is a gradual increase in the contact pressure, a number of dull red luminous contact areas can be observed. If the interfacial temperature is further increased by manipulating the speed or load, the areas become brighter and whiter. Experiments with this cell [9] have allowed the following conclusions to be drawn, which agree with theoretical predictions:

(1) At a given load and speed, the hot spots vary widely in size and temperature.

(2) The larger the size of a hot spot, the longer it takes to reach the maximum temperature commensurate with load and speed.

(3) The higher the thermal conductivity of the rider, the lower is the maximum temperature which is attained, although the rate of attainment of this temperature is greater.

Infra-red sensitive cells can now be obtained with very small time constants. A suggestion [10] is that rather than a transparent glass disc a small hole, about the diameter of the pin, can be provided in a steel disc. The rider can be observed and the flash temperatures recorded by suitably mounting the cell underneath the rotating disc.

Infra-red techniques have shown [9] that junctions achieve temperatures of several hundred degrees even at moderate loads and speeds, although the bulk of the components remains cool. The duration of the hot spot is of the order of 10^{-4} s.

9.4.4 Other methods

Apart from thermocouples, bulk surface temperatures can also be measured by rubbing the interface with wax pencils of varying melting points. The highest melting point wax in the series which fuses is isolated by trial and error and its melting point is taken as the bulk temperature of the sliding interface. Bulk surface temperatures can also be measured by radiation pyrometers but this is not very accurate while the wax pencils only give a crude estimate. We have already pointed out that micro-hardness measurements to assess the temperature at which certain phase change occurred is by no means reliable; certain micro-structural changes, for example, recrystallization of the subsurface may point to some approximate temperature range to which the interface was raised due to frictional work.

9.5 Effect of heat

Although the condition of the interface is not simple, provided the temperature of the surface rises to a high enough value, frictional work should influence the microstructures of metals. Consider the case of a medium carbon steel with a mean thermal conductivity, $k = 0.13$ cal cm^{-1} s^{-1} °C^{-1} and mean specific heat $c = 0.153$ over a tempera-ture range up to 700 °C. Taking the density of the steel as $\rho = 8$ g cm^{-3} and its hardness $H = 25 \times 10^6$ g cm^{-2}, the flash temperatures at various loads and speeds can be calculated from Equation 9.12. This has been done taking $\mu = 1$ and the results are shown in Fig. 9.4. Disregarding the effect of deformation, if we assume that martensite forms if the interface achieves a temperature of 800 °C, we can see that up to a load of 1 kg, the flash temperatures are high but the junctions do not undergo martensitic transformation. Phase change occurs if the speed is increased further and at 400 cm s^{-1}, martensite, according to our assumption, can be expected even at a low load of 70 g.

Wear of steel has been shown to increase with speed initially. We can see this in Fig. 9.5 where, at a load of 1 kg, a maximum wear rate is noted at a speed slightly lower than 100 cm s^{-1}. One explanation for severe wear such as this is that at low speeds surface oxide films do not form effectively to provide their protective effect. However, according to Equation 9.12, at 50 cm s^{-1} and at a load of 1 kg, the flash temperature of this particular steel is 537 °C. The temperature is high enough for protective oxide films to form. Although, at best, the calculation of flash temperature using Equation 9.12 is approximate, the suggestion for future work to understand the mechanism of wear of steels at varying sliding speeds and loads should receive strong support. It is permissible to speculate, however, that although a surface oxide film forms, the metallic substrate shears readily because of its diminished shear stress

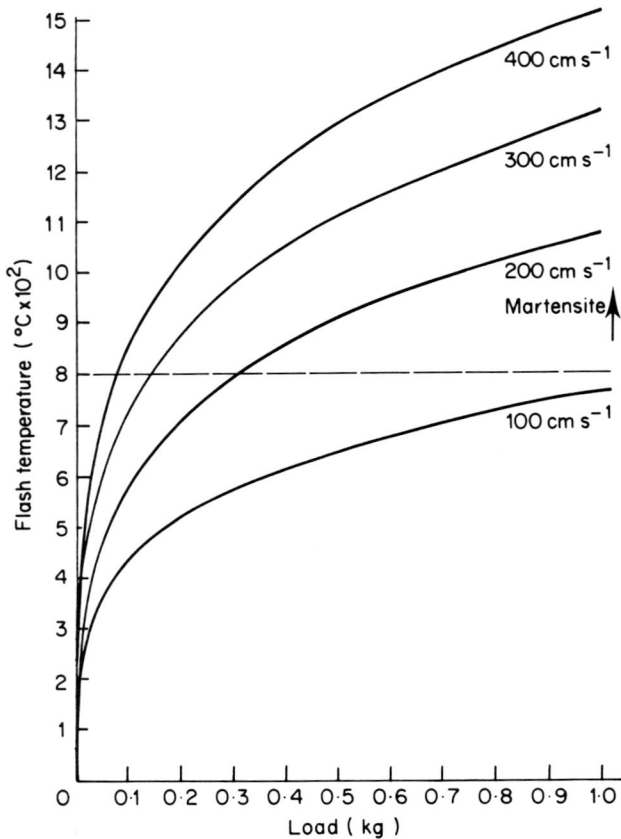

Fig. 9.4   Theoretical flash temperatures at various loads and surface speeds as calculated from Equation 9.12.

at high temperature. This will have the effect of removing both the oxide film and metal.

We can accept that the falling part of the curve in Fig. 9.5 with increasing surface speed may be a result of phase change due to a flash temperature higher than 800 °C. That is, as the temperature rises, the junctions are austenitized as they are sheared and the rapid quenching by high velocity air produces some martensite. This is plausible, particularly since a change in surface speed by a factor of 2 can raise the junction temperature by a factor of 3. The corresponding drop in the wear rate is more than 2 orders of magnitude, which is a significant decrease.

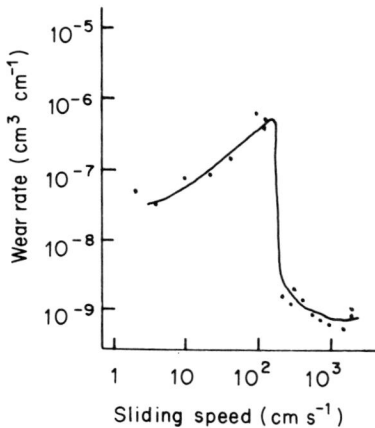

Fig. 9.5   Wear rate of a 0·52% C steel at a load of 1 kg at various surface speeds. (Archard, J. F. (1958). *Wear* **2**, 438.)

The effect of load is to increase the flash temperature as well. This should facilitate martensitic transformation of a steel. Load effect is also known to cause a severely work hardened layer and certain metals are known to become wear resistant at heavy loads because of a hard sliding surface. We should note again, therefore, that although the role of phase change may be positive the dependence of wear rates on load and speed is not exclusive to this.

Figure 9.4 shows that at a moderate load of even 0·2 kg the flash temperature at a speed of 400 cm s^{-1} is quite high for a medium carbon steel. It is clear that further increases in speed and load would cause incipient fusion of the interface. Progressive increase in interfacial temperature may result in melting of the surface layer. This is the reason why large calibre, high muzzle velocity cannons need special metals to withstand the effect of velocity. The combined effect of load and speed shows that at high velocity an initially large value of friction decreases to a low value [11]. It leads at once to the not so unreasonable speculation that the high temperature produces a film of molten metal which acts as a lubricant. Wear, on the other hand, is progressive and severe and the loss of metal probably occurs by direct removal of part of this molten layer. Experiments show that the wear rate q could be empirically related to the melting point T_m of the metal as follows

$$q = \{A \exp (B/T_m)\}(\mu \sigma v)^2 \tag{9.20}$$

where A and B are constants; μ = coefficient of friction; σ = applied contact pressure; and v = sliding velocity.

The relationship in Equation 9.20 was obtained from a surface speed of 54 000 cm s^{-1} and it shows that wear rates of metals depend on their melting points. A plot of q against $1/T_m$ does provide a linear relationship except for metals such as copper and aluminium. These do not fit into a linear pattern so well and they do give significantly lower wear rates than those of nickel or steel. It is fair to state that metals of high thermal conductivity will dissipate frictional heat to the surroundings quickly and will thus impede the formation of a molten layer.

Apart from gun barrels, rocket propelled vehicles running on rails develop high velocities and experimental studies have always been important to obtain basic information. In one apparatus, relative surface speeds of the order of 1 km s^{-1} have been obtained by spinning spheres against flat surfaces [12, 13].

In what has been termed the deceleration technique, a freely suspended sphere is accelerated by a rotating magnetic field. The ball spins close to three flat surfaces (Fig. 9.6) which are placed symmetrically at 120° to one another. By releasing a spring, the three surfaces can be made to contact the spinning ball so that the interfacial friction slows down its rotational speed. The bulk temperature of the interface can be measured with thermocouples and the coefficient of friction can be calculated as follows:

$$\mu = F/W = I\dot{\omega}/3aW \qquad (9.21)$$

where I is the moment of inertia of the sphere of radius a and $\dot{\omega}$ is the angular deceleration.

One of the difficulties with the arrangement shown in Fig. 9.6 is that spring loading becomes unsatisfactory at heavy loads. An alternative apparatus employs a spinning sphere which is made to fall and hit an inclined plate, positioned about 50 mm below the original position of the ball (Fig. 9.7). The angle of inclination of the flat plate to the vertical is $\theta = 30°$ and as the ball hits this plate, rebound occurs sideways. This is obviously due to the frictional force which opposes the spinning motion of the ball as it contacts the inclined plate. The angle of deviation is obtained from the mark A on the wall of a cylinder which is mounted concentrically with the original axis of rotation of the ball. The coefficient of friction is expressed as

$$\mu = \cos\theta \tan\alpha$$

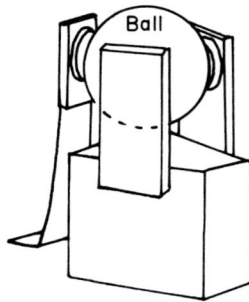

Fig. 9.6   The decelerating technique of measuring friction at high speeds. (Bowden, F. P. and Tabor, D. (1964). "Friction and Lubrication of Solids", Part 2, Oxford University Press, Oxford.)

where α is the angle giving the horizontal displacement of the ball after striking the inclined plate.

The nature of the wear scar of flat surfaces by spinning steel balls shows marked velocity dependence. At a speed of about 1 m s^{-1}, the surface shows abrasion and tearing. The metal shows evidence of flow

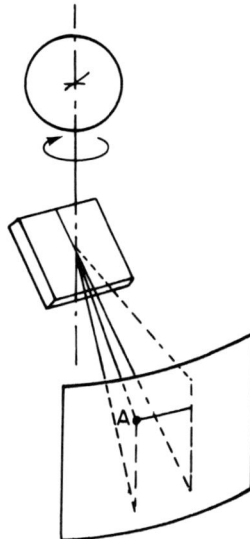

Fig. 9.7   The arrangement of ball and flat inclined plate used to measure friction at a high speed. (Bowden, F. P. and Tabor, D. (1964). "Friction and Lubrication of Solids", Part 2, Oxford University Press, Oxford.)

and behaves like a highly viscous matter at a speed of 150 m s^{-1}. At a much higher speed of 600 m s^{-1}, the interface appears smooth.

Experiments with a polymer such as nylon show that fusion or melting of the material occurs at high speed but only a thin interfacial layer is involved. With metals, friction and wear first decrease with increased speed in the spinning ball experiment but they rise again at a characteristic speed depending on the metal. At very high speeds, there is large scale melting of the metal and the volume of the molten interfacial layer increases with rising speeds.

There are possibly two main factors which decide on the friction and wear mode of materials at very high sliding speeds which are as follows:

  (1) The contact region is subjected to severe heating.
  (2) The interface is deformed at very high rates of strain.

The explanation offered [14] is that surface interactions of a friction couple at these speeds occur under isothermal condition. That is to say that those properties of the solid under study which play the main role in the friction and wear modes become irrelevant at the temperatures generated. Additionally, since the interface is deformed at very high rates of strain, it can be said that sliding is essentially adiabatic. That is, the junctions are sheared so rapidly that the generated heat has inadequate time to be removed and cooling of the hot spots becomes completely inefficient. The net effect is that a solid substrate supports a thin film of viscous metal of very low shear strength. For practical applications, such as for a gun barrel, the suggestion at once is that a high melting point metal is the answer for low wear rates. A refractory coating will probably be acceptable provided it does not fail mechanically, for example, by an abrasion effect. While choosing materials for application at high sliding speeds, a consideration of their thermal conductivity is undoubtedly important.

An analogous situation to high speed interaction with metals and non-metals is the frictional heating effect on ice when the surface layers melt [15]. The low friction values which result are important in the parts of the world where transportation is on ice. The nature of the opposing surface is important and PTFE is fitted to the skids on aircraft in the Antarctic to facilitate landing and take-off. It should be appreciated, however, that speeds involved in the usual vehicles operating on snow are not as great as those suggested in the experiments with a spinning ball. The interface melts, of course, because of the low melting point of the ice.

9.6 Thermal shock

Metal–metal interactions at high velocities force the junctions to be heated and cooled rapidly. The result is that the zones of hot spots undergo thermal shock and the bulk of the solid supporting these junctions is subject to a temperature gradient. A thermal shock effect results in cracking and spalling of surfaces and the resistance, Ω, to thermal shock is dependent [16] on two parameters: $\sigma =$ maximum thermal stress on the surface under study; $\sigma_0 =$ resistance to crack propagation in the body in question. σ_0 and σ are related to Ω by the following expression,

$$\Omega = \sigma_0/\sigma \qquad (9.22)$$

The criterion for complete resistance to failure due to thermal shock is $\Omega > 1$.

It is unlikely that the majority of tribological failures would occur as a result of a single cycle of rapid heating and cooling. If n is the number of cycles which causes failure,

$$n = \exp\left[A(\Omega - 1)\right] \qquad (9.23)$$

where A is constant which depends on such variables as speed and load. The statement by Equation 9.22 simply means that the resistance to crack propagation is a stress acting to oppose the effect of the maximum thermal stress. σ_0, however, is difficult to quantify.

A semi-quantitative measure of thermal shock resistance can be obtained by considering a ratio,

$$\lambda = \frac{k}{\xi E} \qquad (9.24)$$

where k is the thermal conductivity, E the modulus of elasticity and ξ is the coefficient of thermal expansion of the solid. The ratio is used to demonstrate that the higher the value of λ, the greater is the thermal shock resistance of the material. A high value of λ means a shallow temperature gradient in the solid which is provided if k is high. For a high value of λ, it is also necessary for the coefficient of thermal expansion and the modulus of elasticity to be low. A ductile material can withstand thermal shock to a greater degree than a typical brake counterface, viz., cast iron.

References

1. Ingersoll, L. R., Zobel, O. J. and Ingersoll, A. C. (1955). "Heat Conduction", Thames and Hudson, London.
2. Carslaw, H. S. and Jaeger, J. C. (1947). "The Conduction of Heat in Solids", Oxford University Press, Oxford.
3. Bowden, F. P. and Tabor, D. (1971). "The Friction and Lubrication of Solids", Part 1, Oxford University Press, Oxford.
4. Blok, H. (1937). "General Discussion on Lubrication", Vol. 2, Institution of Mechanical Engineers, London, p. 222.
5. Jaeger, J. C. (1942). *J. and Proc. Roy. Soc. NSW* **76**, 203.
6. Archard, J. F. (1958). *Wear* **2**, 438.
7. Moore, D. F. (1975). "Principles and Applications of Tribology", Pergamon Press, Oxford.
8. Bowden, F. P. and Ridler, K. E. W. (1936). *Proc. Roy. Soc. A* **151**, 610.
9. Bowden, F. P. and Thomas, P. H. (1954). *Proc. Roy. Soc. A* **223**, 29.
10. Powell, D. G. and Earles, S. W. E. (1967). *Proc. Inst. Mech. Eng.* **182**(3G), 114.
11. Montgomery, R. S. (1976). *Wear* **36**, 275.
12. Bowden, F. P. and Freitag, E. H. (1958). *Proc. Roy. Soc. A* **248**, 350.
13. Bowden, F. P. and Persson, P. A. (1961). *Proc. Roy. Soc. A* **260**, 433.
14. Bowden, F. P. and Tabor, D. (1964). "Friction and Lubrication of Solids", Part 2, Oxford University Press, Oxford.
15. Bowden, F. P. (1953). *Proc. Roy. Soc. A* **217**, 462.
16. Kragelskii, I. V. (1965). "Friction and Wear", Butterworths, London.

10   Tribology of Polymers

Of the non-metallic materials at the moment, plastics are used for a range of tribological applications such as bearings, gears or as coatings on metallic components. Most plastics are referred to as high polymers and in its simplest form a pure plastic is a mass built up from a large number of molecules. The generic name polymer derives from the Greek term *poly* meaning many and *meros* which translates as part. Thus, a polymer is a large molecule and is built up by putting together simple chemical units, the reaction involved in the process being termed polymerization. For example, a macromolecule of polyethylene results by repeating the methylene or the CH_2 groups thus, $-CH_2-CH_2-CH_2-CH_2-$. A solid block of polymer as in a plain bearing will comprise many of these chains as they are called. An appreciation of the size of these chains can be obtained if one considers that, in commercial polymers, a chain may contain 1000 to 10 000 of these methylene groups. The term resin originally used for natural secretions from coniferous trees is at times used to describe a polymeric material. Polymers can be inorganic or organic, among the former being natural products such as sand and clay or synthetic materials in the form of fibres. The commonly used tribological components are made from organic polymers which can be subdivided into two main groups, viz., natural and synthetic. Some examples of the former are natural rubber, cellulose and starch, all with complex molecular structures. Typical examples of synthetic polymers are adhesives, paints, fibres, plastics and artificial rubber. Apart from gears and dry bearings, polymers are finding increasing uses in the realm of bio-materials. Typical areas are found in dental restorative materials or in total hip replacements. There is an interest in increasing the life of floor coverings, some of which are made from polymers. At the other end, automobile tyres play a very important role in technological societies. All these are discussed later, but in this chapter we shall concentrate on the current state of knowledge regarding the friction and wear modes of polymers in general.

10.1   Laboratory study

As in metals, laboratory studies of polymeric materials are usually carried out on pin–disc machines except when special situations are simulated such as in a tooth brushing machine used to evaluate dental restorative materials. Normally, the disc is made from stainless steel and the pin can be a cylinder, a cone or a hemisphere. Polymeric materials can undergo creep or change in length due to even a slight variation in temperature. Measurements of height loss, therefore, can introduce error if the ambient temperature fluctuates. While measuring wear scars, the load should be removed and the pin allowed to settle a few hours for elastic recovery to be completed before taking any readings. This is impracticable and the wear scar technique is discouraged for polymers with low elastic modulus. Experience suggests that elastic recovery is negligible for materials with a modulus of $20 \, \text{kg mm}^{-2}$ or more and the dimension of a scar can be measured without waiting. For materials with lower values than the limiting modulus, the weight loss technique is the only reliable method.

Two other factors to consider are that polymers absorb moisture continuously in room atmosphere and can pick up dust from the surroundings. One way of overcoming any error due to moisture absorption is to keep a reference sample under the same environmental condition as the pin. If the reference sample is weighed each time the pin is, a suitable correction for moisture absorption can be made. It is safer to work in a clean environment and it is not expensive to put a plastic cover over the testing rig to keep out dust from the atmosphere.

10.2   Counterface roughness

If the counterface is rough, the asperities will plough the polymer as in a situation of two-body abrasion [1]. This will result in a high rate of material removal, but the counterface may also become smoother with time by wear due to rubbing against the polymer. In that case the ploughing component will be less effective, resulting in a diminished rate of wear of the polymer. One attractive feature of a polymeric bearing is that ingress of dust from the atmosphere does not produce a catastrophic effect because polymers generally exhibit good embed-dability. This forestalls potential three-body abrasion, and wear of both the shaft and the bearing is avoided by this mode. The dual problem with three-body abrasion is that in the first instance both the polymer and the steel counterface will wear by abrasion. In the second phase, the roughened counterface will remove material from the polymer by ploughing. It is important, therefore, to have a smooth counterface and to prevent uncontrolled amounts of grit or dust from entering the sliding interface. Otherwise wear rates of both the shaft and the bearing will be accentuated.

The importance of counterface roughness has been demonstrated [2] by rubbing epoxy resin pins reinforced with 60% by volume of carbon fibres on hard stainless steel discs. Two different carbon fibres were used called type I and type II, the latter being more abrasive than the former. The abrasiveness of the pin is relevant, since although a rough counterface will initially cause rapid wear of the polymer, the disc may also be polished, resulting in a reduction of the wear rate of the pin. This wear rate will remain low provided the polymer does not increase the roughness of the counterface by being too severe in its abrasiveness.

It follows that an ideal friction couple will cause polishing of the metal surface but further abrading should cease once a smooth coun-terface commensurate with a low wear rate has been established. Reinforcing the resin pins with type I carbon would appear to be the right choice, but not by itself. Experiments show that the wear rate of the polymer increases as the prior surface finish of the stainless steel disc is made rougher, and being non-abrasive, the pins are unable to carry out any polishing. Introducing an abrasive such as 5% by weight of 8 μm diameter alumina particles in the fibre polishes the disc, resulting in a decreased rate of wear of the pin provided the interface is lubricated with water. The role of the lubricant is to remove the abrasive particles from the interface to prevent their build-up and hence three-body abrasion later in the sliding process. Obviously, the alumina particles remove any sharp protuberances initially and then establish a smooth topography on the disc. Apparently this remains consistent if the interface is lubricated with water. Since continued polishing of the disc results in unnecessary loss of material from the counterface, it is more economical to impregnate a component made from type I fibre with a predetermined thin layer of the abrasive. The object is to allow this layer to polish the counterface and be removed completely from the system once this is accomplished. Thereafter, the interaction is between an abrasive-free fibre reinforced resin and smooth stainless steel counterface only.

Since polymeric components wear more as the roughness of the metal counterface increases, a correlation between the rate of wear and the quantity $(\sigma/R)^{1/2}$ should be possible [3], where σ is the standard deviation of the asperity height distribution of the counterface and R is the mean radius of the asperities. There is no experimental confirma-

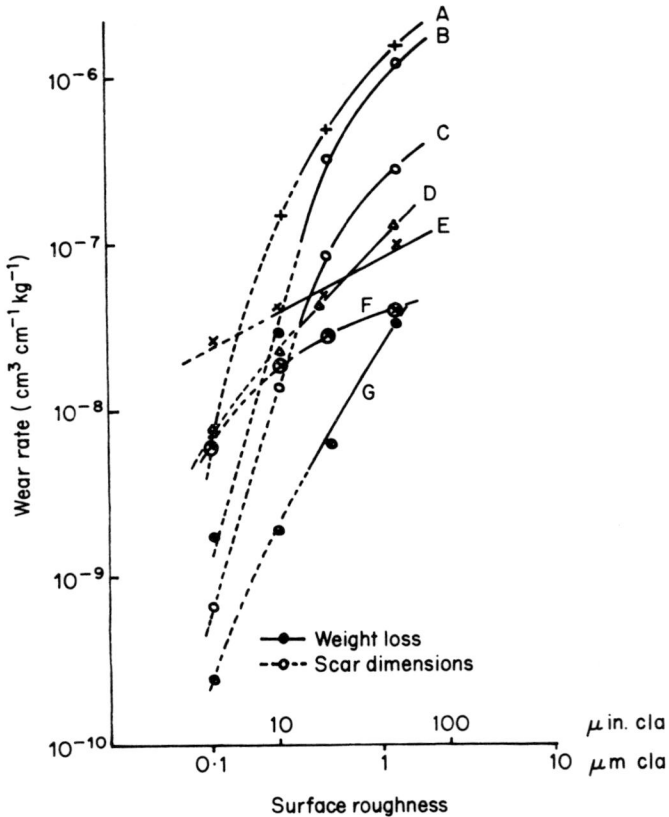

Fig. 10.1   Variation of wear rates of polymers with the surface roughness of a steel counterface. A polystyrene; B polymethylmethacrylate; C polyacetal; D polypropylene; E PTFE; F polyethylene; G nylon 66. (Lancaster, J. K. (1968). *Proc. Inst. Mech. Eng.* **183**(3P), 98.)

tion of this so far, but we should note that the ratio σ/R increases approximately with increasing *cla* values of surfaces. Figure 10.1 shows the variation in wear rates of a number of polymers with increasing cla values of the opposing surface. The interesting point is that the effect of surface finish is the most pronounced with the harder, more brittle polymers, and the ductile materials are less affected. We should note, however, that the results in Fig. 10.1 are from single traversals.

10.3 Transfer films

Although the wear rate is high with a rough counterface, its topography is modified with sliding time as material is deposited from the polymer. Thus in dry rubbing, fibre reinforced pins transfer layers of degraded carbon and resin of sufficient thickness to mask the original topography of the counterface. There is some evidence [4] to suggest that the grain boundaries of the stainless steel act as nucleation sites for these transfer films and the carbon fibres appear as discrete areas in a matrix of epoxy resin of different orientation. It is known that the deposition of these transfer films is impeded in the presence of water lubricant and also if abrasives are trapped at the interface.

PTFE pins sliding on steel show two friction regimes [5]. At high sliding speeds or low ambient temperatures the friction is high, possibly because of the resistance offered by the viscous polymeric film deposited on the counterface. A second reason is that the interfacial interaction is now between the polymer and its film which has been deposited. Polymers are viscoelastic in nature so that the higher the speed of the disc, the greater will be the rate of shear at the interface, which will result in a correspondingly high resistance to shear. Transfer of such films occurs on a massive scale and they can be observed in the form of sheets, ribbons or humps with thicknesses of the order of a few tenths of a μm. At low speeds or moderate temperatures, the friction is low and the thin film of PTFE is drawn over the stainless steel counterface. The film is adherent to the steel with a highly oriented fibrous structure with a thickness between 100 and 400 Å. The low friction is attributed to the easy shear of these thin layers under tangential traction. Obviously, transfer of films means that the pin wears by that amount and attempts have been made [6] to establish the mode by which the polymer layers detach from the bulk. Firstly, the banded structure of PTFE is destroyed with the progress of sliding and long fibres, each with a thickness of about 250 Å, form on the surface of the pin. These connect laterally to produce a film over the contact face and this is then transferred on to the disc. As sliding progresses, the film continues to form and then transfer on to the disc surface, giving the transferred polymer a layer-like deposit, each layer being about 250 Å thick.

The wear rate of PTFE shows an abrupt increase by a factor of about five at its melting point. This is similar to what is expected of metals, but the phenomenology of the interface of a polymer is shown to be analogous at both below and above its melting point. As we have noted, fibres form in the polymer during sliding in an environment below its melting point. The crystalline state is destroyed above the melting point, but fibres 250 Å thick are still formed, except that each of these now comprises a highly dense and disordered aggregate of PTFE molecules. The fibres are still laterally connected to produce a film for subsequent transfer to the counterface, and each fibre is separated from its neighbour by a region of a relatively small number of molecular chains of PTFE. Above the melting point, the fibres are amorphous in nature and the rate of detachment of the films formed is higher than in the case of crystalline fibres at lower temperatures. This is the explanation offered for an abrupt increase in the rate of wear of PTFE at its melting point.

Deposition of a thin film on a metal surface has an important technological application in that a low coefficient of friction of 0·05 or less is obtained for dry bearings when a thin layer of PTFE is applied to a hard substrate such as a metal bearing supporting a shaft [7]. This is a significant gain compared with a value of 0·15 when the polymer is used as a bearing without any metal backing. The object of coating a metallic surface with a polymer such as PTFE is to apply the principle of diversion when the expensive shaft is protected at the expense of the bearing, which is allowed to wear. In this particular case, only the coating of PTFE is lost and the metal bearing is preserved provided recoating is carried out at appropriate times.

An example [8] of such an anti-wear coating on brass bearings is a mixture of 25% PTFE, 10% graphite, 10% lead oxide and 55% polyphenylene sulphide (by weight). The exact mechanism by which the shaft is protected is not known, but it is speculated that the lead ions diffuse into the graphite lattice as sliding commences with the possible formation of an interlamellar compound. The lead ions may also transfer to the steel shaft which is believed to facilitate adhesion of graphite from the coating on to the steel. The PTFE is believed to remain in the coating, thus offering regions of low friction to the graphite now transferred to the steel, the graphite itself possessing low friction due to easy interlamellar slip. Thus the wear of the bearing occurs by the loss of an amount of coating material, all or some of whose constituents migrate to the steel shaft which is unaffected.

10.4 Lubrication

The ability of a steel counterface to retain an amount of transfer film depends on whether rubbing is dry or if a lubricant is introduced into the system. Studies with pin–disc machines using fibre reinforced resin show a complete absence of any deposit on the wear track when a water lubricant is used. Generally, polymers wear only by a low amount when sliding against steel discs in seawater. The reason probably is corrosion of the steel which facilitates its polishing, and thus the smooth profile of the opposing surface is maintained.

Wear rates of fibre reinforced resins are much lower than metal–metal couples when certain organic fluids are used as lubricants. Some of the fluids used with a stainless steel disc in decreasing order of wear rates of the polymer are n-propanol, n-hexane, silicone and n-octanol [9]. Comparing these with a water lubricant, even n-propanol is less harmful but the wear rate is least when rubbing is dry. The problem with water of course is that it disturbs the interface readily. Thus fibre or carbon filled resins reach a steady wear regime in the absence of any lubricant because of a stable transfer film. The deposit, however, is removed rapidly if water is present at the interface and the rate of wear can increase by a factor of 500. On the other hand, PTFE transfer films can be retained partially on the disc in the presence of water and an increase in wear rate, although significant, is not catastrophic. Although it is possible to deposit wear-inhibiting films comprising both resin and a lubricant on a smooth counterface, the polymer–steel couples give maximum efficiency under dry rubbing conditions from the viewpoint of wear resistance.

10.4.1 Conical pins

As a rule, polymers find application in dry rubbing and studies of lubricants have been largely motivated by the necessity for evaluating their undesirable effect in removing the wear-inhibiting transfer film. If such an evaluation is carried out under accelerated tests by using a conical pin of the non-metal, it should be appreciated that a geometric wedge of lubricant forms at the pin–disc interface as a result of wear and the elastic deflection of the load bar assembly. At and above a critical diameter of the generated scar on the pin, wear ceases because the situation at that moment is that of hydrodynamic lubrication. For a conical pin sliding on a steel disc in the presence of a mineral oil, this critical diameter d_c is reached when

$$d_c = K(W/\eta v)^{1/3} \qquad (10.1)$$

where W = applied normal load, v = sliding velocity, η = viscosity of the oil, and K = constant.

If the object of a set of experiments is to compare the wear rates of a material relative to a range of liquid contaminants, the critical wear scar should be large so that the steady state regime of wear is achieved.

10.5   Friction

For most metallic couples the coefficient of friction is constant within the practical range of loads and speeds, but this is not so with plastics [10]. This is because friction of plastics is dependent on the amount of deformation and the elastic hysteresis losses since polymers are visco-elastic in nature. The low coefficient of friction of about 0·05 is attainable only at heavy loads and low sliding speeds or when a film of the polymer is used on a hard substrate. At light loads or high speeds, the coefficient of friction can be as high as 0·3. When a filler is incorporated in the plastic, the coefficient of friction may be entirely that expected of the filler material if it slid itself on the counterface.

The adhesion theory of friction for metals shows that the coefficient of friction μ equals the ratio τ_c/H, where τ_c is the critical shear stress of the softer metal and H is its hardness. An attempt has been made [11] to express μ in terms of S and H where S is the shear strength of the bulk polymer and H is the indentation hardness. Sliding like couples of polyethylene, PTFE, polytrifluorochloroethylene (Kel-F) and perspex (polymethylmethacrylate) at a load of 25 kg in the temperature range -100 to $80\,°C$, shows that the coefficient of friction can be completely described in most cases as follows:

$$\mu = \mu_0 + K(S/H) \qquad (10.2)$$

where μ_0 is the contribution to the coefficient of friction due to some other causes than interfacial adhesion, e.g., surface roughness and K is a constant which is generally greater than unity.

Equation 10.2 is not quite the same as that for metals, and the discrepancy is attributed to the additional strengths which the points of contact attain due to the normal load, or due to molecular structural factors which give a low intrinsic adhesion between couples. For brittle materials such as perspex and Kel-F, the junctions under load can withstand very large applied shear stresses because of high superimposed hydrostatic pressures at the sliding interface. It is evident therefore that the shear strength S of the bulk polymer is not the criterion to describe the frictional resistance. PTFE falls into a different category because of its low intrinsic adhesion to metals at low temperatures. If the temperature is raised to the softening point of the polymer, marked adhesion occurs with the opposing surface and the coefficient of friction rises. King and Tabor [11] suggest that, subject to these qualifications, the adhesion concept can explain the frictional resistance of these polymers over a wide range of service temperatures. However, the details differ from metals in two ways [12]. Firstly, there is no junction growth under a tangential pull as with metals, so that the true area at sliding for polymers is that due to the normal load only. Secondly, the nature of deformation of the interface is intermediate between a plastic and an elastic mode.

Since a layer of polymer is deposited, the property of the transferred film must be taken into account [13]. When rubbing is dry, a polymer should rub on its own transferred film at equilibrium. This has led to the suggestion that the friction of a sliding couple should be governed by two shear strengths, viz. S and S_i of the bulk and the deposit respectively. When $S_i > S$, shearing will occur in the polymer, and at the interface when $S_i < S$. That is, friction at any situation will depend on the smaller of the two shear strengths S_i and S. This suggests that there are occasions when the property of the bulk polymer can be correlated with friction and the bulk hardness is a viable parameter since it is closely related to S. With an amorphous polymer such as polystyrene, there is no bulk transfer of the material to the counterface as sliding commences under load. However, as the ambient temperature is raised, the bulk strength of polystyrene decreases and there is an onset of wear near its glass transition temperature. The wear process of polymers then is governed by their mechanical strengths in the sense that an intrinsically soft material will readily provide transfer films to the opposing surface. The topography of the counterface then changes and rubbing is between like materials so that the frictional resistance of the couple is decided by the orientation of the deposited film. PTFE gives a substantial amount of static friction at room temperature but filaments of the polymer are deposited on the wear track as sliding commences. The orientation of the film is in the direction of sliding and kinetic friction is then low because these layered deposits have a low value of S_i. According to Pooley and Tabor [13], friction remains low but the thickness of the deposit does not change. This implies the unlikely case when the polymer ceases to lose material after an amount of initial wear. We note that when brass slides on steel an equilibrium thickness of the former is deposited on the counterface when the rate of transfer equals the rate of removal from the deposit in the form of wear debris.

For all friction couples the effort continues to see if the rate of wear shows any dependence on some easily measurable parameters. A reasonable correlation of the wear rate of polymers is obtained with $1/\sigma_b\varepsilon$ where σ_b is the stress to break the polymer at an elongation ε. Unfortunately, such a correlation is shown to be possible only for experiments where the polymer is traversed once only [3]. The suggestion is that, provided the counterface had no deposit of the polymer, its wear rate could be correlated with $1/\sigma_b\varepsilon$. This is not a practical situation, but apparently correlation with the breaking strength and elongation is possible under three-body abrasion, especially if the intervening particles are coarse [1]. We may speculate that if a suitable lubricant can continuously remove any transfer film, a similar correlation may be possible.

Three main types of wear are recognized, viz., adhesive, abrasive and fatigue wear. Adhesive wear is manifested in transfer of polymeric films and abrasive wear occurs either by a rough counterface or by loose adventitious particles trapped at the interface. The model [7] of fatigue wear starts with the relationship $n \propto (\sigma_0/\sigma)^t$ where n is the number of cycles to failure, σ_0 is the stress to failure by a single application of the load, σ is the applied cyclic stress and t is a material constant. The rate of wear $\propto (1/n) \propto \sigma_0^{-t}\sigma$. The stress σ on a polymer will be governed by the geometry of the indenting asperity if the encounter is elastic. For a hemispherical asperity of radius R, penetrating a softer polymer, $\sigma \propto R^{-2/3}$ when the interaction is elastic. In that event,

$$\text{Wear rate} \propto \sigma_0^{-t}R^{-2t/3} \qquad (10.3)$$

10.6   Speed and temperature

High local pressures at the contact zone do not alter the viscoelastic characteristics of a polymer substantially during rolling [14]. In sliding, however, high shear strains develop in the surface causing appreciable tearing and transfer of polymer to the other surface. The coefficient of friction is seen to increase with load and to fall after reaching a maximum (Fig. 10.2). The reason for the rise in friction is attributed [7] to the increase in the contact area when the load is increased. The fall in friction after the maximum is possibly due to extra heat which is generated during sliding at loads higher than a critical value. If the temperature is high enough, a layer of low shear strength material will be expected to form at the interface which should provide low values of the coefficient of friction. There is some indirect evidence to support this view in that the peak in the friction curve appears at a lower load as the velocity of sliding is increased.

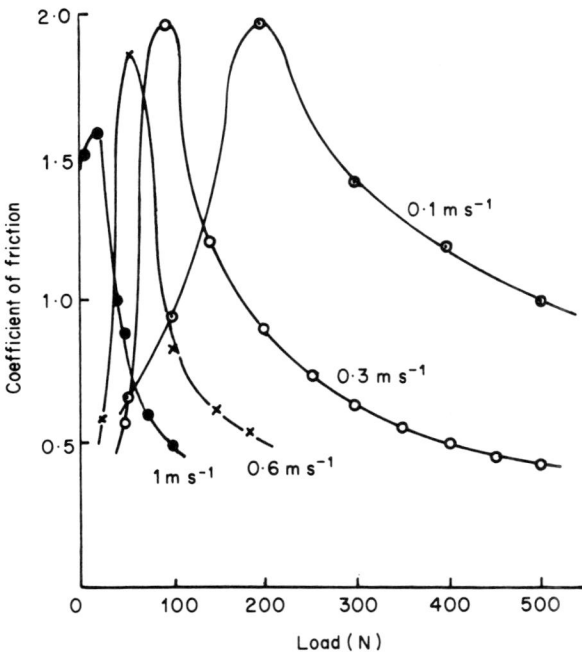

Fig. 10.2   Variation of the coefficient of friction with load at various speeds for nylon 6, sliding on steel. (Lancaster, J. K. (1973). *Plastics and Polymers* **41**, 297.)

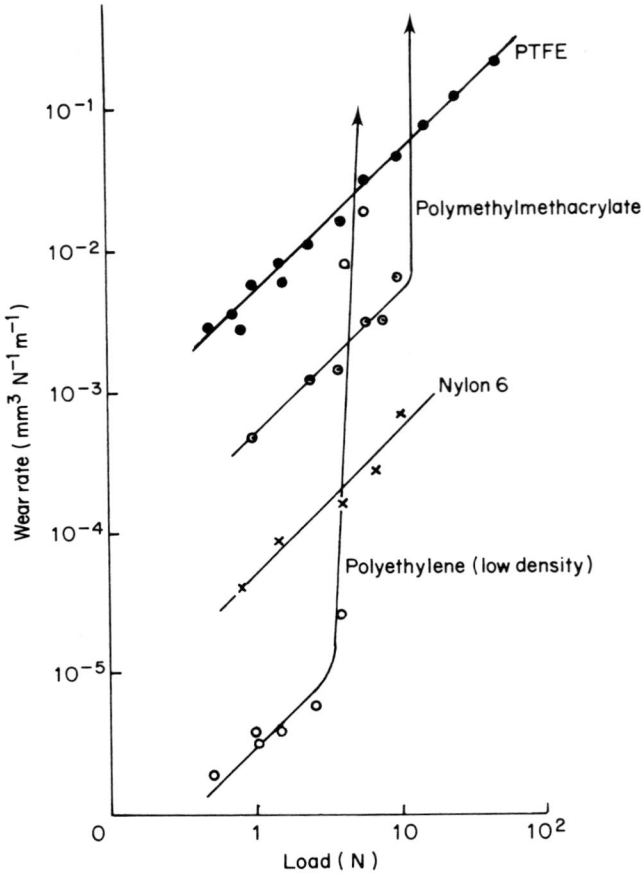

Fig. 10.3   Variation of wear rates of polymers with load sliding against steel.
(Lancaster, J. K. (1973). *Plastics and Polymers* **41**, 297.)

The rate of wear, if plotted against log (load), follows a similar pattern to that of some metals, showing an initial linearity followed by a transition load when the rate of wear accelerates (Fig. 10.3). The transition load, of course, varies according to the nature of the polymer and probably again depends on the generation of high interfacial temperatures.

Usually, there is no correlation between wear rates and the apparent area of contact. The true contact area is load dependent and the number of junctions increases [15] with load but not infinitely. The effect of a

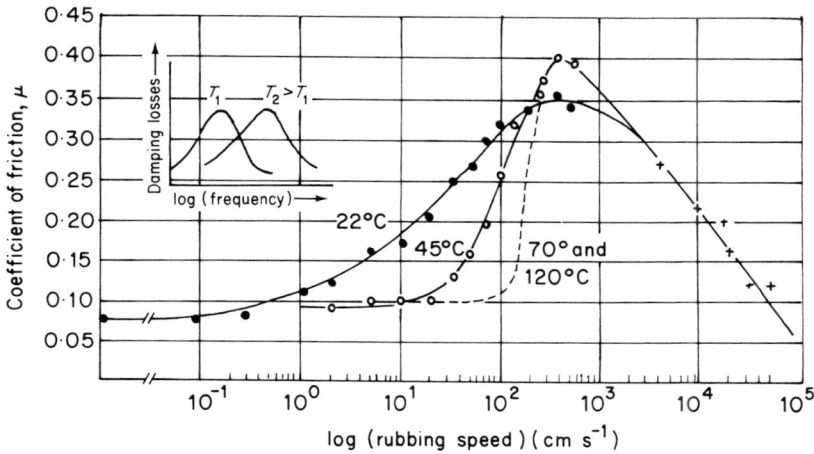

Fig. 10.4 Friction of 49% crystalline PTFE on itself as a function of speed. The inset shows that the damping loss peak for a viscoelastic material is shifted to a higher frequency if the ambient temperature is raised. (McLaren, K. G. and Tabor, D. (1963). Proceedings: 1st Lubrication and Wear Group Convention, Institution of Mechanical Engineers, London, paper 18.)

tangential pull is to deform the asperity, whereas with metals there is usually lateral plastic flow. There does not appear to be any attempt to obtain a basic relationship between the true area of contact and either the frictional resistance or the wear rate of a polymer.

The effect of sliding velocity on the coefficient of friction, μ, is shown [16] in Fig. 10.4. The frictional resistance increases with speed and at the lower speed range, μ decreases as the ambient temperature is raised. At high speeds, at or above 250 cm s^{-1}, the coefficients tend to be similar and the effect of raising the ambient temperature is to displace the curve towards the higher speed range. Figure 10.4 shows the variation of μ with log (speed) and the pattern of these curves is similar to the damping loss characteristic of the polymer as shown in the inset. The losses are dependent upon the frequency of deformation and the bulk temperature. For PTFE at 22 °C, a maximum viscoelastic loss occurs at a frequency of 10^3 MHz. Comparing this with the corresponding friction maximum which occurs at 500 cm s^{-1}, it is calculated [16] to show that the polymer undergoes a unit displacement of about 50 Å. The conclusion is that friction of polymers as a result of sliding occurs

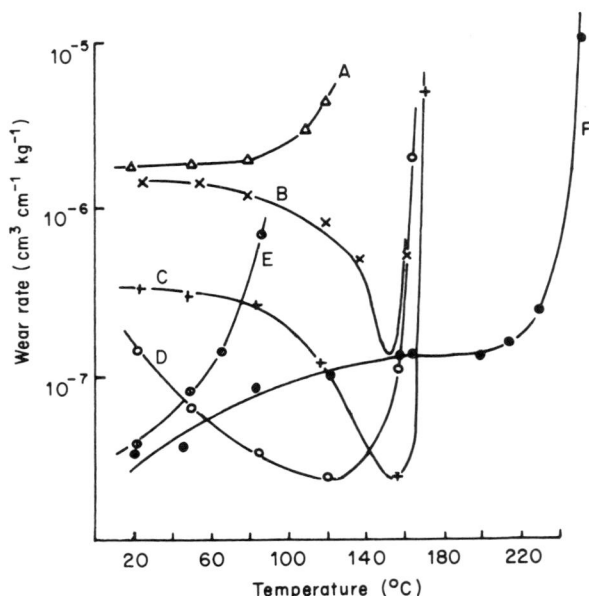

Fig. 10.5   Variation of wear rate with temperature on a lubricated 1·2 μm cla
steel counterface. A polystyrene; B polymethylmethacrylate; C
polyacetal; D polypropylene; E polyethylene; F nylon 66. (Lancaster, J. K. (1968). *Proc. Inst. Mech. Eng.* **183**(3P), 98.)

due to a process at a molecular level. We note that metallic friction
involves interaction at a much larger scale where asperities deform and
the junctions grow.

Results [3] from a pin–disc machine where the worn volume was
measured after sliding a distance of 10^3 cm using a spiral track to avoid
rubbing on transferred films are shown in Fig. 10.5. It shows that, as the
ambient temperature is increased, the rate of wear falls initially. The
suggested reason for this is that there is a decrease in the modulus of
elasticity with a rise in temperature. If the mode of wear is by fatigue,
this means that the number of cycles to initiate fatigue increases and
wear is low because the running time is kept constant. At a characteristic temperature, however, thermal softening results in a rapid rise
in the rate of wear. In a similar manner [17], the wear rate rises
markedly at a characteristic speed (Fig. 10.6). The experiments were
carried out for about 300 revolutions only at a relatively low load of 1 kg
to avoid any excessive rise in bulk temperature. The idea is that in this

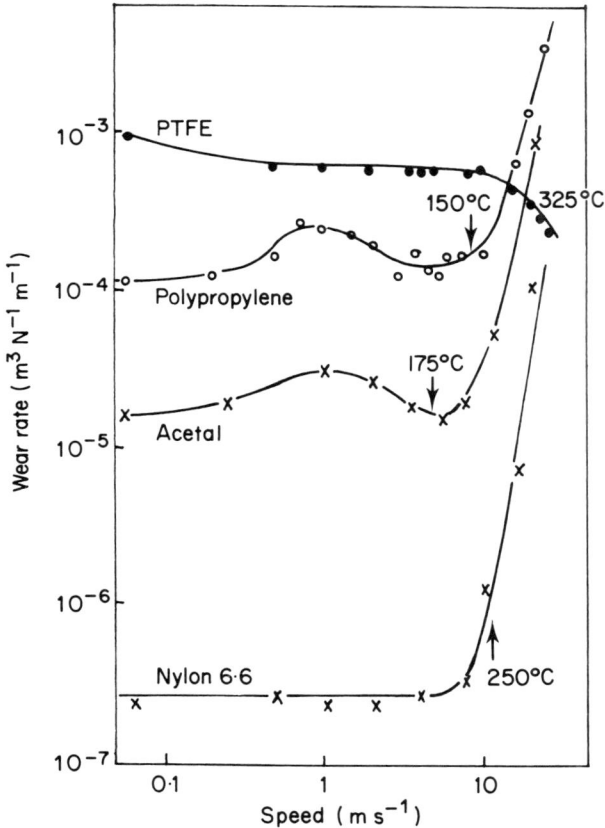

Fig. 10.6   Variation of wear rate with speed for polymers sliding against mild steel. (Lancaster, J. K. (1971). *Tribology* **4**, 82.)

way only the localized rise in temperature at the points of contact, that is the flash temperature of the interface, will affect the wear rate. The effect of a high flash temperature should be to increase the wear rates by facilitating transfer of polymer to the counterface. However, Fig. 10.6 shows that not all the polymers show a dramatic increase in wear with speed and, in fact, PTFE shows a decrease in the rate of wear.

No doubt, the surface temperature developed during sliding plays a dominant part in the friction and wear modes of polymers. Unfortunately, it is not possible to measure these temperatures easily and recourse is often taken to calculating the flash temperatures as we have already discussed in Chapter 9. Using the assumptions enumerated in Chapter 9, expressions for flash temperatures at the polymer interface have been deduced [17]. It is first assumed that the counterface is steel so that the deformation is confined to the polymer only. Although polymeric deformation involves a substantial amount of an elastic component, for the sake of simplicity, it is assumed that deformation is plastic. Two extreme speed ranges are considered as before so that the flash temperature θ_m varies as follows:

For low speeds, $\theta_m = 10^{-2}\,\mu H^{1/2} W^{1/2} v$ (10.4)

For high speeds, $\theta_m = 5{\cdot}7 \times 10^{-5}\,\mu H^{3/4} W^{1/4} v^{1/2}$ (10.5)

where μ = coefficient of friction, H = hardness of the polymer, W = applied normal load, and v = sliding velocity.

We have already discussed that, unlike high speed sliding, thermal equilibrium is reached at low speeds and the heat generated due to frictional work is expected to be shared between the polymer and the counterface. The implicit assumption in Equations 10.4 and 10.5 is that the hardness of the polymer does not vary with temperature and hence sliding speed. This is another source of error in calculating flash temperatures and hence a correction is applied. Experiments show that the hardness of polymers tends to decrease almost exponentially with temperature, so that to a close approximation

$$H = H_0 \exp{(-a\theta_m)}$$ (10.6)

where H_0 = hardness of the polymer at a room temperature of about 20 °C, a = material constant and not much error is introduced if it is given a value of $0{\cdot}005$, and θ_m = flash temperature attained at the points of contact.

The effect of hardness variation is the most pronounced at high speeds and, in that situation, the flash temperature θ_m is shown to be

$$\theta_m = \theta_0 \exp{(-0{\cdot}0038\theta_m)}$$ (10.7)

where θ_0 is the flash temperature calculated from Equation 10.4, assuming that there is no change in hardness.

The mean surface temperature θ is given by

$$\theta = \theta_a + c\mu W v,$$ (10.8)

where θ_a is the ambient temperature and c is a constant. The total temperature rise θ_t of the polymer is the sum of the mean surface temperature and the flash temperature. That is,

$$\theta_t = \theta + \theta_m \tag{10.9}$$

For design purposes, materials are rated according to the operating service temperature and generally surface temperatures are not calculated while deciding on the application of dry rubbing bearings which are manufactured from polymeric materials. The usual assumption at the design stage is that the bearing will wear more at temperatures higher than the ambient temperature and an appropriate factor is used while estimating loss of life due to wear.

10.7 Crystalline polymers
Unlike metals, most polymers are amorphous and they possess a limited amount of short range order. A truly amorphous polymer has a random distribution of chains in its main molecular structure and this gives rise to the viscoelastic nature of many of these solids. A crystalline polymer shows considerable three-dimensional order and there is a dramatic change in such properties as density, elasticity and mechanical response in general when compared with the amorphous variety. We should attempt to seek a comparison of these polymers with metals, but it should be appreciated that perfect crystallinity is rare in the former. Some typical crystalline polymers are polyethylene, polypropylene, nylon and polyacetal, and it is useful to outline the nature [18] of these materials before collating the available information on their tribological behaviour.

10.7.1 *Polyethylene*
The polymer polyethylene is popularly called polythene, and other well-known trade names are Alathon, Alkathene and Carlona. It is a long chain hydrocarbon of the type $-CH_2-CH_2-CH_2-CH_2-$ and its intermolecular forces are not strong. Most of its strength arises out of the close molecular packing, resulting due to crystallization. Its tribological use, if any, is not known but its good chemical resistance and toughness should make it a candidate material for a tribological application. It is worth noting that polyethylene becomes brittle when exposed to ultra-violet light. High energy irradiation results in a decided loss of crystallization due to cross-linking. Progressive radiation eventually renders the polymer amorphous and results in surface oxidation in the presence of air. This is degradation, and the effect of

irradiation on the tribological properties of polymers is being studied.

10.7.2 *Polypropylene*

Like polyethylene, polypropylene is a linear hydrocarbon and hence they have certain properties in common. Polypropylene is more prone to oxidation and has a lower density, but possesses a melting point some 50 °C higher than polyethylene. Polypropylene becomes brittle under cryogenic conditions and will probably wear readily if used at temperatures below 0 °C. Certain polypropylene products show good abrasion resistance and find application in ropes and brush tufting. If used as sheets for covering floors, it needs development by adding fillers to resist scratching and wear.

10.7.3 *Nylons*

The term nylon is a generic name for the many synthetic polyamides and is designated by numerals e.g. nylon 6, to specify the chemical structure. Some polyamides such as nylons 6 and 66 are thermoplastic. Because of the characteristic chemical structure, polyamides crystallize with inherent high intermolecular attraction, but the structure also incorporates amorphous zones which give a manufactured component the welcome degree of toughness. On the other hand, a high molecular attraction gives these polymers a high melting point which is a tribological advantage in terms of the consequences of frictional heating. The physical and mechanical properties of nylons are considerably affected by the degree of crystallization, which can be controlled by altering the production technique. As an example, a component in nylon 6 if cast and slowly cooled followed by annealing will probably be about 60% crystalline, especially, if the component is thin-walled. Solidification of polymers is controlled by forces somewhat analogous to those in metals. Thus small spherulites can be attained in a polymer, in a similar manner to small grains in metals, by a heterogeneous nucleation process if about 0·1% fine silica dust is added to the melt. The resulting solid is known to show high tensile strength and hardness. There is some reduction in the ductility and impact strength but this is compensated for by an increase in the abrasion resistance. It is important to appreciate that the morphology of a cast polymer may vary because of the differences in the cooling rate from the surface to the centre. Thus the outside of a casting will be less crystalline because of a rapid solidification rate and may be less resistant to wear as a result of this. While carrying out experimental work, therefore, the variation in properties between the surface and the centre must be taken into account. Fortunately, nylon manufacturers have produced materials which do not show this gradient in crystallinity from the surface to the centre [18]. Mechanical properties of nylons are also affected by the operating environment such as the ambient temperature or the humidity. The field of application of nylons is wide and the main properties which make them attractive are the rigidity, toughness, and heat resistance. The material is expensive and typical components in the area of tribology are bearings, valve seats, cams and gears.

10.7.4   Polyacetals

Polymerization of formaldehyde results in the product known as polyformaldehyde, also known as polyoxymethylene and polyacetal. Like polyethylene, polyacetal is linear structurally and is hence thermoplastic. The acetal resins show favourable stiffness and have high fatigue endurance and resistance to creep. All these properties are important from the tribological point of view particularly since the acetals are known to provide low interfacial friction. Compared with nylons the acetals are superior in such properties as fatigue and creep, but inferior from the point of view of wear resistance. It is not known if a comparative study of nylons and acetals with respect to wear has been made, but there is scope for carrying out a tribological study of acetals. As an example, the crystallinity of acetals increases considerably if annealed at high temperatures, and there is some evidence that we can expect improved wear resistance with an increase in the degree of crystallinity of a polymer.

10.7.5   Counterface

All resins are modified from the point of view of appearance and mechanical, physical and chemical properties by incorporating various additives. Of particular interest to a tribologist is the use of such fillers as glass. The crystalline acetals are good bearing materials because of an absence of stick–slip motion and the lowest coefficient of friction is experienced with a steel counterface. Other metals, in particular aluminium, increase the friction, and as Table 10.1 shows nylon sliding on itself is not a desirable friction couple. A high coefficient of friction due to nylon sliding on itself is attributed to a thermal effect but we shall

Table 10.1   Kinetic friction of nylon 66 [18]

Counterface	Nylon (moulded)	Nylon (machined surface)	Mild steel
Nylon (moulded)	0·63	0·52	0·31
Nylon (machined surface)	0·45	0·46	0·33
Mild steel	0·41	0·41	0·60–1·00

discuss this aspect later. Table 10.1 shows that the nature of the nylon surface is not sensitive to frictional resistance and similar metals are incompatible. Apart from sliding mechanisms, polymers are employed as rolling elements. In general acetals are used as pump impellers and fan blades, although there exists the possibility of erosion in a chemical medium.

10.7.6   Friction and wear
Experimental work suggests that the frictional behaviour of polyethylene is sensitive to temperature and a theory incorporating the relaxation characteristic of the material has been outlined [19]. It is suggested that, in a similar manner to metal couples, strong junctions are formed when polyethylene slides on itself. A junction formed in this manner is under imposed normal and horizontal forces, and it thus extends until it fails under a tensile stress. This being the case, if A_t is the true area of contact and σ_a is the average tensile stress on a junction causing fracture, the coefficient of friction is proportional to $A_t\sigma_a$. Since the hardness of a polymer varies markedly with temperature, it is more convenient to measure the amount of penetration h and say $A_t = K_1 h$, where K_1 is a constant. Using a new constant K, we can express the coefficient of friction μ as

$$\mu = Kh\sigma_a \qquad (10.10)$$

σ_a is obtained from the stress–strain curve of the bulk polymer but the rate of strain of the junctions will be orders of magnitude greater in a sliding situation. Since polyethylene is viscoelastic, the junctions being strained at a higher rate will fail at a lower extension when compared with the failure strength in a tensile test. The average tensile strength of a junction will also be low, the degree of loss depending on the relaxation characteristic of the polymer under study.

Applying a simple Maxwell model, the stress σ at a time t, assuming the rate of extension to be constant, is

$$\sigma = \sigma_Y [1 - \exp(-t/t^*)] \tag{10.11}$$

where σ_Y is the yield stress and t^* is the relaxation time of the polymer. If $\bar{t}$ is the time required to break a junction, the average tensile stress $\bar{\sigma}$ necessary to do this is,

$$\bar{\sigma} = \sigma_Y(1+c) \tag{10.12}$$

where

$$c = -\frac{\bar{t}}{t^*} [1 - \exp(-\bar{t}/t^*)]$$

Replacing σ_a by $\bar{\sigma}$ in Equation 10.10,

$$\mu = Kh\sigma_Y(1+c) \tag{10.13}$$

The authors [19] of this theory have incorporated a surface energy term as well, but for simplicity, Equation 10.13 is expressed in terms of yield stress, the time of relaxation, and the fracture characteristic only. The shear stress of the junctions rather than the concept of a tensile pull should also provide a plausible expression for polymeric friction.

Table 10.2 Values of the product $h\sigma_Y$ at various temperatures for polyethylene [19]

Temperature (°C)	20	30	40	50	60	70	80	90
$h\sigma_Y$ (kg mm^{-2} μm)	15·3	16·7	16·5	16·56	15·1	14·2	12·5	13·0

One difficulty with Equation 10.13 is that the parameters h and σ_Y are not easily obtainable unless the yield stress is taken from a tensile test. Some typical values at 20 °C have been quoted [19] as 17 μm and 0·9 kg mm^{-2} for h and σ_Y respectively. As Table 10.2 shows, the product $h\sigma_Y$ for polyethylene is relatively insensitive to temperature. The parameter c is known to decrease with temperature and the relaxation time is independent of temperature. On the basis of these observations, it is concluded that it must be the time to break a junction which decides on the net frictional resistance. This is interesting because the postulate with metals is that the contact area increases with time in the direction of motion and the junction then breaks.

Sliding speed has a decided effect [20] on the friction and wear of polyethylene. A low density polyethylene shows a small drop in μ but the wear rate reaches a minimum and then rises rapidly with sliding speed. It is interesting that while the counterface is an insulating material such as glass, the friction falls steadily albeit in a shallow manner but wear is very high when compared with the results for the same polymer running on a thermally conducting steel counterface. A high density polyethylene, on the other hand, shows a similar increase in wear while rubbing on glass but a wear rate which is nearly independent of the sliding speed when the counterface is steel. The coefficient of friction also remains constant while rubbing on steel but a glass counterface gives a maximum value of friction at a characteristic speed. A maximum coefficient of friction is also noted with nylon 6, but while rubbing against steel. The reason for these maxima and the inconsistent influence of the thermal conductivity of the counterface is not clear, but a maximum friction is also observed at a particular sliding speed while rubbing polypropylene on steel.

The friction and wear mode of these crystalline polymers are thus shown to depend on the rubbing speed, which facilitates frictional heating. If the counterface is prepared from an insulating material, heat loss from the interface is impeded and the temperature of the polymer increases. Overheating results in softening and incipient fusion occurs at a critical speed which is lower for a glass than for a steel counterface. Melting is known to occur to a distinct depth in the polymer, the degree depending on its nature, the rubbing velocity and the thermal conductivity of the counterface. For a steel disc, this molten depth is consistent over a range of sliding speeds, but a counterface with a low thermal conductivity such as glass can produce a viscous layer on a polypropylene pin. This layer is about 10^{-2} cm thick at a sliding speed of 100 cm s^{-1} when the normal load is 1 kg and the corresponding depths for polyacetal, nylon 6 and high density polyethylene are about $0 \cdot 15 \times 10^{-2}$, $0 \cdot 45 \times 10^{-2}$ and $0 \cdot 12 \times 10^{-2}$ cm respectively. We have already discussed that this viscous layer is removed from the pin to be transferred to the counterface. It would appear that materials which transfer as large lumps must produce viscous layers which are deeper than those produced by these crystalline polymers, except for low density polyethylene, which only smears a thin layer of material on the counterface.

10.7.6.1   *Friction*

As the machine starts, the value of the coefficient of friction increases with time. This has been explained [20] in terms of a bedding-in process. It is first assumed that the frictional resistance F equals τA_t where τ is the shear stress of the polymer and A_t is the true area of contact. The value of A_t is low initially but is believed by Lee [20] to increase with continuous running so that F should increase since τ, being a material property, can be assumed to be a constant. The true area of contact can increase by creep or it can grow as a result of a tangential pull. This contradicts our earlier statement that there is no junction growth with polymers but melting will cause flow of material, when the true area of contact should increase. The dependence of A_t on time thus becomes credible because a little time is necessary for the rubbing process to generate heat and establish an equilibrium temperature at the interface.

It is logical to think that frictional resistance of polymers should be defined in terms of the sliding velocity v and h, the depth of the molten layer measured from the sliding interface. Ignoring what may be termed the running-in friction, the steady state frictional resistance has been expressed in terms of v, h and the apparent area A of the interface as,

$$F = \eta(v/h)A \qquad (10.14)$$

where η has been called the effective viscosity of the polymer layer. This qualification is necessary since the portion of the pin which is directly in contact with the counterface will attain the highest possible temperature as a result of frictional heating. The temperature falls away progressively from the interface, the effect being a viscosity gradient in the polymer. Furthermore, a polymeric melt behaves in a non-Newtonian manner. That is, the viscosity of the melt varies with the rate of shear which probably changes significantly from the interface to the cooler part of the polymer. Examination of transverse sections of polypropylene pins suggests that the rate of shear is the highest at the sliding interface. The ratio (v/h) in Equation 10.14 is the shear rate and experiments show that it has a low value if the effective viscosity is high, as it should be. As expected, experimental plots of η against (v/h) show that the polymer interface is low in effective viscosity when a thermally conducting counterface such as steel is used.

Although Equation 10.14 provides an expression for the frictional resistance, we should consider the true area of contact A_t. It has been observed [20] that $A_t = (A/2)$ with polymers so that Equation 10.14 can be modified to the following form:

$$F = f\eta(v/h)A \qquad (10.15)$$

where $f = A_t/A$ and is less than unity.

10.7.6.2   Wear

Of the crystalline polymers included in this chapter, observation of the wear process tends to suggest that the low density polyethylene wears by a slightly different mechanism to the others. Generally, at a low sliding speed, the polymer is removed by a two-body abrasion mode due to the hard asperities of the metal counterface. At high speeds, the polymer loses material by depositing the viscous layers which form due to heating.

A peculiarity of the other crystalline polymers is that at high sliding speeds, the viscous layer is not completely transferred to the counterface and the bulk of it is drawn out towards the rear edge of the pin. We could say that the extraneous sheet thus formed, analogous to a burr in metals, is wear and would be thus recorded if height loss of the pin was taken as the criterion. Wear therefore occurs by the lateral expulsion of the viscous layer and its magnitude increases if a glass counterface is used.

The model of a parallel plate plastometer has been cited [20] to explain this tail with the associated high rate of wear. Thus consider a viscous liquid being held between two parallel circular plates, each of diameter D. Suppose that a normal load W is placed on the top plate causing the liquid to be forced out equally in all directions so that the separation distance h between the plates decreases with time, t. For Newtonian flow,

$$\frac{dh}{dt} = \frac{2Wh^3}{3\pi D^4\eta} \qquad (10.16)$$

Applying Equation 10.16 to a pin of diameter D for the case where the lateral outflow of polymer occurs,

$$\frac{dh}{ds} = \frac{2Wh^3}{3\pi D^4\eta v} \qquad (10.17)$$

where s is the sliding distance and v is the velocity. Taking appropriate values from experiment, the wear rate is seen to be some four orders of magnitude higher when calculated from Equation 10.17. There may be a number of reasons for this discrepancy, viz., the outflow being nonsymmetrical and the value of η being uncertain at the interface. Examination of Equation 10.17 shows that if h is large the rate of wear is high, while if η is large, the rate of wear is low. The higher the normal load, the greater is the amount of wear. The velocity term is more difficult to justify since the relationship suggests that the higher the value of v, the less the rate of wear, which is not true at all speeds for any polymer.

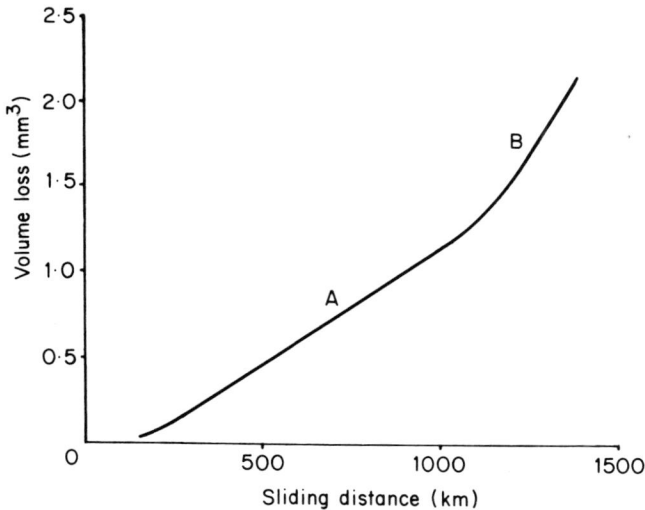

Fig. 10.7   Three regimes of wear of a high molecular weight polyethylene. (Dowson *et al.* in Lee, L. H. (Ed.) (1974). "Advances in Polymer Friction and Wear", Plenum Press, New York and London.)

Work [20] on High Molecular Weight Polyethylene (HMWP) with an average molecular weight $\bar{M}_w$ between 3·5 and 4·0 million shows three regimes of wear when sliding in the form of pins against steel discs (Fig. 10.7). Up to a sliding distance of about 150 km, the volume loss is unpredictable, possibly as a result of bedding in of the pin surface. Beyond this, there are two distinct regimes of wear A and B, both producing linear patterns when the volume of the polymer removed is

plotted against the sliding distance. In the first linear regime, the pin
transfers material on to the counterface, the deposit appearing as dark
patches. The effect of continual transfer is to decrease the effective
surface roughness of the steel disc, initially by filling in the valleys. With
further transfer, however, the heights of the deposits appear to become
greater than the highest peaks of the steel surface, as shown by an
increase in the values of cla (Fig. 10.8). That is, the nature of the transfer
process is discrete, and in spite of a tangential motion at the interface,
the deposit is not polished to give low cla values.

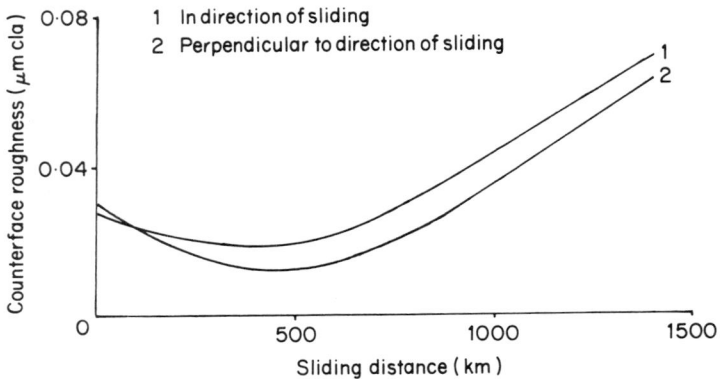

Fig. 10.8   Variation of counterface roughness with sliding distance. (Dowson
et al. in Lee, L. H. (Ed.) (1974). "Advances in Polymer Friction and
Wear", Plenum Press, New York and London.)

Being a viscoelastic material, the asperities on the pin are likely to
undergo alternate elastic deformation and relaxation as they encounter
the protuberances of the counterface. This is a potential condition for
fatigue and it is likely that this will result in transverse cracks in the pin
surface relative to the direction of motion. Such cracks have actually
been observed when the slope of the wear curve changes (B, Fig. 10.7)
with a higher rate of wear. The suggestion thus is that wear of this
HMWP occurs in a complex manner, the modes being a combination of
abrasion, adhesion and micro-fatigue.

Current studies [20] with crystalline polymers show that wear by the
abrasive mode becomes low as the hardness, tensile yield strength,
resistance to tear and the density of the material increases. If a

polyethylene sheet is slid against an abrading surface such as a grinding wheel, the first effect is for the grits to indent and deform the polymer. The polymer is pulled and ploughed as the wheel starts to rotate and strips of the polymer are removed continuously by stretching them until they tear. A high molecular weight is associated with a good tear strength and hence with an increased resistance to abrasion. A high degree of crystallinity means an increase in hardness, tensile and tear strength, all of which will favour increased resistance to abrasion.

10.8 Composites
Polymers have poor mechanical and thermal properties which can be limiting when the materials are being considered for use in stressed engineering applications. Inadequate mechanical properties have the consequence that even at loads which are considered low and safe for metals, the polymer may deform plastically or break under tensile loads. Because of inherently low thermal conductivity, heat build-up can occur readily resulting in structural degradation of the polymer. The role of additives is in the first place to improve the mechanical and thermal properties. Additives can also act as solid lubricants and resist degradation of a polymer by ultra-violet light.

10.8.1 Fillers
Polymer composites incorporate fillers, some examples being china clay, calcium carbonate, talc and barium sulphate. The properties of composites cannot be reproducible without controlling the types of fillers and their sizes and shapes. Coarse particles make a component brittle and the plate-like shape of china clay makes the product aniso-tropic in mechanical properties. It is quite common to use fibrous fillers such as wood, flour and cotton flock to improve the impact strength, rigidity and toughness of the resin matrix. Apart from enhanced strength, addition of asbestos and glass fibres to certain polymers has the beneficial effect of improving the heat resistance of moulded components. The production technique can be controlled to embed fibrous fillers in a resin matrix in such a manner that the service strain acts in the plane of the fibres. These fillers have higher moduli of rigidity than the matrix and in this way a high strength composite is produced. Use of laminar fabrics as strengthening fillers is illustrated in a later chapter while discussing bearings lined with woven materials for use in highly stressed mechanisms.

Of the additives fulfilling the role of a lubricant, the common ones are graphite and molybdenum disulphide used in quantities up to 2% in thermoplastic resins. Solid lubricants are known to exude from the interface and coat the counterface. An interface thus supplied with a low friction material during the life of a friction couple can be created with other additives, such as stearates of calcium, barium, cadmium and lead. Hydrocarbons such as paraffin wax, low molecular weight polythene and ethyl palmitate also have possibilities as dry lubricants to be incorporated in a plastic.

Hydrocarbon molecules can degrade if exposed to ultra-violet radiation and the effect on tribological interactions has received limited attention. Degradation of polymers in this manner is undesirable and can be minimized by incorporating ultra-violet absorbers. The most important additives which absorb ultra-violet light are *o*-hydroxybenzophenones, *o*-hydroxyphenylbenzotriazoles and salicylates.

10.8.1.1   *Applications*

The possibility of using iron powders is strong when one considers that the heat transfer rate of low density polyethylene can be increased by a factor of four by using iron powder as a filler. The particles should be fine, elongated and dense but they adversely affect the flow properties of the composite during injection moulding to manufacture the component.

A large number of fillers such as glass, carbon, MoS_2, bronze, polyphenylene sulphide and poly-*p*-benzoates are incorporated in PTFE for tribological applications. PTFE belongs to the group known as fluoropolymer and the majority of tribological components with a PTFE matrix are produced by pressing and sintering. A more recently developed group employs the casting technique to manufacture components and is based on organic fillers added to such bases as ethylene-tetrafluoroethylene (ETFE), vinylidine fluoride-tetrafluoroethylene ($VF_2 - TFE$) and fluorinated ethylene-propylene copolymer (FEP).

PTFE by itself wears readily, but like other fluorocarbons exhibts low wear rates if suitable fillers are incorporated. The lowest wear rate is observed when bronze is added as a filler in PTFE and other additives used are carbon or graphite, glass and ceramics. Incorporation of a suitable filler can reduce the wear rate of PTFE by a factor of 1000. There is, however, an optimum volume of the filler between 20 and 30%

in the composite, beyond which the PTFE matrix loses its integrity and the wear rate increases substantially.

One reason for the reduction in wear rate is quite likely the improved compressive strength and thermal conductivity of the composite. The other reason would appear to be that, at the initiation of sliding, the PTFE portion of the composite begins to transfer to the counterface, but this is interrupted as soon as the fillers protrude. The counterface is protected by the PTFE films and the composite wears at a low wear rate because loss of material is governed by the wear rate of the fillers. An ideal filler should, therefore, be able to interrupt the transfer of the resin to the counterface, but should not roughen it since a rough counterface will increase the wear rate of the composite. It seems that the fillers in the PTFE matrix are only held mechanically but carbon fibre in a matrix of $VF_2 - TFE$ exhibits adhesion to the polymer. This adhesion is believed to inhibit rather than interrupt the transfer of the resin to the counterface, giving rise to a composite with a low wear rate. The coefficient of friction remains low and similar tribological properties are experienced by reinforcing $VF_2 - TFE$ with glass fibre or by using ETFE filled with carbon and glass fibre. Nylons have also been reinforced with glass spheres, glass fibres and alumina particles.

References

1. Lancaster, J. K. (1969). *Wear* **14**, 223.
2. Lancaster, J. K. (1972). *Wear* **20**, 315.
3. Lancaster, J. K. (1968). *Proc. Inst. Mech. Eng.* **183**(3P), 98.
4. Bonfield, W., Edwards, B. C., Markham, A. J. and White, J. R. (1976). *Wear* **37**, 113.
5. Makinson, K. R. and Tabor, D. (1964). *Proc. Roy. Soc. A* **281**, 49.
6. Tanaka, K. and Ueda, S. (1976). *Wear* **39**, 323.
7. Lancaster, J. K. (1973). *Plastics and Polymers* **41**, 297.
8. West, G. H. and Senior, J. M. (1973). *Tribology* **6**, 269.
9. Lancaster, J. K. (1972). *Wear* **20**, 335.
10. Lancaster, J. K. (1973). *Tribology* **6**, 219.
11. King, R. F. and Tabor, D. (1953). *Proc. Phys. Soc. B* **66**, 728.
12. Pascoe, M. W. and Tabor, D. (1956). *Proc. Roy. Soc. A* **235**, 210.
13. Pooley, C. M. and Tabor, D. (1972). *Proc. Roy. Soc. A* **329**, 251.
14. Ludema, K. C. and Tabor, D. (1966). *Wear* **9**, 329.
15. Ohara, K. (1976). *Wear* **39**, 251.
16. McLaren, K. G. and Tabor, D. (1963). Proceedings: 1st Lubrication and Wear Group Convention, Institution of Mechanical Engineers, London, paper 18.
17. Lancaster, J. K. (1971). *Tribology* **4**, 82.

18. Brydson, J. A. (1975). "Plastics Materials", Newnes–Butterworths, London.
19. West, G. H. and Senior, J. M. (1972). *Wear* **19**, 37.
20. Lee, L. H. (Ed.) (1974). "Advances in Polymer Friction and Wear", Plenum, New York and London.

11 Rubber

The natural product cis-polyisoprene, commonly known as rubber, is viscoelastic and can exhibit recoverable deformation of as much as 1000 per cent. In its raw natural state, the material is tacky with a low resistance to abrasion and is sensitive to oxidative degradation. However, resistance to the latter and to abrasion and creep can be imparted together with an improvement in tensile strength and elasticity by vulcanization, which involves cross-linking the natural product. One example of the process of vulcanization is to heat the natural rubber with sulphur. There are many other ways of cross-linking rubber and a large number of polymers is available which exhibit similar properites to vulcanized rubber. The generic term elastomer is now used to describe this group of polymers which demonstrate the following common characteristics:

(1) The materials have glass transition temperature well below room temperature.
(2) They exhibit a high degree of elasticity and retract rapidly after stretching.
(3) They show a high modulus when elongated.
(4) They are largely amorphous.
(5) They have large molar masses and are readily cross-linked.

Among the many tribological applications of rubber are automobile tyres, sealing rings, flexible pad thrust bearings and soft lined journal bearings. There are statistical data which tend to suggest that up to 90% of automobile tyres have to be withdrawn from service because of wear in the tread. Even a small increase in the wear resistance of these components will result in a large saving of money and material. An understanding of the mechanisms of friction and wear of rubber is of practical value. There have been many attempts to provide experimental results and theoretical models to improve our understanding of the tribology of rubber. Although great progress has been made in ensuring a long life of many rubber products working under sliding or rolling conditions, controversy remains regarding mechanisms.

11.1   Adhesion and hysteresis

Being a viscoelastic material, the nature of the interfacial interaction between a rubber and another solid is different from metals and similar to many polymers in general [1–3]. Thus consider a sheet of rubber loaded against a hard counterface when the pressure generated at the true area of contact A_t is elastic. If W is the applied normal load and n is the total number of contacts, each being under a mean elastic pressure $\bar{\sigma}$, it can be assumed that,

$$W = \bar{\sigma} A_t \tag{11.1}$$

and

$$A_t = Kn(\bar{\sigma}/E)^m \tag{11.2}$$

where K is a constant; E is the modulus of elasticity of the rubber; and $m \to 1$ for most practical situations.

The establishment of this true area of contact gives rise to a finite amount of static friction, which is part of the resistance offered to motion under a horizontal pull. This resistance is the force F_a due to adhesion of the elastomer to the counterface and is believed to be the result of interaction between surface molecules at the contact area. If now the tangential pull is increased so that the rubber sheet just begins to slide, an additional resistance F_h due to hysteresis is experienced. The hysteresis phenomenon is as a result of the delayed recovery of the rubber after it has been dragged across the protuberances of the opposing surface. The kinetic friction F is,

$$F = F_a + F_h \tag{11.3}$$

The adhesion component is confined to the surface but hysteresis is a bulk effect and depends on the viscoelastic property of the elastomer. Although the amount is considerable, only a layer of the rubber a few thousandths of a centimetre thick undergoes deformation [3].

11.2 Velocity

The frictional resistance of rubber is influenced by both sliding velocity and temperature [4], the phenomenon being interrelated with the temperature and rate dependence of the viscoelastic properties of the elastomer. If the coefficient of friction μ is plotted against sliding velocity, a maximum value of μ is reached at a characteristic velocity in a similar manner to that shown in Fig. 10.4. The effect of increasing the temperature is to provide a similar maximum value of μ but the peak is shifted to a higher velocity. It has also been established that the maximum is symmetrical when the rubber is made to slide on a smooth surface but shows marked asymmetry when the counterface is rough. The reason probably is that when the counterface is very smooth, the sole mechanism for friction is molecular adhesion at the interface. That is, the flexible polymeric chains of the rubber form local junctions with the opposing surface, and as sliding commences, the bonds stretch and then break and relax before forming new junctions elsewhere [5]. When the surface is rough, the elasticity of a rubber permits it to drape the hills and valleys of the counterface. Sliding causes part of the rubber to accumulate on the positive flanks of the asperities [2] because of hysteresis but the negative slopes are not draped by any material. This type of deformation effect causes dynamic energy losses giving rise to the additional hysteretic component of friction. The concept of molecular adhesion suggests that a bond breaks by a molecular jump. This can only happen if the prevailing activation energy barrier is overcome, which is lowered progressively in the direction of sliding by an amount proportional to the applied tangential force. The formation and disruption of a frictional bond are separate thermally activated processes and the total number of bonds formed is modified according to the velocity and temperature of the sliding interface.

The velocity and temperature dependence of elastomeric friction can be expressed [6] by a master curve (Fig. 11.1). In Fig. 11.1, a_T is defined by the Williams–Landel–Ferry equation [7] as follows:

$$\log a_T = \frac{-8 \cdot 86(T - T_s)}{101 \cdot 5 + (T - T_s)} \tag{11.4}$$

where $T_s = T_g + 50$ and $T_g =$ the glass transition temperature of the rubber; $T =$ ambient temperature. v is the sliding velocity and a_T is also called the horizontal shift factor.

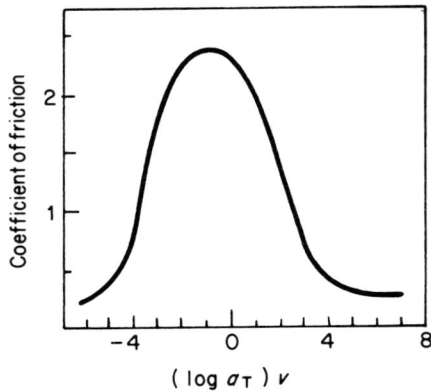

Fig. 11.1 Variation of the coefficient of friction with $(\log a_T)v$, v being the sliding velocity (see Equation 11.4 for $\log a_T$). (Sarkar, A. D. (1976). "Wear of Metals", Pergamon, Oxford.)

11.3   Mechanical models

There are several theories [1] of elastomeric friction incorporating both the adhesion and the hysteresis terms. In the *mixed theory*, part of the mechanism considers molecular adhesion giving rise to stick–slip and the remainder presents the information by using a mechanical model. The mixed theory propounds the component F_a in Equation 11.3 but there are also two theories of hysteretic friction, viz., the *unified theory* and the *relaxation theory*. The relaxation theory uses a Maxwell model but the unified theory is the simpler of the two.

11.3.1 *Mixed theory*

Suppose a sheet of rubber is dragged across a hard surface with an imposed velocity v (Fig. 11.2). Consider a small element of the rubber, shown at A in Fig. 11.2(a), which sticks with the rubber sheet on to the counterface in spite of the imposed horizontal pull. Let it stick for a small distance λ (Fig. 11.2(b)) before it separates at A' to assume a new position A as shown in Fig. 11.2(c). Assume that the maximum stress on the elemental area δA is σ_m so that the work done in stretching the rubber a distance λ is $(\sigma_m \lambda\, \delta A)/2$. As the rubber is relieved from its stretched position, an amount of energy is returned to it which equals $(\sigma_m \lambda\, \delta A)(1-\alpha)/2$, where α is the fraction of energy lost from the system during the process. The fraction α can be computed by using a

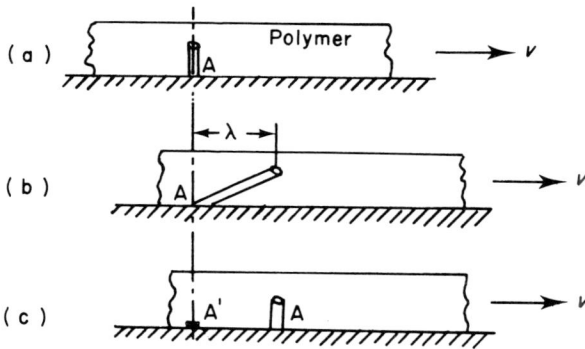

Fig. 11.2 A model for the mixed theory of elastomeric friction. (Moore, D. F. (1975). "Principles and Applications of Tribology", Pergamon, Oxford.)

mechanical analogy, e.g., the Voight model (Fig. 11.3). Thus it can be shown that the work done to extend the spring during a quarter cycle of sinusoidal vibration with a frequency ω is $(EL\delta_m^2)/2$ and the energy dissipated during this period is $(\pi/4)\eta\omega L\delta_m^2$. In these expressions, E is the spring modulus, η is the dashpot viscosity, L is a characteristic length and δ_m is the maximum spring extension. The loss fraction α is,

$$\alpha = \frac{\text{Energy dissipated in the dashpot}}{\text{Energy stored in the spring}}$$

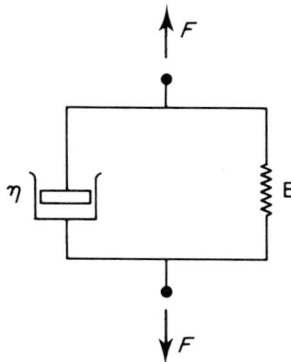

Fig. 11.3 Voight model. The elastic behaviour of the rubber is represented by a spring, E, and the viscosity by an ideal dashpot, η. This is a mechanical model and E and η are in parallel. The external force is F.

Therefore

$$\alpha = \frac{2\pi\eta\omega L\delta_m^2}{4EL\delta_m^2} = \pi\left(\frac{\eta\omega}{2E}\right)$$

The relaxation time t of the Voight element is $t = \eta/E$, therefore,

$$\alpha = \tfrac{1}{2}\pi t\omega = \tfrac{1}{2}\pi \tan\delta$$

where $\tan\delta$ is the tangent modulus or the damping factor. Thus the energy loss ΔG per contact area is

$$\Delta G = (\pi/4)\sigma_m\lambda\,\delta A \tan\delta$$

The total energy loss G for n bonds is

$$G = (\pi/4)n\sigma_m\lambda\,\delta A \tan\delta$$

If F is the frictional force, the external work done is $F\lambda$ and $F\lambda = G$. Thus

$$F = (\pi/4)n\sigma_m\,\delta A \tan\delta \qquad (11.5)$$

The total interfacial area formed due to molecular adhesion is $n\delta A$ and

$$n\delta A \simeq W/H \qquad (11.6)$$

where W is the applied normal load and H is the hardness of the rubber. Substituting this value in Equation 11.5,

$$F = K\sigma_m(W/H) \tan\delta \qquad (11.7)$$

where K is a constant.

The coefficient of adhesional friction is $\mu_a = F/W$, so that

$$\mu_a = K(\sigma_m/H) \tan\delta \qquad (11.8)$$

We note that although it is postulated that the frictional resistance is due to adhesion, μ_a depends also on the damping factor of the rubber, which is a viscoelastic property.

11.4   Hysteretic friction

The hysteresis effect is manifested by the deformation of the polymer by macro-asperities such as are encountered by the tyres of auto-mobiles on road surfaces. Consider again a sheet of rubber draped on an asperity (Fig. 11.4). In the static state, an elastic pressure distribution will be established as shown in Fig. 11.4(a). As sliding begins due to a horzontal pull, the rubber piles up at the leading edge of the hill and the

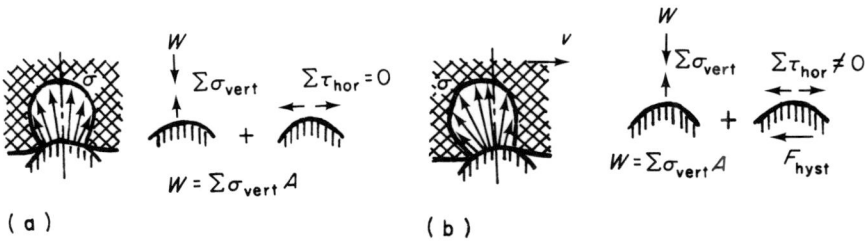

Fig. 11.4 A sheet of rubber draped around an asperity. Pressure distribution at static condition, (a), and during sliding, (b). W = normal load; σ = normal stress, τ = tangential pull. (Moore, D. F. (1975). "Principles and Applications of Tribology", Pergamon, Oxford.)

contact is broken at a higher point on the downward slope of the asperity (Fig. 11.4(b)). This results in an unsymmetrical pressure distribution and a force due to hysteretic friction is established which opposes the attempt by the rubber for forward motion. It has been suggested [2] that the hysteresis term in the total elastomeric friction can be viewed as the inertia which is inherent in a sheet of material wrapped over the hills and valleys of a hard surface. The effect of this inertia is that there is a delayed reaction in the bulk of the sheet material when an external force endeavours to induce motion.

Considering a rubber sheet attempting to slide over two asperities on a hard surface (Fig. 11.5). The static situation is represented in Fig. 11.5(a) for the elastomer under a normal load W. The pressure profile over one asperity only is shown, which is distributed symmetrically. Hysteretic friction at this stage is zero. As argued before, a finite horizontal velocity results in an asymmetric pressure distribution and the contact area takes up a new position cd (Fig. 11.5(b)) and then to position ef (Fig. 11.5(c)). Note that the dimensions ef < cd < ab and in this manner, the rubber contracts or stiffens, the contact area decreasing with increasing speed. As Fig. 11.5(d) shows, the hysteretic friction rises to a maximum at a characteristic speed but then falls.

The hysteresis which develops due to the encounter of a rubber block with a rough surface has been termed macro-hysteresis. Hysteresis also results when the micro-asperities interact and the term micro-hysteresis is also used. This latter component, however, decreases progressively as the surface finish of the opposing solid gets finer, until at a characteristic smoothness of the counterface, frictional resistance can be attributed entirely to adhesion at the interface.

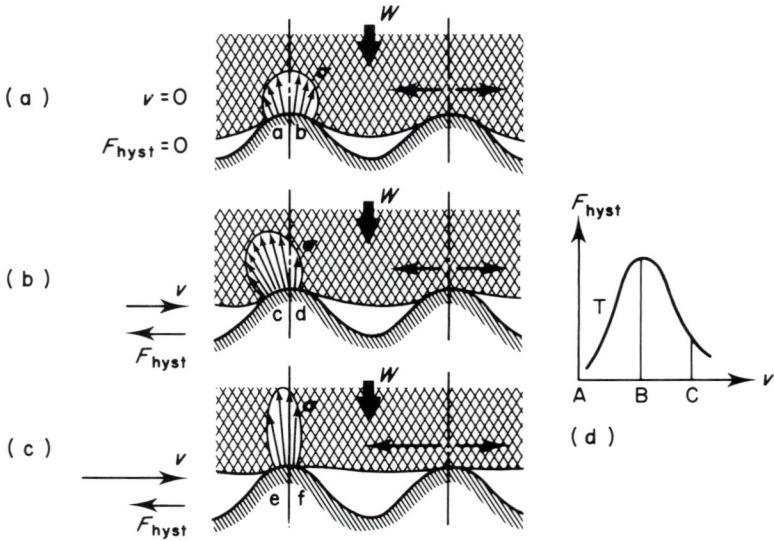

Fig. 11.5   Hysteresis force due to asperity interaction. (Moore, D. F. (1975). "Principles and Applications of Tribology", Pergamon, Oxford.)

The two theories of hysteretic friction [1], viz., the *unified theory* and the *relaxation theory* provide expressions which are broadly similar and the coefficient of hysteretic friction μ_h is shown to depend on the damping factor, $\tan \delta$ as well. A generalized coefficient of hysteretic friction is defined as $(\mu_h/\tan \delta)$ which is related to the modulus of elasticity E and the mean normal pressure $\bar{\sigma}$ as

$$(\mu_h/\tan \delta) = \text{constant}(\bar{\sigma}/E) \tag{11.9}$$

11.5   Schallamach waves

By a suitable experimental technique, wrinkles can be seen to develop in the contact region when a soft rubber pad loaded on a hard track is subjected to a tangential force [8]. These have been called Schallamach waves (Fig. 11.6) which cross the region of contact at high speed from the front to the rear. There is no true sliding between the rubber and the track and the relative motion is in these waves. The areas between the Schallamach waves and the counterface are the regions where interfacial contact is temporarily lost, the movement being similar to that of a caterpillar moving over a leaf.

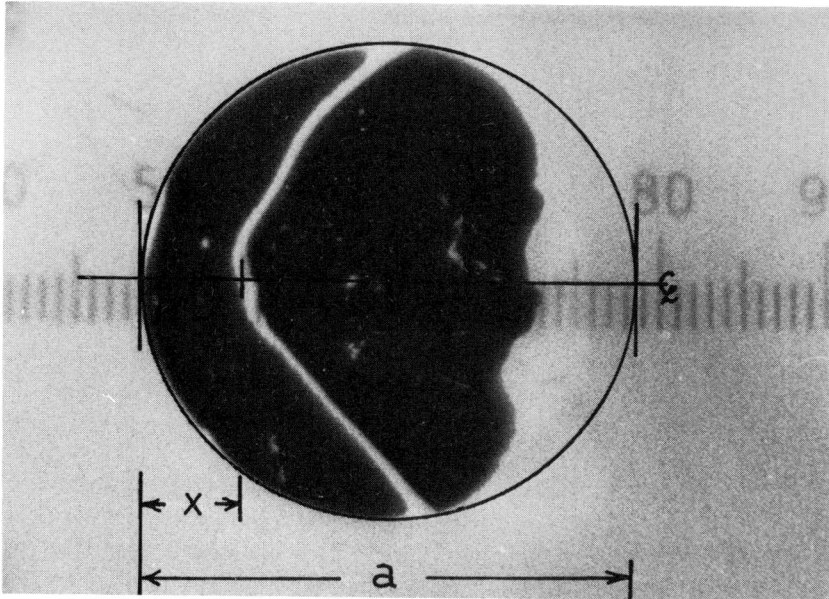

Fig. 11.6 The contact diameter *a* between a rubber and a glass counterface. *X* is the position of the Schallamach wave. The rubber specimen was held stationary while the glass counterface was moved from right to left at $0 \cdot 8 \text{ mm s}^{-1}$. The wave moves from left to right. (Briggs, G. A. D. and Briscoe, B. J. (1975). *Wear* **35**, 357.)

The Schallamach waves arise due to the buckling of the rubber pad as a result of elastic instability in certain regions of the contact area induced by tangential compressive stresses. Thus, consider a hard slider moving under load against a soft rubber block (Fig. 11.7). The normal stress inside a diameter 2*a* is parabolic as expected of an elastic solid and the strain distribution in the contact region is non-uniform. At the front end of the rider the rubber is in a state of compression, while the rear end is subjected to tensile stretching (Fig. 11.8). Because of this strain distribution the actual sliding speed v_a will vary according to the following relationship:

$$v_a = v(1 + \varepsilon) \tag{11.10}$$

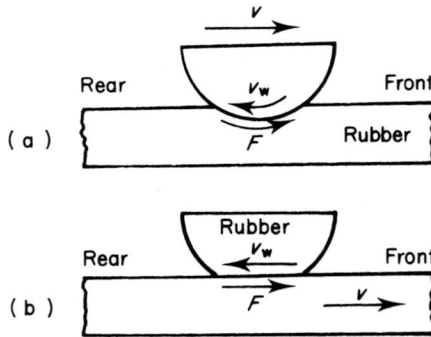

Fig. 11.7   Relative directions of sliding velocity, v, wave velocity, v_w and frictional force F. (a) Hard rider on rubber; (b) rubber slider on a hard solid. (Schallamach, A. (1971). *Wear* **17**, 301.)

where ε is the longitudinal strain distribution in the contact region and v is the overall velocity applied to the slider externally. The strain distribution in Fig. 11.8 is for the case where the normal stress in the contact zone is symmetrical. The strain in that event is positive and negative about the centre of the contact. Since $\varepsilon = 0$, $v_a = v$ at the centre

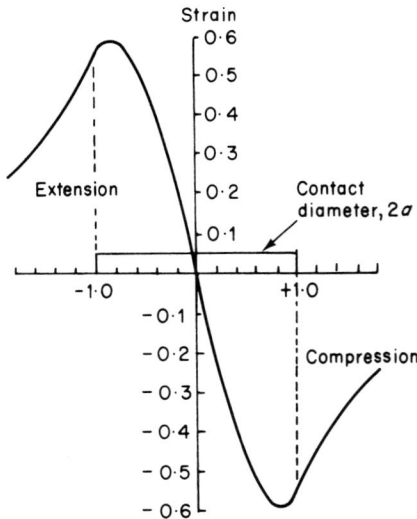

Fig. 11.8   Distribution of surface strain due to a hard rider indenting a circular area of radius *a* in a rubber surface. (Schallamach, A. (1971). *Wear* **17**, 301.)

of the contact. $v_a < v$ in the front of the slider and $v_a > v$ in its rear.

Cine films of contact areas of soft rubber loaded against perspex counterfaces show that, in general, the Schallamach waves originate from the front end and become thinner and break up into smaller components as they approach the rear (Fig. 11.9). A high imposed velocity gives coherency to the waves as they appear in the front but break up into fine ripples in the rear end of the contact area. The ripples start earlier at the front end if the external velocity is increased further.

For the friction couple shown in Fig. 11.7(b), the compressive stress is lowest in the front but the tangential compressive strain is the highest. The net result is that the front end buckles by a maximum amount and there is a stress relaxation effect. When this happens, the whole slider moves forward by an amount determined by the size of the fold in the rubber during buckling. We can, therefore, attempt to explain the difference between the static and the dynamic friction of elastomers contacting a hard counterface. When the imposed tangential force is low, the rubber is held to the opposing surface by an adhesive bond and sliding begins as the external effort is increased to a value at which the compressive induced strain in the front of the rider becomes large enough for the rubber to buckle. There is a critical speed v_c which must be exceeded before Schallamach waves appear [9]. If a hard slider on a rubber block (Fig. 11.7(a)) is subjected to a tangential force, adhesive forces keep both surfaces together and the total area of contact moves as a whole. Micro-sliding during this movement, however, can take place on an annular region of the interface where the normal stress is zero or very small. Experiments suggest that friction of natural rubber at low speeds and temperatures is probably due to adhesive dragging and to a continuous interfacial relaxation on a macroscale. At a critical speed, Schallamach waves appear resulting in a lowering of the frictional resistance. The higher the temperature of the surroundings, the greater is the value of v_c and the waves traverse the interface faster and more frequently [10] without any change in the coefficient of friction. A high imposed sliding speed increases the wave velocity and a lowering of the frictional resistance at the interface.

Schallamach waves do not appear even with very soft rubbers when the surfaces are contaminated. Hard rubbers fail to produce these waves even when completely clean. On the other hand irradiating a rubber

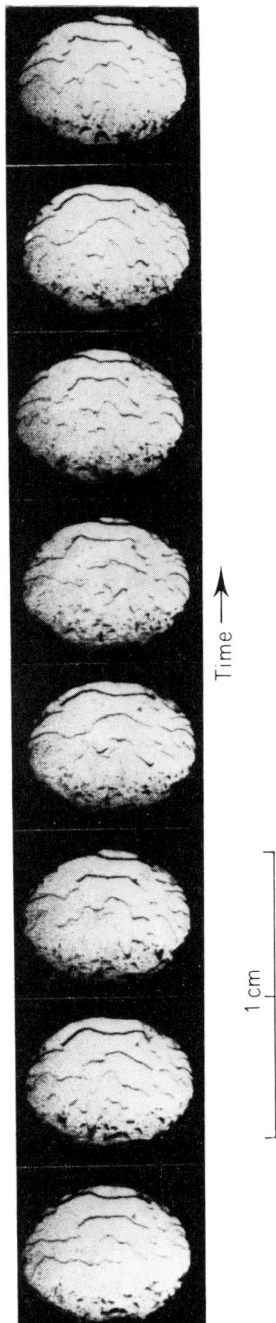

Fig. 11.9 Eight frames from a film shot at 32 frames a second of the contact area between a perspex plate and a natural rubber sphere which is sliding over it at 0·043 cm s⁻¹ . (Schallamach, A. (1971). *Wear* **17**, 301.)

with ultraviolet light modifies its outer layer which is hard and Schallamach waves are noted.

11.6   Work of adhesion

A review of published literature points to some broad generalization [11]. These are that the coefficient of elastomeric friction increases with the imposed sliding velocity, reaching a peak with a fall thereafter. Since it is an elastic solid, friction of rubber is load dependent, being low as the contact pressure is increased. A fall in the coefficient of friction is noted when the interface is lubricated or is contaminated with dust. Nevertheless some of the conclusions are anomalous and it has been pointed out as recently as 1976 [11] that even after three decades of concentrated work, the debate on the mechanism of elastomeric friction continues.

The molecular theories provide another approach to explain some of the experimental results observed during sliding of elastomers. These postulate that rubber bonds are formed with the counterface which are then strained before being broken. The net frictional force is a compromise between the increase in bond strength with speed and a decrease in the number of sites occupied by the rubber molecules.

Experiments show that the coefficient of friction decreases with contact pressure at low loads but becomes sensibly constant at high loads [12] and the coefficient of friction μ can be related [13] to the normal contact stress as follows for a wide range of rubbers and gelatine compounds:

$$(\mu)^{-1} = A + B\sigma \qquad (11.11)$$

where A and B are constants.

The usual explanation for the nature of the μ–σ relationship is that the true area of contact increases with load until at and above a limiting load the hills barrel out and further deformation is prevented. We could liken this to the imaginary situation with metals when at a characteristic load, the true area of contact equals the apparent area. A steady coefficient of friction at an unchanging area of contact is acceptable but a fall in μ with increasing area defies certain basic postulates. The mechanism of friction *vis-à-vis* normal load is probably quite complex and may be tied up with the viscoelasticity of a rubber.

Another theory [10] attributes the phenomenon of elastomeric friction to the work of adhesion at the interface. Thus if a pad of an elastomer is pressed against a clean surface, the energy to separate the two solids is considerably greater than that needed to bring them together. A work of adhesion term is defined which is the difference in energy between a complete cycle of peeling and sticking. Note that for a Schallamach wave to traverse, the rubber buckles and then lifts vertically off the counterface; that is, the contact peels and then adheres again. Such peeling and sticking continues as long as conditions are conducive for Schallamach waves to emanate from the front of the rider. The work of adhesion should be some function of the speed of the Schallamach waves. If the frictional phenomena are explained by Schallamach waves, for an elastomer sliding over smooth surfaces, energy should be lost viscoelastically during the peeling process in a region of the rubber close to the edge of the contact area which is in a state of high strain. There may be other sources of friction, e.g., energy loss in the formation of the waves, but the bulk of the friction is accounted for by the work of adhesion in this theory. It should be noted that Schallamach bonds are broken in a direction normal to the interface and not tangentially, as is believed to happen with metallic junctions.

The surface energy or the peel energy is rate dependent and the concept has been incorporated in the phenomenology of Schallamach waves [14] to explain how energy is lost giving rise to friction. Thus, consider a rubber pad under a tangential velocity v over a hard track. Let the frictional resistance be F while Schallamach waves, spaced apart by a distance λ, move at a velocity v_w. It can be shown that

$$F = \gamma v_w / \lambda v \qquad (11.12)$$

where γ is the rate dependent surface energy per unit area of the contact interface and is the peel energy required by the rubber to be lifted off the contact. Equation 11.12 shows that the frictional resistance falls with increasing sliding speed. We have seen that this is only true after a limiting speed depending on the ambient temperature so that the idea of peel energy probably is not a complete explanation of elastomeric friction.

11.7   Wear

The quasi-empirical laws with metals suggest that there is a lack of correlation between wear and the frictional resistance. It is known that a material with a high hysteresis loss tends to undergo low wear and exhibit high friction, prompting the corollary that the higher the frictional resistance the lower the wear. An examination of the wear laws of rubber suggests such a correlation and hence it is important to have a good background knowledge of elastomeric friction before undertaking a study of wear. Rubber has been shown to exhibit three types of wear, viz.,

(1) abrasive wear;
(2) fatigue wear;
(3) wear by roll formation.

11.7.1 Abrasive wear

This is defined as the wear caused by an abrasive projection when an elastomer slides against it under a normal load. This is a case of two-body abrasion and the abradant should remove material by ploughing. Both the theories and experimental results with metals show that wear rates depend on the normal load together with the shape, size and the hardness of the particles acting as abrasives. For elastomers, however, the following three main criteria have been used to describe abrasive wear:

(1)
$$q_E = \frac{\text{Volume of layer removed}}{\text{Work of friction}} = \frac{\Delta V}{FS}$$

where q_E = energetic wear rate or the energy index of abrasion; F = frictional resistance; and S = sliding distance.

(2)
$$q_A = \frac{\text{Volume abraded}}{\text{Work of friction}} = \frac{\Delta V}{FS} = \frac{\Delta V}{\mu WS} = \frac{(\Delta V/WS)}{\mu}$$

or

$$q_A = \frac{A'}{\mu}$$

where q_A = abradability; W = normal load; and A' = abrasion factor = $\Delta V/WS$.

(3)
$$\beta = \frac{\text{Work of friction}}{\text{Volume abraded}} = \frac{1}{q_A} = \frac{1}{q_E} = \frac{\mu}{A'}$$

If an elastomer is abraded over a rough surface, a set of parallel ridges forms on the sliding surface in a direction transverse to that of motion (Fig. 11.10). As expected, the spacings between these ridges are the

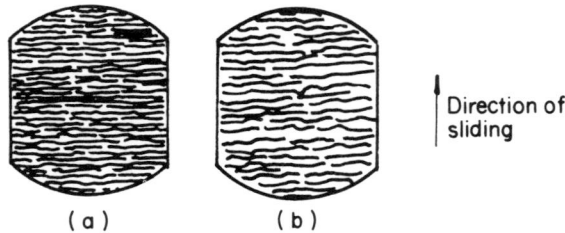

Fig. 11.10   Abrasion patterns on rubber. (a) Fine abradant; (b) coarse abradant. (Moore, D. F. (1975). "Principles and Applications of Tribology", Pergamon, Oxford.)

larger, the greater the coarseness of the abrading surface. These ridges are called *abrasion patterns* and a vertical section cut along the direction of motion through the rubber shows the pattern of the profile to be sawtooth in nature (Fig. 11.11) and its peaks lie in a manner which is opposed to the motion of the counterface. In a situation such as in a car tyre, it is the rubber which moves on the track. Either way, the peaks will be stretched and loss of material will be expected. Wear by a rough counterface will occur by a tearing action and the process will continue to maintain a saw-tooth profile. The question to be answered is whether the transverse ridges develop by a fatigue process in the elastomer.

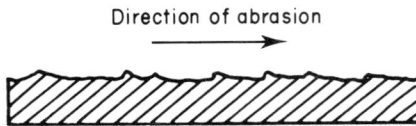

Fig. 11.11   Profile of a typical abrasion pattern on rubber. The arrow shows the motion of the counterface. (Moore, D. F. (1975). "Principles and Applications of Tribology", Pergamon, Oxford.)

Whether the road surface is rough or smooth, it would appear that the two-body abrasive wear involves stretching and tearing of the rubber rather than removal of material by ploughing, as in metals. Photoelastic techniques have revealed that if a cylindrical rider moves slowly on a rubber pad, the rear end of the contact shows the isochromatic fringes to be the closest, giving evidence of the intensity of stress concentration there. The rear end is stretched and shows a series of tears across the surface of the track. The nature of the tears is periodic in that they appear as a discontinuous series of transverse cracks and suggests a macroscopic stick–slip phenomenon. Thus at the beginning of the interaction, a layer of the rubber drapes around a hill, moves forward until the elastic restoring force exceeds the frictional force, and the rubber strip snaps back. This happens repeatedly, giving rise to tears at right angles to the direction of motion. These originate in the rear where the stress concentration is a maximum. With increasing number of cycles, the cracks grow and rubber is detached from the main body in the form of a wear particle.

11.7.1.1   *A wear model*
A simplified theory of abrasive wear of an elastomer starts with a surface asperity of radius R supporting a mass of rubber under a normal load, δW (Fig. 11.12). The elastomer produces a contact area of width $2a$ and there are regions of tension and of compression as shown. If E is an

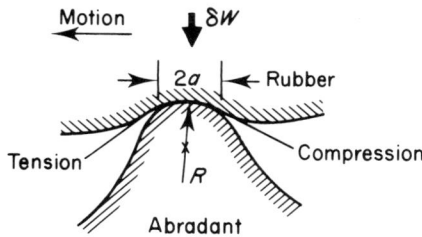

Fig. 11.12   An elastomer interacting with an abradant during sliding. (Moore, D. F. (1975). "Principles and Applications of Tribology", Pergamon, Oxford.)

elastic constant of the elastomer,

$$a = \phi(R, \delta W, E) \tag{11.13}$$

It can be written that

$$(a/R) = c_1(\delta W/ER^2)^\alpha \tag{11.14}$$

where (a/R) and $(\delta W/ER^2)$ are dimensionless quantities. c_1 and α are constants and they can only be evaluated by experiment. If there are n^2 abrasive protuberances per unit area, the total load W is given by,

$$W = \Sigma\, \delta W = n^2\, \delta W \tag{11.15}$$

We have seen that the abrasion factor $A' \propto \Delta V$, therefore

$$A' = n^2 a^3 \qquad (11.16)$$

Substituting n^2 from Equation 11.15 and a from Equation 11.14 into Equation 11.16, and taking $\alpha = \frac{1}{3}$,

$$A' = c_2(WR/E) \qquad (11.17)$$

where $c_2 = c_1^3 = $ constant.

The abrasion resistance has been defined as

$$\beta = \mu/A'$$

Therefore substituting for A' from Equation 11.17,

$$\beta = K(E\mu/WR) \qquad (11.18)$$

where $K = 1/c_2 = $ constant and μ is the kinetic coefficient of friction.

Equation 11.18 shows that a sharp protuberance or heavy service load decreases the abrasion resistance of an elastomer. The rubber is wear resistant if it has a high elastic modulus and if the kinetic frictional resistance is high. Obviously, a good tear resistance will improve the life of a component but a rubber which shows a high hysteresis loss gives a large frictional resistance.

11.7.1.2 Degradation

Although difficult to quantify, wear rates of most materials should be influenced by a change of mechanical and physico-chemical properties of the surface layers of the interface. Chemical degradation of the sliding interface of rubber has been observed [15] and is attributed to a thermal effect. Degradation of rubber molecules is known to occur whether the counterface is rough or smooth, and apart from temperature, the chemical bond is attacked by light and oxygen. Severe mechanical loads assist degradation possibly by raising the temperature of the surface layer, but we have seen that the friction of rubber tends to decrease with increasing normal load. High sliding speeds may increase the surface temperatures, and if excessive, resinification of the surface layer can occur, producing typical wear products described by some as crumbs. Anti-oxidants such as phenyl-β-naphthylamine are reported to effect a reduction in the amount of wear. Of the additives, vulcanizates containing an active carbon black such as anthracene have greater wear resistance compared with those containing inactive lamp black.

11.7.2 *Fatigue wear*

We should note that although the obvious mode in the foregoing is two-body abrasion, the appearance of transverse cracks indicates fatigue. Although in practice dust particles must enter the interface, three-body abrasion has not received much attention for rubber. Fatigue wear in rubber, of course, has been recognized and the idea of a critical shear stress has been used to explain the various types of wear. Thus denoting the intrinsic critical shear stress of the rubber by τ_c, fatigue wear predominates when $\tau < \tau_c$ where τ is the applied tangential stress. At $\tau > \tau_c$, the rubber will wear by a two-body abrasion mode if the counterface is rough. The total amount of wear by fatigue is believed to be low for rubber although we have seen that even an obvious two-body situation probably has a fatigue component.

11.7.3 *Wear by roll formation*

If a pad of rubber slides on a smooth surface so that hysteretic deformation is low but the adhesional friction is high, a severe form of wear occurs which has been termed wear by roll formation. This mode of wear is augmented by a rubber of inherently low tear strength. Suppose a rubber block with a projection is subjected to a normal load W and a horizontal velocity v (Fig. 11.13(a)). If the tangential force is inadequate, there will be no slipping of the rubber but the projection will undergo deformation in a complex manner, increasing in severity with rising friction at the interface. If the imposed velocity is such that the tangential force at the interface exceeds the frictional resistance, the contact spots will slip and sliding will commence. This assumes of course that the rubber layer draping the hills is strong enough not to tear. The projection shows a cut perpendicular to the direction of the tensile pull (Fig. 11.13(b)) and this is the initiation of wear. If the imposed horizontal force is maintained, the cut will grow, allowing relative movement of the contact zone (Fig. 11.13(c)) without total slipping. This is only possible if the rubber forms into a roll. Subsequently, the condition is that of rolling friction (Fig. 11.13(d)), with roll formation beginning at other projections on the rubber. The rolled rubber sheets detach, as wear debris and roll formation is

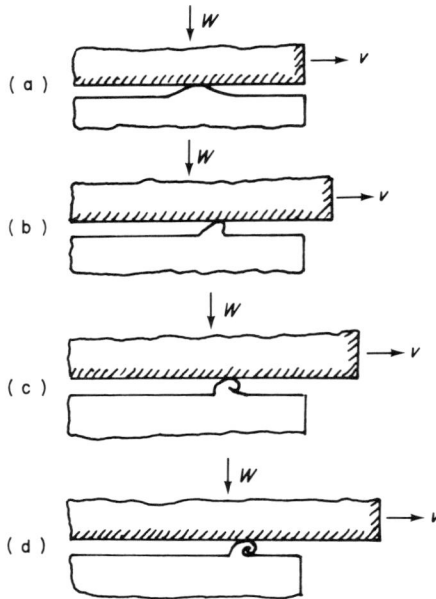

Fig. 11.13   Wear by roll formation, showing the sequence from the moment a normal load is applied. (James, D. I. (Ed.) (1967). "Abrasion of Rubber", Rubber and Plastics Research Association of Great Britain, London.)

facilitated if frictional heating is high, since this will lower the tear strength of the elastomer.

Wear by roll formation occurs under conditions of adhesional friction so that the condition to be satisfied is that

$$\tau = \mu\sigma \qquad (11.19)$$

where τ = imposed shear stress, σ = normal applied stress, and μ = coefficient of friction.

Experiments show that there is a critical value of the coefficient of friction, μ_c, below which the wear rate is low. For a constant normal load, the value of μ_c can be altered by changing the nature of the elastomer. Typically, a soft rubber wears readily by roll formation but the mechanism changes to fatigue wear, which is mild by nature, if the hardness of the rubber is increased by incorporating sulphur in the rubber.

11.8 Vehicle tyres

A common example of application of rubber is found in tyres for vehicles. For a rolling tyre, the interaction with the road surface is elastic so that the pressure distribution in the contact region is parabolic (Fig. 11.14). It is seen that the pressure distribution along A_1A_2 is such that the pressure generated is greater at the point A_1 where the tyre meets the road than at A_2 where the contact is broken. The general loss of pressure along A_1A_2 is attributed to the hysteresis loss of the tyre during deformation. The pattern of pressure distribution in the width direction BB is irregular, giving a minimum at the centre and is due largely to the unequal radial stiffness of the tyre across the contact area. It is seen that the contact pressure is not the highest in the centre of the tread path. When the vehicle is static, both the longitudinal and transverse forces are directed towards the centre of the contact area and arise as a result of the internal elastic forces in the tyre developed due to deformation under a normal load.

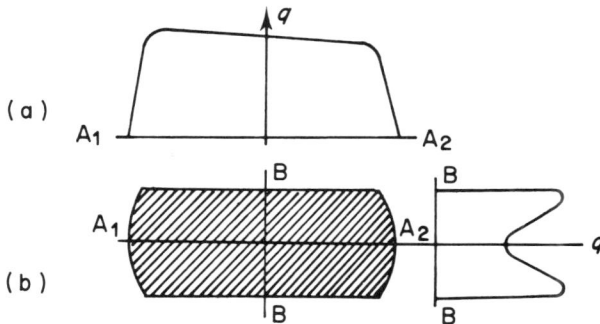

Fig. 11.14   Pressure distribution over the contact area, shown shaded, between a vehicle tyre and the road. (James, D. I. (Ed.) (1967). "Abrasion of Rubber", Rubber and Plastics Research Association of Great Britain, London.)

Wear of vehicle tyres occurs largely as a result of slip of the interface between the elastomer and the road surface. It will therefore be useful to understand the mode of slip of a wheel as it rolls. Figure 11.15 represents a tyre rotating on a flat surface, and in so doing, the tyre tread undergoes an amount of angular deformation ε which is expressed as

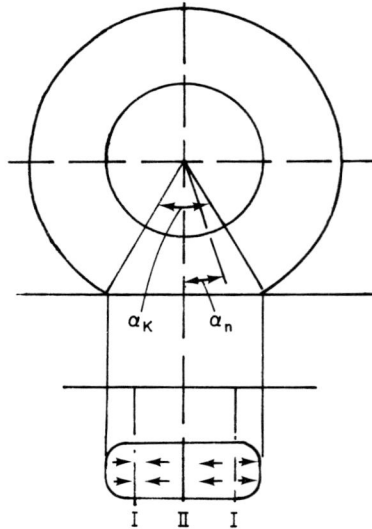

Fig. 11.15   A tyre rotating on a flat surface. (James, D. I. (Ed.) (1967). "Abrasion of Rubber", Rubber and Plastics Research Association of Great Britain, London.)

follows:

$$\varepsilon = \left[\arcsin \left(\cos \frac{\alpha_K}{2} \tan \alpha_n \right) - \alpha_n \right] \qquad (11.20)$$

where, as shown in Fig. 11.15, α_K = angle of contact of the wheel with the road and α_n = the angular displacement of the tread element.

The tangential stress pattern in the contact zone will have a longitudinal component τ_g which determines the slip of the contact region such that,

$$\tau_g = K\varepsilon \qquad (11.21)$$

where K is a constant.

As the wheel rotates, additional tangential resistances arise due to rolling (τ_c) and shearing of the tyre tread elements (τ_y) while they pass through the contact zone. τ_c is about 2–3% of the total tangential components and thus can be neglected. The contact area in Fig. 11.15 shows three zones. In the two outside zones marked I, the tyre tends to slide in the direction in which the wheel is rotating, while in the central contact area marked II, the trend is for slip to occur in the opposite direction. It is argued [15] that slipping of a driven wheel occurs mainly in the direction of wheel rotation, but in the opposite direction during braking.

An automobile tyre can be out of service due to the failure of the carcass and by separation of the tread pattern. If these causes are inhibited, the life of a tyre is determined by and large in accordance with its wear resistance.

References

1. Moore, D. F. (1972). "The Friction and Lubrication of Elastomers", Pergamon, Oxford.
2. Moore, D. F. (1975). "Principles and Applications of Tribology", Pergamon, Oxford.
3. Schallamach, A. (1953). *Proc. Phys. Soc. B* **66**, 817.
4. Schallamach, A. (1963). *Wear* **6**, 375.
5. Schallamach, A. (1953). *Proc. Phys. Soc. B* **66**, 386.
6. Grosch, K. A. (1963). *Proc. Roy. Soc. A* **274**, 21.
7. Williams, M. L., Landel, R. F. and Ferry, J. D. (1955). *J. Amer. Chem. Soc.* **77**, 3107.
8. Schallamach, A. (1971). *Wear* **17**, 301.
9. Barquins, M. and Courtell, R. (1975). *Wear* **32**, 133.
10. Briggs, G. A. D. and Briscoe, B. J. (1975). *Wear* **35**, 357.
11. Roberts, A. D. (1976). *Tribology Int.* **9**, 75.
12. Denny, D. F. (1953). *Proc. Phys. Soc. B* **66**, 721.
13. Thirion, P. (1946). *Rev. Gen. Caoutch* **23**, 101.
14. Roberts, A. D. and Thomas, A. G. (1975). *Wear* **33**, 45.
15. James, D. I. (Ed.) (1967). "Abrasion of Rubber", Rubber and Plastics Association of Great Britain, London.

12 Other Non-metals

Our understanding of the mechanisms of friction and wear of metals, plastics and rubber is by no means clear, but these materials are well known for their usefulness in tribological applications. Some other non-metals such as wood or cement are less obvious to us as materials for tribological components, but are used all over the globe. For example, wooden and cement floors undergo sliding friction in the presence of abrasives. Textile fabrics need replacing because of wear, as do the heels of shoes. We shall discuss these in this chapter because a civilized world must recognize the need to design components for as long a life as possible. Other non-metals are ceramics, which are being studied for high temperature application but also as implants in human bodies. Materials such as diamond are well known for their wear resistance. We shall collate available information on these in the following sections.

12.1 Wood

The use of wood has played a large part in human development. Some examples of commonly used wooden friction elements are wheels, chutes, gears, pinions and shoes. Apparently, wheels were not known in the Americas until introduced by the marauding Europeans only recently [1]. The use of the chariot is recorded as being introduced in India some five millenia ago by the savage hordes of Aryans. The early Egyptians during the same period used wooden sledges on wooden rollers to transport stone building blocks and statues. Wooden sledges are also used today for transportation on snow and ice in the polar regions. Another application where wood underwent sliding under load was the railway system in metal mines in medieval Europe where wooden trucks transported ores, being drawn by men or horses. The trucks moved over wooden rails whose resistance to wear was of primary importance. This led to the selection of hardwoods like oak and beech for the rails. An interesting aspect in the historical perspective was the employment of softwood rails in some cases with replaceable strips of hardwood nailed to these in the same manner as wear resistant linings which are used in certain cases to-day. The wheels on the trucks were manufactured from a dense compact wood such as elm so that both the rails and the wheels would appear to have been of similar hardness. A well-known modern application of wood is in marine bearings.

12.1.1   The nature of wood

Wood is produced by the activity of a single cell layer called the *vascular cambium*. Growth of the trunk occurs as the cambial cells divide, forming the xylem or the sapwood, which is on the inside of the tree, and phloem or bark on the outside. The non-bark portion is the wood. Of the dry weight of wood, cellulose, holocellulose and lignin constitute 99% of the total constituents [2]. Holocellulose is the term given to components related to cellulose and the remaining 1% of the wood comprises resins, silica and gums among others. Cellulose is a linear polymer and may constitute 50–60% of the total weight of dry wood. The molecule consists [3] of chains of up to 5000 glucose rings joined by oxygen atoms. The cellulose structure shows crystallinity separated by amorphous regions. The microstructure is complex but a simplified picture is that wood is composed of a number of cells called tracheids (Fig. 12.1). The tubular tracheids are hollow and the centres are called lumens, which are full of sap when the cells are alive. The hydroxyl groups in a cellulose structure are polar by nature and form strong attractive fields between the cellulose molecules so that molecular bundles of fibres, or fibrils as they are called, are formed. As Fig. 12.1(b) shows, a cell has various layers and the fibrils, which are arranged in these walls, show varying orientations.

There are four types of xylem elements:
(1) parenchyma cell;
(2) fibre;
(3) tracheid;
(4) vessel.

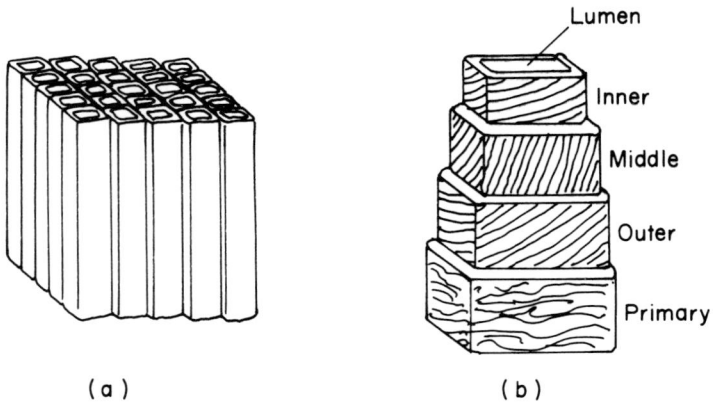

Fig. 12.1   Structure of wood (schematic). (a) Shows the assembly of tracheids, and (b) layers in the cell walls. Note the differing orientations of the cellulose fibrils. (Wyatt, O. H. and Dew-Hughes, D. (1974). "Metals Ceramics and Polymers", Cambridge University Press, Cambridge.)

Not all the above elements will be found in every type of wood although the parenchyma cell is invariably present and acts as storage for food reserves or as waste disposal units. Fibres are mechanical elements and act to support a tree while the vessels and tracheids convey water and dissolved mineral salts from the roots to the leaves and branches. Although a piece of wood comprises essentially axial elements, sheets of cells extend in a radial direction as well and these are referred to as rays. Rays are uniseriate when they are one cell wide. Sometimes large structures with a height of several hundreds of cells and a width made up of twenty or more cells are formed and the rays are then said to be multiseriate. The minerals in wood are usually referred to as white deposits and consist chiefly of silica and salts of calcium and are frequently found in vessels. The term wood resin is used for any gummy material in softwoods and is referred to as gum when present in hardwoods. Certain woods such as teak give an oily feel and the oil has been identified as oleo-resin. Living cells are shown to house starch whose role in tribological applications of wood is not known, but the low friction of wood bearings has been attributed to gums and resins.

Gums are carbohydrates formed by the degradation of cell walls and are soluble in water but not in alcohol. Resins on the other hand are insoluble in water but soluble in alcohol and are oxidation products of certain oils. There are hard resins which are devoid of any oil while the oleo-resins are liquid and contain appreciable amounts of oil. A mixture of true gum and resin is called a gum resin and a well-tried tribological material lignum vitae contains this, usually described as gum guaiacum. Gums and resins, among others, are held in special ducts in the wood. The ducts are found both in the axial elements and in the rays, being of two types:

(1) Schizogenous ducts: these are formed when two contiguous cells split apart.
(2) Lysigenous ducts: these are produced when the cells collapse, thus leaving a cavity.

Schizogenous ducts are lined with parenchyma cells but this is not so with lysigenous ducts. Schizo-lysigenous ducts exist which are large in size and form by enlargement of schizogenous ducts by a process of breakdown of the surrounding cells.

In softwoods, the bulk of the woody tissues is in the form of small axial tracheids. The growth of a tree by addition of new wood occurs progressively during the favourable season of the year. The first-formed wood in the year is known as early wood and the portion formed later as the late wood. The other names of these growth rings are spring wood and summer wood respectively. The early wood is soft and wears more readily than the late wood.

12.1.2 Lignum vitae

The wood *lignum vitae*, also known as *Guaiacum officinale*, can be translated as wood of life named so because it was used for medicinal purposes. It has a high resin content which gives it a self-lubricating property when used as a bearing. Lignum vitae is grown mainly in the West Indies and South America and is supplied in the form of billets about 3 m long and 75–300 mm in diameter. The wood is greenish black in colour and is about 80% heavier than oak. It is strong, hard and possesses a fine texture.

Lignum vitae has been used in many situations where it undergoes sliding and rolling but the most commonly quoted example is its use as a lining material for plain bearings in ships. Metal bearings for the underwater portion of the propeller shaft pose a difficult maintenance problem because of corrosion by seawater. Wood, in particular the genus *Guaiacum*, has proved to be the most successful material and has been used since the middle of the nineteenth century [4, 5]. It is quite common to line the plain bearings, or stern tubes as they are called, with lignum vitae so that sliding is between the steel shaft and wood.

Although wear rates vary between trees, possibly because of the likely variations in composition or physical and mechanical properties, it seems certain that variations in the amount of wear occur depending on whether the steel shaft runs along or across the grain (Fig. 12.2). The terms often used are end-grain strips and long-grain strips, and laboratory studies have been carried out where mild steel shafts have been rotated on bronze bearings lined with lignum vitae [6]. The wood wears more when the long grain is against the shaft, as in Fig. 12.2(b). Experience seems to confirm this although it must be emphasised that being a natural material the properties vary between batches. Lignum vitae resists attack by a number of acids and alkalis [7]. Carefully conducted experiments using the heartwood of a billet of lignum vitae, show a substantial reduction in wear resistance, in boiling water, although the wood is known to be used under such a condition. The wear resistance is not impaired if used at sub-zero temperatures.

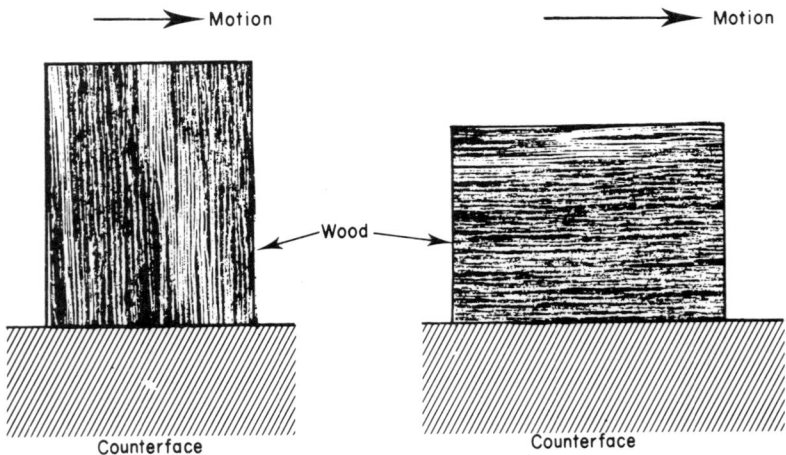

Fig. 12.2   Sliding of wood against a counterface. (a) Across the grains; (b) along the grains.

Lignum vitae contains resin and wax which are distributed in the body of the wood making it self-lubricating. This gives a low coefficient of friction, average values being 0·1 and 0·03 for sliding and rolling contacts respectively [8]. Since the shaft will run on wet interfaces in a marine environment, it is useful to know that the wood does not absorb any significant amount of water even after prolonged soaking. The frictional resistance may be high at a very low load due to the presence of interfacial surface tension of the water film. This is obviously squeezed out of the way at heavier loads and the coefficient of friction is then comparable to the value of the dry wood.

12.1.3 Balsam wood

Resistance to sliding offered by a hardwood such as lignum vitae is believed to be due to interfacial adhesion. For a softer wood such as *Abies balsamea*, or balsam fir as it is known popularly, the coefficient of sliding friction μ is composed of μ_h and μ_a, the contributions due to hysteresis and adhesion respectively. Balsam wood contains a small amount of fatty substance and resin and if these are extracted and the wood is then slid against balls of steel and PTFE, the coefficient of friction is seen to increase with load (Fig. 12.3) [9]. Figure 12.3 also shows that the value of μ becomes higher as the ball diameter decreases. Sliding disrupts the wood fibre in the track but no permanent deformation is noticeable either by visual or by microscopic examination of the section of the track. During the process of sliding, however, the wood shows deformation. The magnitude of the coefficient of friction as

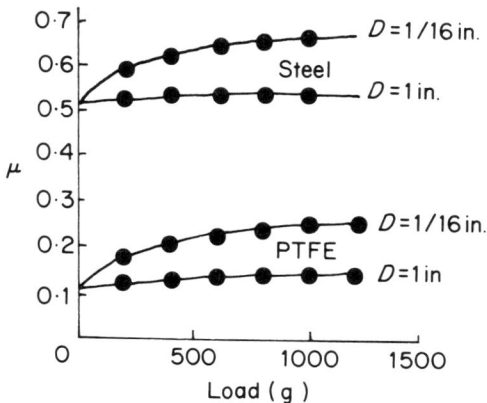

Fig. 12.3   The variation of the coefficient of friction, μ, with load for PTFE and steel sliding on wood. D is the diameter of the slider. (Atack, D. and Tabor, D. (1958). *Proc. Roy. Soc. A* **246**, 539.)

a result of hysteresis is independent of the opposing material and is a function of the bulk property of the wood. Adhesion of wood is strong with a steel counterface. Dry wood also shows strong adhesion with PTFE which falls to a very low value if the interface is wet.

A study of balsam wood is interesting because of its simple structure. It is not known, however, if it has found many applications in sliding or rolling.

12.1.4 *Wood flooring*

An interesting use of wood is for flooring. For either domestic or industrial use, hardwoods are preferred for both light and heavy pedestrian traffic, the classification being up to 100 and over 500 persons per day, respectively [10]. Examination of floorings after being in service for 10 years or so and laboratory abrasion tests show that the mode of material loss varies. Some floorings wear progressively leaving a smooth surface, but others develop a roughness, the nature of which can be classified under four categories:

(1) the fibres along the grains of the wood are lifted slightly but there is no splintering;

(2) surface cracking around the rays;

(3) severe splintering of the wood;

(4) severe surface breakdown providing a scabbed appearance.

Taking transverse sections of worn surfaces and examining them under the microscope provides a valuable insight into the mechanisms of wear. It would appear that the size of the pores and their distribution in the wood are the main parameters which influence the wear resistance. The first effect of the load is to compress the pores. When the pores are small, the main effect is the consolidation of the wood tissue. For a pore with a large diameter, the expected effect is collapse under load with accompanying rupture of the connecting walls of adjacent pores. The walls of large pores are generally thin, and hence weak under radial loads. The eventual result is splintering which is regarded as wear possibly by the combined effect of fatigue and abrasion. In certain types of wood such as eucalyptus, large to medium sized pores are distributed at an angle to the normal load so that these define planes of weakness of the wood. As traffic moves, the pores are compressed and relaxed. The surface pores fail first and cracks usually initiate at the surface. For light pedestrian traffic, the cracks are confined to very thin surface layers and the wood wears producing small flakes of wear debris. If traffic is heavy, especially in the form of trucks, the cracks tend to extend deeper and open up. The opening up of the crack is facilitated by the abrasive particles inevitably present in industrial traffic. Examination of the appearance of a wear product about to separate makes the delaminated nature of the failure obvious. Since the total quantity and density of the fibres decide on the mechanical strengths of wood, their role in wear resistance has been examined. The possibility of a correlation is there, provided the large pores are well separated from one another. Resins and gums at the interface should protect the wood from wear due to abrasion. Although not investigated, the fatigue mode may still remove flakes of material from the surface of the wood and a quantitative study of pore size, fibre density etc. in terms of wear of wood would be of great value. Disregarding certain anomalies, we could cautiously accept the recommendation that for heavy traffic a hardwood with large pores is unsuitable. We should also note that laboratory studies show that a fine-textured wood of low density will wear more than a similar wood with a higher density. Wear appears to be progressive without marked surface damage when the pore size is fine. There are many other variables to consider, for example, whether the sliding component is made from early wood or late wood. Fast grown hardwoods have a large proportion of late wood and hence they provide an overall high resistance to wear. Chutes made of pine to guide cereals are seen to wear quickly if made from the soft early wood.

12.2 Fibres

A fibre can be considered [11] as the unit from which cloth and various fabrics are produced by weaving. Natural fibres can have their origin in animals, vegetables or minerals, a common example [12] of the latter being asbestos. Mammalian hairs and wool are typical fibres of animal origin, while an example of vegetable fibre is coir from the shell of a coconut. There are also man-made fibres which are subdivided into two types. A regenerated man-made fibre is produced from natural high polymers such as cellulose, while the raw material for synthetic man-made fibre is an artificially produced polymer. An obvious tribological application of certain types of fibres is the manufacture of sweeping brushes where the aim is a long life at a minimum purchase price. Other areas are clothing and floor coverings.

A fibre has a length which is many times larger than the dimension of its cross-section [13]. Typically [14, 15], the transverse dimension, also

called the fibre fineness, can vary from a few microns up to 100 μm
while its length falls between a few mm and 1 m. A short length of fibre,
about 100 mm, is called a staple fibre, examples being wool and cotton.
Man-made fibres are produced as continuous filaments, so called
because of their infinitely large length. Wool has a wavy form, called
crimp, and this results in its possessing softness and elasticity. This
waviness can be imparted to man-made fibre as well, when they are
known as bulked or textured.

In the textile literature [16], a distinction is made between wool and
hair. Wool refers to the fine undercoat of a sheep while the coarse fibre
forming the outer coat is hair. However the chemical structure of all
hair fibres is closely related to that of wool. Chemically, wool consists of
a protein called keratin which has the empirical formula of
$C_{72}H_{112}N_{18}O_{12}S$. Microscopically, a wool fibre shows three distinct
structures (Fig. 12.4):

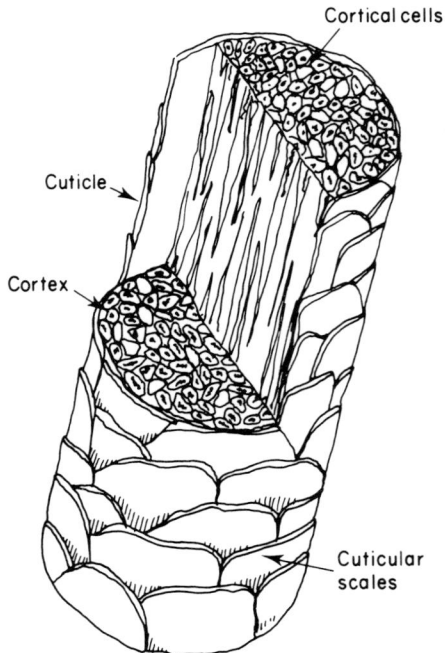

Fig. 12.4   Structure of wool fibre. (Goswami, B. C., Martindale, J. G. and
Scardino, F. L. (1977). "Textile Yarns, Technology, Structure and
Applications", John Wiley and Sons, New York.)

(1) There is an outer sheath, not shown in Fig. 12.4, called the epicuticle and it repels moisture.

(2) Underneath the epicuticle are scale-like cells called cuticles.

(3) The cuticle encloses the cortex which forms the bulk of the fibre.
A yarn is an aggregate of fibres and can be one of two types:

(1) continuous filament;

(2) spun yarns.

Fig. 12.5 Idealized diagram of plain weave in plan view and in cross-section. (Robinson, A. T. C. and Marks, R. (1967). "Woven Cloth Construction", Butterworths, London.)

The continuous filament yarn is smooth and lustrous. The second type is spun from small fibres and the strength of this yarn depends on the inter-fibre cohesion and the twist given to the fibres during manufacture. Because of small length, the ends of the fibres protrude from the body of the yarn, thus producing a fibrous surface. A fabric is woven by interlacing two sets of yarns known as the warp and the weft (Fig. 12.5). The former is the direction of length of the cloth and the weft is at right angles to it. Individual warp and weft yarns are termed ends and picks respectively. Interlacing produces a crimp at each junction. The crimp, the spacing between the warp and weft yarns, the physical and mechanical properties of the fibres and modification imparted to the woven fabric by any finishing technique are known to influence the wear resistance of fabrics.

12.2.1 Testing methods

There is much interest in the wear of fabrics but laboratory test rigs have not always provided a correlation with wear rates experienced in service. This is true of most wear tests and, of course, the purpose of laboratory study is to obtain basic information. One condition that must be satisfied in simulated studies is to appreciate that a fabric is subjected to repeated stretching under small forces and to intermittent rubbing. There have been various attempts to standardize tests which will simulate service condition and there are more than sixty types of laboratory machines [11]. Wear of a fabric will depend on stretching forces, method of laundering and the nature of the abrading medium.

Abrasive wear has been classified as follows:

(1) Plane or flat abrasion when a flat area of the fabric undergoes wear.
(2) Edge abrasion as evidenced at folds.
(3) Flex abrasion when rubbing occurs as the cloth is subjected to flexing and bending.

While designing a laboratory apparatus, the following factors should be considered:

(1) Standardization of the fabric and the testing equipment and conditions such as cleanliness, etc.
(2) Direction of abrasion must be known because abrasion rates are known to vary between the warp and the weft.
(3) The tension in the fabric, the nature of the abradant and the interfacial pressure must be standardized.
(4) The hardness of the backing material upon which the fabric is held affects the wear results. Some of the many backing materials used for testing wear resistance of fabrics are stainless steel, felt, foam rubber or inflated rubber diaphragms.

The motion can be unidirectional, reciprocating or oscillatory and assessment of total wear can be made by the weight loss technique plotted against the sliding distance or the number of cycles. Other methods of assessing wear are comparison with an unabraded sample, appearance of a hole, change in pile height, loss in strength of an abraded sample compared to an unabraded sample, and change in lustre. A microscopic examination of fibres and yarns during and after abrasion provides basic information. Of the various abrasion testers [11, 14], the Taber abraser uses abrasive wheels which rub on flat circular samples and the ball toughness test in the B.F.T. machine uses a 5 mm diameter ball which rubs on the fabric mounted on a hard backing.

A cantilever apparatus has been used [18] to obtain basic information on friction between fibres. The lower fibre is held taut in a carriage capable of a linear speed of 2 mm min^{-1}. The upper fibre is held at right angles to it and pressed onto it and the load at the interface is known. As the lower surface moves the upper specimen is dragged with it because of interfacial friction which can be evaluated by noting the deflection with a microscope. The whole experiment can be carried out in a controlled atmosphere or in vacuum and the material combinations can be changed as necessary. The nature of damage can be observed with a microscope although, at very light loads, the torn portions of a fabric are difficult to locate.

12.2.2   *Wear*

Examination of worn shirts shows that the collar wears along the folds and the fibre ends have the appearance of a brush because the cuticle disappears, causing fibrillation [19]. It seems clear that firmly held fibres are vulnerable to wear if rubbed and the presence of abrasives aggravates the damage. For example, with boiler suits, abrasion occurs where the garment is held under tension and pressed against a hard surface. Shoe laces wear against one another because they are tied under tension and abrasion is encouraged by dust which settles on them. A source of tension is the use of starch so that bed sheets, pillow slips, etc. can undergo sliding wear. Examination of various parts of a garment shows that the portions which are loosely held undergo gentle wear and only the cuticle seems to be affected.

There is evidence to show that the geometric form of a fabric is important [20]. For example, the face and the back of woven fabrics may resist abrasion to different degrees. The effect of the direction of sliding for a rayon satin fabric is shown in Fig. 12.6. Wear is seen to be low when rubbed in the weft direction.

An idea of the nature of the interface can be obtained by microscopy but only at relatively heavy unrealistic loads. A study of the nature of surface damage has been made [21] by sliding a 1 mm long flat metal rider on a nylon fibre 50 μm in diameter at normal loads in the range 40–280 g. The interesting point is that at low loads wear is low, but the scar is uniform in appearance over the apparent contact area. At heavy loads of 160 g or over, the central region is torn more than the surrounding. The explanation is that contact is at least partly elastic and the Hertzian stress is the greatest at the centre. Since it has been established that fibres wear more with increasing load, the centre should be damaged excessively.

The effect of mechanical wear of wool fabrics is that the cuticular scales are removed so that flat areas of exposed cortex are revealed [21].

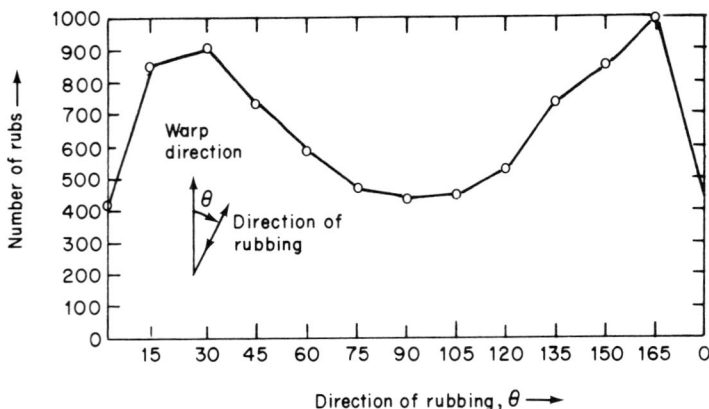

Fig. 12.6   Effect of direction of rubbing on the durability of a satin fabric. (Booth, J. E. (1964). "Principles of Textile Testing", Temple Press Books, London.)

Another effect of sliding under load is the fraying of the ends of the fibres. A technique has been described which examines vulnerable areas of garments such as elbows and collars using a scanning electron microscope [21]. This can then be compared with non-used fibres rubbed on a laboratory apparatus. If the load, speed etc. of the laboratory evaluation could be established, this could be used to rank new materials and to elucidate mechanisms.

12.3 Shoes

The intrinsic pattern of human locomotion suggests that the underside of a shoe is subjected to a cyclic compressive stress. As the stress is relaxed when the foot is lifted for the next step, slip is expected at the appropriate areas of the sole and the heel. The impacting action, particularly, at the heel has also to be considered. Evidently, forces acting on the probable surface which undergoes wear are diverse with the added variations imposed by peculiarities of individual locomotion. It is reasonable to suppose that the effect of repeated impact should be to produce fatigue wear, whereas sliding will result in material loss by the adhesive and abrasive modes. The degree of adhesion should depend on the physical and chemical nature of the road and that of abrasion will be affected also by loose grits, if any, on the road. These variables being equal, wear of shoes should depend on the material from which the wearing surfaces of footwear are manufactured.

A method of obtaining data on the loss of material for footwear is to divide the sole and the heel of each shoe into arbitrary squares and follow the amount of wear of the affected zones of the surface [23]. For a particular time of use, wear of a specific zone can be expressed by observing a large number of shoes. If the edge of the heel shows a maximum wear in n_1 shoes out of a total number n which have been assessed, the percentage of shoes showing wear of the edge is $(n_1/n) \times 100$. Figure 12.7 shows that between 76 and 100% of the shoes examined [23] show wear of the edge of the heel. If a particular zone is followed and its amount of wear plotted against walking time, the area of damage first increases with time as expected (Fig. 12.8). The sole wears less than the heel but the interesting aspect is that the damaged area does not expand and loss of material is confined then largely to the height direction. There does not appear to be much published information on the wear of shoes. The reason for this paucity of data for the general readership is possibly because individual firms carry out their own investigation which remains confidential. Of course, built-in obsolescence has been a feature with footwear for guaranteed sales revenue and also to provide scope for changes in design. Heels and soles have been manufactured variously from wood, leather and synthetic non-metals, whereas design of shoes has concentrated on improving aesthetic appeal and to offer more comfort to the subject. Conservation of material is important and published information on wear of shoes would be welcome.

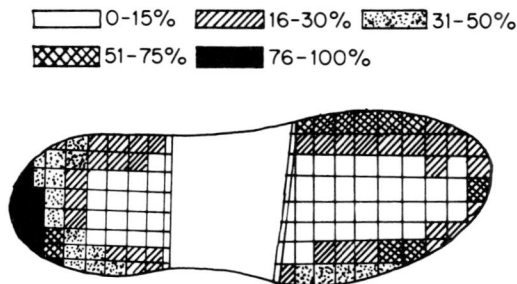

Fig. 12.7   Topology of wear of a shoe. Between 76 and 100% of the shoes examined showed wear of the edge of the heel. (James, D. I. (Ed.) (1967). "Abrasion of Rubber", Rubber and Plastics Research Association of Great Britain, London.)

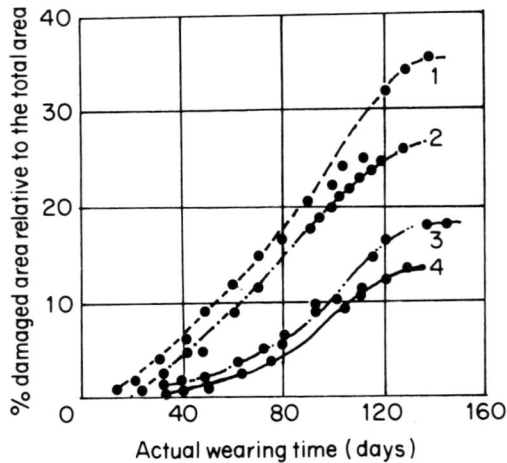

Fig. 12.8   Changes in the areas of wear of two different shoes. 1 and 2 are heels. 3 and 4 are the soles corresponding to 1 and 2 respectively. (James, D. I. (Ed.) (1967). "Abrasion of Rubber", Rubber and Plastics Research Association of Great Britain, London.)

12.4 Ceramics

The term ceramics has its origin in the Greek work *keramos*, meaning burnt matter and is probably associated with cerami, an ancient district in Athens. The natural raw material, clay, essentially hydrated alumino-silicates of varying compositions, produces such domestic articles as crockery, tiles and statuettes. Apart from clay based articles, the word ceramics is also used to describe a group of synthetic compounds. Like pottery, they involve shaping in the green state and then firing at elevated temperatures. More specifically of tribological interest are the so called new ceramics, viz., graphite, nitrides, cermets, concrete and glass.

The inclusion of concrete and glass is frowned upon in certain quarters largely because they do not entail sintering treatment at high temperatures for manufacture. However, glass is a useful counterface for laboratory studies of friction and wear of materials and it is used in nose cones of missiles where it may undergo erosion by rain. Concrete is a popular material for building roads and paths and their tribological success is of interest. Cermets have not been studied, but carbides and graphite are well known in such applications as cutting tools and bearings and there is a growing interest in nitrides. We shall collate the

Other non-metals   305

available information regarding the tribology of some of the ceramics in this section. First, however, it will be instructive to have an insight into the nature of this family of ceramics.

Broadly speaking, the binding force between atoms of most ceramics falls somewhere between ionic and covalent bonding. The bonds are very stable and provide the materials with a high chemical stability and melting points. For example, the melting point of titanium carbide, TiC, is 3120 °C while tantalum carbide, TaC, is capable of withstanding a temperature of more than another 700 °C before it fuses. The sublimation temperature of boron nitride, BN, is 2730 °C and that of graphite is 3800 °C. Obviously, these ceramics make an attractive proposition for application at high temperatures. Unfortunately, most ceramics lack plasticity. They fail at very low tensile loads but are hard and possess high compressive strengths. Graphite and carbides are well-tried materials but studies of ceramics generally in sliding or rolling contacts are not widespread.

12.4.1 Graphite

Graphite has a hexagonal layer structure (Fig. 12.9). The carbon atoms in a layer are 1·42 Å apart and are held together by strong covalent bonds. The interlayer attraction is low, the magnitude being decided by weak van der Waals forces and the separation distance between the layers is about 3·40 Å. Figure 12.9 shows how the basal arrangement of

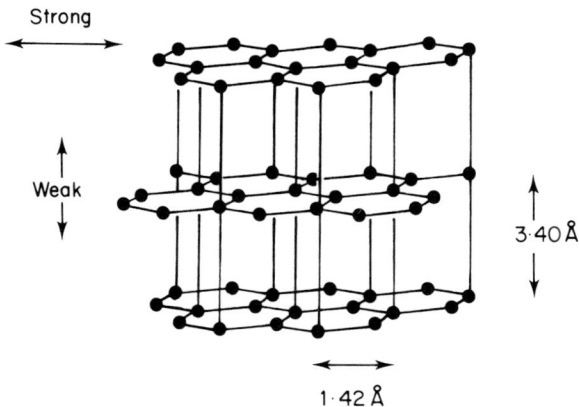

Fig. 12.9   The structure of graphite. (Bowden, F. P. and Tabor, D. (1964). "The Friction and Lubrication of Solids", Part 2, Oxford University Press, Oxford.)

atoms is repeated in alternate layers, leading to an anisotropic structure. Since the surface energy of the sheets is at least one order of magnitude lower than that for the edges, external forces cause separation of blocks of carbon atoms along the basal planes. Failure of a graphite crystal thus occurs by cleavage along the hexagonal planes. In a polycrystalline solid, the crystals will be variously oriented and failure is believed to occur by cleavage from an edge of a grain or from a defect within a grain and not by plastic flow [24].

There are three main types of graphite, viz., commercial, pyrolytic and graphite fibres. Commercial polycrystalline graphite is produced by heating coke in the region of 3000 °C. Pyrolytic graphite is produced by passing methane over a mandrel kept at a temperature of up to 3000 °C. Atomic carbon is deposited and the c-axis of the graphite is orientated normally to the mandrel surface. Commercial poly-crystalline graphite has about 30 per cent porosity. This can be virtually eliminated by a further graphitization process when a fine-grained microstructure is produced.

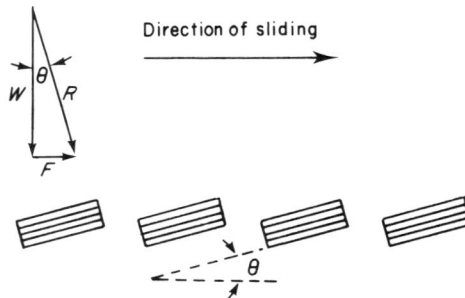

Fig. 12.10   Tilting of the basal planes of polycrystalline graphite after sliding. (Bowden, F. P. and Tabor, D. (1964). "Friction and Lubrication of Solids", Part 2, Oxford University Press, Oxford.)

Commercial graphite is noted for its strength at high temperatures and good resistance to thermal shock. Pyrolytic graphite is dense and finds application in rocket nozzles. Graphite fibres have a high Young's modulus but are expensive. Examination of polycrystalline graphite samples which have been slid shows that some of the grains are orientated in such a manner that their basal planes are tilted to the sliding direction by an angle $\theta = 5°$ (Fig. 12.10). If the sliding is done in the opposite direction, the angle of tilt is reversed. The value of tan 5° is 0·1, and this is the value of the kinetic friction of graphite whether it slides on itself or on steel. It has been concluded that the tilted basal planes become normal to the resultant force R of the normal load W and the frictional resistance F as shown in Fig. 12.10.

Graphite has a high interlamellar binding energy when de-contaminated of air, moisture and some other vapours. Thus if a graphite surface is allowed to rub on steel or copper in vacuum, friction is high because the basal planes are unable to shear readily. The reason for low friction in the presence of certain vapours is believed [25] to be due to the neutralization of some of the charge mosaics on the graphite. It does not follow that a high binding energy in vacuum means a low rate of wear. In fact the reverse seems to be the case if wear runs are conducted in a clean environment [26]. The coefficient of friction can be as high as 0·8 but fragmentation of the graphite crystals occurs, as evidenced by loose carbonaceous wear debris. Practical situations provide similar results, as is known for aircraft flying at a high altitude when the carbon brushes of electric motors undergo excessive wear. As graphite slides on steel or copper, a layer of carbon is deposited so that the graphite slides on itself. The high wear rate of the material in vacuum is probably due to the interfacial interaction forming junctions and increasing the probability of wear under shearing stresses. The presence of certain gases results in an interfering film so that wear due to adhesion or abrasion is minimized. Friction is low because the parti-tioning gas films act as lubricants.

Water vapour is an effective lubricant as are ammonia, acetone, benzene, ethanol and diethyl ether. They are all useful in lowering wear and the explanation lies in the efficacy of the gas to partition the graphite from the counterface. The *condensation–evaporation* theory suggests that a gas continues to settle on a surface and also to evaporate from it. If the rate of condensation is greater than that of evaporation, a monolayer will remain on the surface which will tend to prevent contact between the opposing asperities. The condensation life, as it is called, is very short for gases, being of the order of 10^{-6} s and has been shown to depend on their boiling points. Hydrogen has a low boiling point so that it evaporates rapidly and hence is unable to stay on a surface long enough. Like hydrogen, the probable cause for the ineffectiveness of nitrogen and carbon monoxide from the point of view of friction and wear is a low condensation life. The efficacy of the lubricating gas depends on the saturation pressure p of the gas over its liquid at the operating temperature and p has been correlated [27] with the ratio p_1/p where p_1 is the vapour pressure of the gas at which the rate of wear is zero. The pressure in question varies according to the gas involved. For example, for comparable wear rates, the gas pressure has to be increased by two orders of magnitude when dealing with oxygen compared with water vapour.

The gas films are not very effective if the graphite itself is impure [28] and, not unexpectedly, the chain lengths of the condensible gases play a role in the effective maintenance of a partitioning film. The higher the chain length, the lower the pressure necessary to maintain the lubricating gas at the interface. With some vapours, the pressure necessary may be as low as 10^{-4} mm Hg. Thus for n-heptane with a chain length of about 14 Å, this is 8×10^{-4} mm Hg and this is equivalent to it being present in amounts of only a few parts per million in ordinary air. This being the case, even a small amount of n-heptane in dry air should reduce both friction and wear of graphite when it is sliding.

Although the reason is not clear, the coefficient of friction μ of polycrystalline graphite sliding on itself or on gold, silver or copper diminishes with temperature [29]. However, when the counterface is tantalum, iron or nickel, the friction again falls up to 1000 °C but then increases to a value of about $\mu = 2 \cdot 5$ at 1500 °C. One suggestion [24] is that the metals react with graphite to form carbides which increases the strength of the interface and hence friction. This is not convincing since the friction of the interface in this case should be decided by the weaker graphite and not the carbides of the counterface.

12.4.2 *Diamond*

An important tribological application of diamond is in the form of dies which are used for high speed wire drawing when the final diameters are less than 0·075 mm. The speeds are high, being in the range 50–200 m s^{-1}, and the die entry area is known to wear. Diamond, of course, is a well known material for polishing metals and its friction and wear modes are of interest.

Diamond is notable for its extreme hardness and when pure is composed entirely of carbon atoms. Whereas graphite is hexagonal in structure, diamond possesses two interpenetrating face-centred cubic lattices (Fig. 12.11). The friction of diamond is sensitive to the orien-

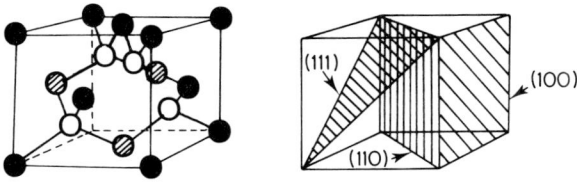

Fig. 12.11 Structure of diamond. (Bowden, F. P. and Tabor, D. (1964). "Friction and Lubrication of Solids", Part 2, Oxford University Press, Oxford.)

tation of the plane relative to a slider and as a rule it is the lowest when sliding occurs along the direction of the most closely packed atomic planes. The directions of low friction are also those which are difficult to abrade [30] and these are known as hard directions. Conversely, the soft directions abrade more readily. If plotted against load, the coefficient of friction decreases with increasing load, suggesting an elastic encounter. The anisotropic friction and wear of diamond would be difficult to exploit in practice but some information on the wear of diamond, e.g., in the form of dies, is available. If a metal wire is drawn through a die, the former softens due to frictional heating as a result of high sliding speed which increases the interfacial temperature. The metal yields and transfers onto the diamond so that the latter is protected since the metal now slides on itself. There is, however, a critical speed below which the diamond wears due to carbonization of the interface as evidenced by the nature of the wear debris, which is mainly amorphous carbon. It is known that carbonization does not occur below 1000 °C and hence it is speculated that sliding creates hot spots and the temperature there will be above 1000 °C due to very high speeds of the process. These true areas of contact will graphitize, weaken and then detach under the combined action of radial and longitudinal forces. The higher the melting point of the metal, the more difficult it is to transfer a metallic film on to the diamond. Consequently, a high melting point metal will result in a large amount of wear of the diamond. Typical wear rates of diamond in $(g\,m^{-1}) \times 10^{-12}$ are about 500 and 2 for chromium and copper respectively in the speed range 100–200 m s^{-1}. Copper has a melting point of 1083 °C and should therefore deposit a protective coating on the diamond. Chromium melts at 1615 °C and a higher speed or load will be necessary to lower its yield point sufficiently for metal transfer to be possible.

12.4.3   Nitrides

Nitrides can be grouped under ceramics generally but their application in tribological areas is not as yet very common, although both boron nitride and silicon nitride have possibilities as friction materials. The former exists in two crystallographic forms which resemble, respectively, the structures of diamond and graphite. The latter type is known as lamellar BN (Fig. 12.12). It differs from graphite in detail in that its cleavage face contains two elements, viz., boron and nitrogen whereas in graphite the plane comprises carbon atoms only. Boron nitride is produced from boric oxide by the powder metallurgy technique and gives a coefficient of friction against metals of about $0·4$ when slid in air. The friction is not much affected by water vapour but is lowered to below $\mu = 0·2$ in an atmosphere of organic gases such as heptane. Like graphite, the friction falls with increasing temperature, but boron nitride shows an overall high friction compared with graphite.

It is unlikely that boron nitride as such is a proven tribological material but much attention has been paid to silicon nitride. Essentially, the component in the latter material is produced by shaping the product in silicon powder and then reaction sintering it in nitrogen to produce silicon nitride. Early work [31] showed that the material had the potential for application as bearings at light loads up to a temperature of 400 °C. A ground finished bearing gave a coefficient of friction of $0·3$ which could be reduced to $0·1$ by lapping the bearing area, indicating the importance of surface finish of a component produced from silicon nitride. More recent appraisal [32] shows that hot pressed silicon nitride

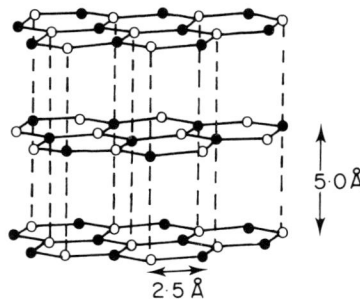

Fig. 12.12   Structure of boron nitride. • boron; ○ nitrogen. (Bowden, F. P. and Tabor, D. (1964). "Friction and Lubrication of Solids", Part 2, Oxford University Press, Oxford.)

will give a high fatigue resistance provided its surface has a fine topography and the component is free from defects such as microcracks and subsurface voids. Unfortunately, the friction of silicon nitride against itelf rises at high temperature and the wear rate is high. Silicon nitride, in common with all ceramics, is stronger under compression than under tensile forces. Components, therefore, should be designed so that compressive loading is used in service [33]. It is very important that stress concentration is minimized so that sharp corners are smoothed with generous radii. The compressive load should be well distributed over the supporting area, and there should be no impact forces.

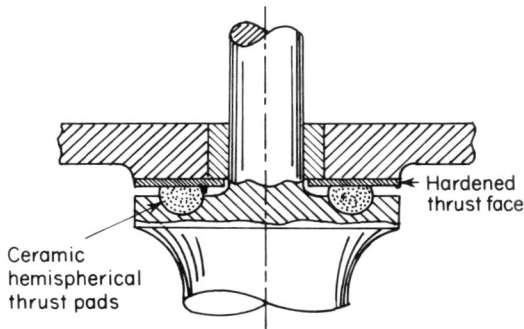

Fig. 12.13 Use of hemispherical thrust pads made in silicon nitride to avoid corrosion products. (Parr, N. L. (1971). Tribology Convention, Institution of Mechanical Engineers, London, p. 112.)

One advantage of silicon nitride is its chemical inertness. It does not react with metal such as a steel shaft so that a friction couple will not generate corrosion products. The absence of corrosion products will eliminate high starting torques and a probable use is in marine applications (Fig. 12.13) where hemispherical thrust pads can be employed under compressive loading for the thrust bearing of a propeller shaft. Although there is an absence of corrosion products, dry rubbing of a plain bearing against a metal journal produces wear debris and this can give rise to three-body abrasion. A suggested remedy [33] is to line the journal or the bearing with a graphite bush which, apart from preventing any wear debris from attacking the silicon nitride, acts as a solid lubricant. If a graphite lubricant of some sort is introduced at the interface, self-glazing of the bearing surface occurs which has the welcome effect of reducing friction. The probable mechanism is that the surface layer of the silicon nitride bearing decomposes to silicon in the presence of carbon and frictional heat and then oxidizes to form a surface film of silicate. The beneficial role of graphite is confirmed if silicon nitride is run against a cast iron counterface in internal combustion engines when wear is negligible.

Some other ceramics with a potential for use as bearing materials are shown in Table 12.1 but silicon nitride is the most promising at the moment. It is possible to improve the performance of silicon nitride bearings by incorporating additives. For example, an addition of boron may facilitate the formation of the glassy phase to improve friction and wear characteristics of the components.

One difficulty with ceramics is their inability to relax localized stresses by plastic yielding, and this brittleness makes them an uncertain group for use under dynamic loading. However, they provide low friction and wear under dry conditions of sliding, oscillatory or rolling motion. The generous use of single crystals of alumina as jewels for watches and other appliances against steel counterfaces has been motivated by the low friction and wear of these combinations. A worthwhile advantage of ceramics and cermets is that they retain their hardness at high temperatures and resist oxidation. It is not always easy to find a lubricant which will be effective at high temperatures. Therefore, the possibility of developing friction couples capable of operating at high temperatures under dry contact makes any evaluation programme of ceramics and cermets attractive.

12.4.4 *Carbides*

Many carbides show a fall in the coefficient of friction, μ, with increasing temperature when slid against themselves [20]. The actual value of μ depends on the type of the carbide, being 0·8 and 0·4 at room temperature for tungsten carbide and titanium carbide respectively. Both these values are halved if like couples are slid at low speeds at a high temperature of 900 °C. This is a positive advantage, especially since the wear rates of these carbides remain low at a range of elevated temperatures.

Carbides are the principal constituents of cermets, an example being tungsten carbide crystals embedded in a metallic matrix of cobalt. One of the coveted properties of these materials is their low wear resistance and hence they are used in cutting tools. Since they provide low friction

Table 12.1 Potential ceramics for use as bearing materials [33]

Material	Specific gravity	Coefficient of thermal expansion $(10^{-6}\,°C^{-1})$	Temperature limit of oxidation stability (°C)	Route	Cost relative to metal equivalent	
					Complex shape	Simple shape
Alumina (Al_2O_3)	3·9	8·5	2000	Sintering	Very high	High
Beryllia (BeO)	3·2	8·5	2400	Sintering	Very high	High
Zirconia (ZrO_2)	5·9	11·0	2400	Sintering	Very high	High
Magnesia (MgO)	3·6	13·0	>2500	Sintering	High	High
Silicon nitride (hot pressed) (Si_3N_4)	3·2	2·0	1600	Hot pressing	High	High
Tantalum carbide (TaC)	13·9	6·5	1000	Hot pressing	High	More
Silicate glasses	2·5–3·0	0–1·0	1200	Fuse melt and press	Similar	Cheaper

and wear, they are used in arduous service conditions, for example, for high speed machining when the surface velocity is 300 m min^{-1} or so. The most popular cermet for machining or for drilling rocks is the WC–Co system containing up to 20% by weight of cobalt. The cermet is produced by the powder metallurgy technique and multicarbide cermets containing WC, TiC, TaC and MoC with a cobalt binder are used if service conditions become more severe. One of the newly employed ceramics, alumina, is more attractive because of its high hot hardness, oxidation resistance at high temperatures and low wear rate when used as cutting tools.

12.4.4.1   Tool wear

The role of carbides in wear resistance can be appreciated by following the wear process of a cutting tool such as in a lathe. Assuming orthogonal cutting, two types of metal removal can be identified [35], viz., flank wear and crater wear (Fig. 12.14). Both types of wear are progressive.

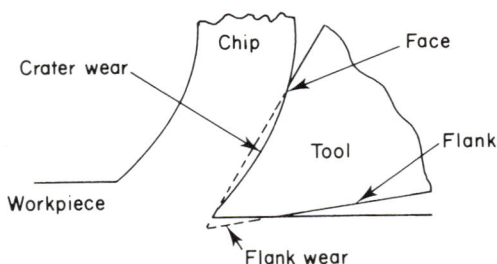

Fig. 12.14   Flank and crater wear of a cutting tool. (Boothroyd, G. (1975). "Fundamentals of Metal Machining and Machine Tools", McGraw-Hill, New York.)

A crater forms due to sliding of the tool material against the underside of the chip and is aggravated by the considerable amount of frictional heat generated. Local temperatures of the order of 1000 °C are not uncommon at high cutting speeds and crater wear may be so severe that fracture of the tool edge occurs due to the impact effect during machining. The area of the tool flank in direct contact with the workpiece (Fig. 12.14) loses material. This is flank wear, which is expressed in terms of the width of the wear land (Fig. 12.15). Plotting

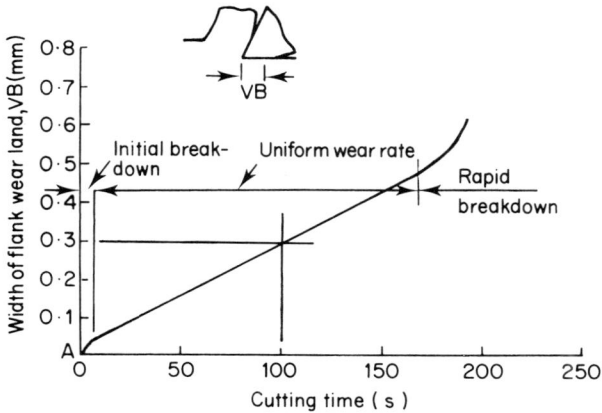

Fig. 12.15   Flank wear with time. (Boothroyd, G. (1975). "Fundamentals of Metal Machining and Machine Tools", McGraw-Hill, New York.)

the flank wear against the cutting time, a steady state of wear is observed once an initial breakdown of the sharp cutting edge has been achieved. As the steady state wear progresses, the apparent area at the flank–workpiece interface becomes larger, the effect of this being an increase in the cutting force at the tool. This results in a rise in temperature of the wear land, which is believed to cause a sudden rise in flank wear (Fig. 12.15). Crater wear can also be expressed by measuring the wear land.

Five types of wear have been identified [36, 37], viz.;
(1) superficial shearing of the tool surface;
(2) diffusion wear;
(3) abrasion;
(4) attrition;
(5) sliding wear.

Superficial shearing is frequently observed in the crater region while machining steel, titanium or nickel based alloys at high cutting speeds. Wear by this mode occurs when the interfacial temperature is high but can be prevented by using cemented carbide tools where the carbide is in high concentration. Removal of small microscopic layers from the tool surface due to rubbing suggests that the mechanism is that of solid solubility so that this may be regarded as a mild form of what has been termed *diffusion wear*. Diffusion wear, as defined, involves loss of tool

Fig. 12.16   Section through the crater surface of a cutting tool showing adhering steel. The interface shows the characteristics of diffusion wear. (Trent, E. M. (1978). In "High Speed Steel Symposium", September 13–15, Fagersta, Sweden.)

material without a serious rise in interfacial temperature. Diffusion involves migration of material across an interface so that atoms from the cutting tool are transferred to the chip or the workpiece. A form of solubility criterion may be pertinent when one considers that TiC and TaC, the important constituents in cemented carbides, diffuse into steel much more slowly than WC and the WC–Co based cutting tools wear more readily. There is evidence [37] to show that if the mode is truly by atomic diffusion, the worn surface shows a smooth interface with the workpiece material which normally adheres to the cutter. The carbide particles protrude (Fig. 12.16) indicating their non-participation in the process so that it is evident that the metallic constituent only of the tool is removed by diffusion. It is likely that during steady state wear, the temperature of the flank interface is low and it wears by a diffusion mechanism.

With certain workpieces, e.g., those made from austenitic stainless steel, hard Ti(C, N) phases in the steel may abrade a tool steel although the overall hardness of the cutter may be as high as 800 HV. Hard particles such as inclusions may act as abradants as well, but abrasive wear in this manner is not a very common situation with cutting tools.

Under interrupted cutting condition or where the interface is under vibration due to a lack of rigidity of the machine, microscopic fragments of grains detach from the body of the tool and this has been termed attrition wear. Attrition wear is expected at low cutting speeds and, although the bulk surface temperature is low, interfacial adhesion is predominant. Tungsten carbide reamers with about 6·5% cobalt show that wear occurs by mechanical detachment of individual or groups of tungsten carbide grains [38]. This has been termed abrasive wear [39] but the mode is probably that of attrition.

A typical sliding action is that by a chip rubbing against the toolpiece while escaping. Sliding in this manner can give rise to a considerable amount of wear, the worn surfaces showing a smooth background with fine scratches aligned in the direction of flow of the escaping particles. Wear probably occurs by both adhesive and abrasive modes, the oxide particles acting as abradants since the tool material, workpiece and the chip are expected to oxidize. Crater wear may be regarded as a severe mode of sliding wear but in the context of cutting tools a typical situation is drilling of metals. The chip is expected to slide at high pressure over the body of the tool but at a low velocity. By definition, there is no seizure in bulk during sliding wear but the amount of material removed can be very high. It is quite common to find deep grooves due to sliding by the chip on the rake and the clearance faces of a cutting tool.

References

1. "The International Book of Wood" (1976). Mitchell Beazley, London.
2. Jane, F. W. (1970). "The Structure of Wood", Adam and Charles Black, London.
3. Wyatt, O. H. and Dew-Hughes, D. (1974). "Metals Ceramics and Polymers", Cambridge University Press, Cambridge.
4. Penn, J. (1856). *Trans. Inst. Mech. Eng.* Jan, p. 24.
5. Penn, J. (1858). *Trans. Inst. Mech. Eng.* July, p. 81.
6. Hide, W. T. (1956). *The Shipbuilder and Marine Engine Builder*, **63**, 644.
7. Greene, S. (1959). *Forest Products J.* Septembr, p. 303.
8. McLaren, K. G. and Tabor, D. (1961). *Brit. J. Appl. Phys.* **12**, 118.
9. Atack, D. and Tabor, D. (1958). *Proc. Roy. Soc. A* **246**, 539.

10. Armstrong, F. H. (1948). "Flooring Hardwoods Their Wear and Anatomical Structure", Forest Products Research Bulletin, No. 21.
11. Booth, J. E. (1964). "Principles of Textile Testing", Temple Press, London.
12. Stoves, J. L. (1957). "Fibre Microscopy", National Trade Press.
13. Chapman, C. B. (1974). "Fibres", Butterworths, London.
14. Meredith, R. and Hearle, J. W. S. (1959). "Physical Methods of Investigating Textiles", Textile Book Publishers.
15. Robinson, A. T. C. and Marks, R. (1967). "Woven Cloth Construction", Butterworths, London.
16. Goswami, B. C., Martindale, J. G. and Scardino, F. L. (1977). "Textile Yarns, Technology, Structure and Applications", John Wiley and Sons, New York.
17. Hamburger, W. J. (1945). *J. Text. Res.* **15**, 169.
18. Pascoe, M. W. and Tabor, D. (1956). *Proc. Roy. Soc. A* **235**, 210.
19. Clegg, G. L. (1949). *J. Text. Inst.* **40**, T449.
20. Weiner, L. I. and Kennedy, S. J. (1953). *J. Text. Inst.* **44**, P433.
21. Chapman, J. A. and Menter, J. W. (1954). *Proc. Roy. Soc. A* **226**, 400.
22. Dowson, D., Godet, M. and Taylor, C. M. (Ed.) (1978). "The Wear of Non-Metallic Materials", Mechanical Engineering Publications Ltd, Bury St. Edmunds.
23. James, D. I. (Ed.) (1967). "Abrasion of Rubber", Rubber and Plastics Research Association of Great Britain, London.
24. Bowden, F. P. and Tabor, D. (1964). "The Friction and Lubrication of Solids", Part 2, Oxford University Press, Oxford.
25. Bryant, P. J., Taylor, L. H. and Gutshall, P. L. (1963). Transactions of the 10th National Vacuum Symposium, pp. 21–6.
26. Savage, R. H. (1948). *J. Appl. Phys.* **19**, 1.
27. Savage, R. H. and Schaefer, D. L. (1956). *J. Appl. Phys.* **27**, 136.
28. Rowe, G. W. (1960). *Wear* **3**, 274.
29. Rowe, G. W. (1960). *Wear* **3**, 454.
30. Wilks, E. M. and Wilks, J. (1959). *J. Phil. Mag.* **4**, 158.
31. Godfrey, D. J. and Taylor, P. G. (1968). "Inorganic non-metallic bearings, with special reference to the use of silicon nitride", Special Ceramics 4, British Ceramic Research Association, Stoke-on-Trent.
32. Godfrey, D. J. (1978). Private communication.
33. Parr, N. L. (1971). Tribology Convention, Institution of Mechanical Engineers, London, p. 112.
34. Mordike, B. L. (1960). *Wear* **3**, 374.
35. Boothroyd, G. (1975). "Fundamentals of Metal Machining and Machine Tools", McGraw-Hill, New York.
36. Trent, E. M. (1977). "Metal Cutting", Butterworths, London.
37. "High Speed Steel Symposium" (1978). Fagersta, Sweden, 13–15 September.
38. Turley, D. M. (1974). *Wear* **27**, 259.
39. Bailey, S. G. and Perrot, C. M. (1974). *Wear* **29**, 117.

13 Friction Materials

Slow moving vehicles such as bullock carts in India rely on the inertia of the system to decelerate or to stop so that all that is necessary is to bring the bullocks to a standstill. Society, however, has always demanded greater speed to progress to what is considered to be a higher state of civilisation. Early fast driven vehicles were the horse driven carriages, which need external assistance to make the decelerating force reliable and reproducible. This was how brakes became popular, in which an external component rubs against a suitable surface connected in series with the rolling wheel. The function of a brake is to produce a controllable amount of frictional resistance so that the vehicle can be stopped quickly. A brake system in a modern automobile or in a railway engine has to perform an arduous task with a reasonable amount of precision and predictability. In the former, a disc or a drum which rotates with the rolling wheels is made to rub against a pad so that the interfacial friction causes the desired retardation. The pad constitutes the component manufactured from a suitable friction material and is suitably held and actuated so that it can be made to rub under pressure against the disc or drum. There may be more than one braking pair in a vehicle. Considering a disc brake, the most commonly used mechanism is the spot type where two pads are used so that both sides of a disc are pressed. The disc is manufactured from high grade pearlitic grey cast iron and the friction pads are produced from metals and non-metals, the latter being the most common friction material today.

13.1 Types of friction materials

A friction material once shaped into the required design must give a high coefficient of friction, μ, when pressed against the moving counterface. An ideal friction material should also be durable and not be abrasive to drums or discs. The material itself and the components manufactured from it should be cheap and the latter must be dimensionally stable throughout their active lives. The friction pad or a lining made of a friction material should have adequate mechanical strength and not give objectionable fumes when hot.

Various types of friction materials in brake lining application for all kinds of vehicles have been used, some typical examples being wood, leather and metal rubbing against wood and metal. Wood against wood or against wood lined with leather have been used for horse drawn carriages. Modern technological society demands satisfactory service under high speed transportation of heavily laden vehicles. Wood or leather became unsuitable for such onerous duties and modern friction materials have been developed so that they can perform at high speeds and loads. For automobile applications, the brake is lined with a non-metallic base multi-composition material which rubs against a cast iron or medium carbon steel counterface, giving a high coefficient of friction. Being proprietary, compositions of these materials are closely guarded but analyses of typical linings show [1] that most of them contain 50% by weight of chrysotile asbestos with rubber or resin or a combination of both as the binder. Many metallic compounds, graphite and coal are also found as constituents of brake lining materials.

The most economical way of producing a pad and disc is by casting but metal–metal couples become unsuitable for heavy duty application because of the generation of frictional heat which inevitably occurs during braking. The complicated compositions of current day friction materials involve the only choice of manufacturing method viz., pressing and sintering. Sintered components are known to exhibit improved performance, possibly because the manufacturing technique can accommodate additives which improve mechanical, thermal and tribological properties of the linings. The basis material for these linings is usually copper or iron [2] and the additives can be adjusted such that the end product becomes a cermet. A friction pad is pressed against a disc by a piston which is actuated by an hydraulic fluid under pressure. An advantage of metal and to a lesser extent of cermet is that brake pads produced from them have high thermal conductivities. This prevents any serious heat build-up at the interface. There is, however, the possibility of heat being conducted readily towards the fluid via the piston, causing vaporization. Sintered metal and cermet pads are used against cast iron and steel counterfaces and should be acceptable for heavy duty application because of their thermal advantages. Under similar service conditions, the cermets are known to give a longer life than resin-based materials. However, production of sintered components is not cheap because expensive dies and powder must be used to produce dimensionally accurate components.

13.2   Frictional heat

A pad or a brake lining rubs against a rotating disc or drum and the interfacial friction retards the rotation of the wheels of the vehicle. The frictional work is dissipated as heat, resulting in a rise in temperature of both the pad or the lining and the opposing surface. An effect of heat build-up in the friction material is to give rise to brake fade, a term used to describe the loss of frictional resistance due to high temperatures. A temperature rise is liable to cause distortion of the friction component which will change the torque output from the brake, reducing its efficiency. As a rule, the higher the temperature of the sliding surface, the greater is the wear of the component. Apart from these difficulties, the thermal stresses developed in the lining may cause it to crack and to fail prematurely. Temperature rise in a brake system is, therefore, of interest and can be measured by pyrometry. Heat generated at the braking interface due to frictional work can also be calculated [3] quite accurately.

13.2.1   Calculation of heat generated at the interface

Suppose a vehicle of weight W is brought to rest on a level road under a uniform deceleration of f. If the vehicle had a speed u at the instant the brake was applied, its velocity v after a time t counted from the instant of braking is,

$$v = u - ft \qquad (13.1)$$

Only a fraction z of the rotational energy of the wheels, drums and the discs is contributed to the kinetic energy of the vehicle, a typical value of z being 0.05. If g is the acceleration due to gravity, the total braking force $= (Wf/g) + z(Wf/g)$, or

$$\text{Total braking force} = (1 + z)Wf/g \qquad (13.2)$$

Neglecting losses such as resistance of the tyres against the road, the total instantaneous energy dissipated at the interface due to work done by the friction component against the metal counterface is given by,

$$(1+z)Wfv/g \qquad (13.3)$$

Therefore, the heat generated at the interface is given by

$$(1+z)Wfv/gJ, \qquad (13.4)$$

where J is the mechanical equivalent of heat.

Generally, the loads carried by the front and the rear wheels of a vehicle are not equal, and hence, the braking effort is divided between the front and the rear in the same ratio as the distribution of the vehicle weight. The braking effort is described as a braking ratio and is decided at the design stage, although error is experienced due to the weight distribution of passengers, luggage, etc. in a car. To take account of this, if y is the amount of braking effort transmitted to the front axle, the heat generated at the rubbing path is,

$$(1+z)Wfvy/gJ \qquad (13.5)$$

If, as is usually the case, there are two disc brakes in the front, the heat generated per disc is,

$$\frac{(1+z)Wfvy}{2gJ} \qquad (13.6)$$

Suppose α is the fraction of heat entering, say, a disc with a rubbing path area A_1; the heat inflow H_1 is

$$H_1 = \frac{(1+z)Wfvy\alpha}{2gJA_1} \qquad (13.7)$$

Substituting for v in Equation 13.7 from Equation 13.1,

$$H_1 = \frac{\alpha y W(1+z)f(u-ft)}{2gJA_1}$$

which can be rewritten as

$$H_1 = \frac{\alpha y W(1+z)fu}{2gA_1J}\left(1-\frac{ft}{u}\right) \qquad (13.8)$$

If t_s is the time taken to stop the vehicle due to braking,

$$u-ft_s = 0 \quad \text{or} \quad f = u/t_s$$

Substituting this value of f in Equation 13.8,

$$H_1 = \frac{\alpha y W (1+z) u^2}{2gA_1 J t_s} \left(1 - \frac{t}{t_s}\right) \qquad (13.9)$$

or

$$H_1 = \frac{E_1}{t_s} \left(1 - \frac{t}{t_s}\right) \qquad (13.10)$$

where $E_1 = \alpha y W (1+z) u^2 / 2gA_1 J$. Denoting $(E_1/t_s) = N_1$ and $(1/t_s) = (f/u) = M$, Equation 13.10 can be written as

$$H_1 = N_1 (1 - Mt) \qquad (13.11)$$

Calculation of heat generated at the interface due to frictional work is thus possible from a knowledge of load and velocity, among others. The amount of heat generated increases with speed and load, as expected. If t_s, the time required to stop the vehicle by braking is high, the amount of heat generated is low, as is the case of a vehicle stopped by gentle repeated braking. It follows that emergency braking or a bad driver who is unable to regulate his driving will cause excessive heat evolution at the interface. The fraction of heat α which enters the drum or the disc depends largely on the thermal conductivity, density and the specific heat of each member of the friction couple. As an average figure, for a drum brake, $\alpha \approx 0.95$ so that about 5% of the heat generated at the rubbing interface enters the lining. For a spot-type disc brake $\alpha = 0.99$ so that only 1% of the heat generated enters the friction component.

13.2.1.1 Non-level road

A truly level road is seldom found and the effect of gradient must be taken into account while carrying out calculations. This is particularly important while driving down a steep slope. Thus consider a vehicle being driven down a hill of inclination θ to the horizontal. If the engine is switched off, the force developed by acceleration is Wf/g, where $f = g \sin \theta$. Thus the accelerating force down the incline is $W \sin \theta$. Neglecting deceleration due to drag losses, a constant braking force of $W \sin \theta$ must be applied to the vehicle if some steady speed v is to be maintained. At that velocity, the instantaneous energy dissipated is $Wv \sin \theta$. Therefore, the net heat evolved H_i per unit area of the front drum or disc is,

$$H_i = \frac{\alpha y W v \sin \theta}{2 A_1 J} \tag{13.12}$$

Note that in Equation 13.12, the z term is omitted.

13.2.2 Single and repeated braking

A single, very violent brake application is necessary for an emergency stop. If the initial speed is high, this results in excessive rise in the interfacial temperature, unlike the situation where the vehicle has been brought to rest by repeated gentle braking. For a brake drum whose weight is W_1, shortly after the application of brakes, the drum attains a uniform temperature ΔT, given by

$$\Delta T = \frac{\text{Total heat received by the drum}}{\text{Thermal capacity of the drum}}$$

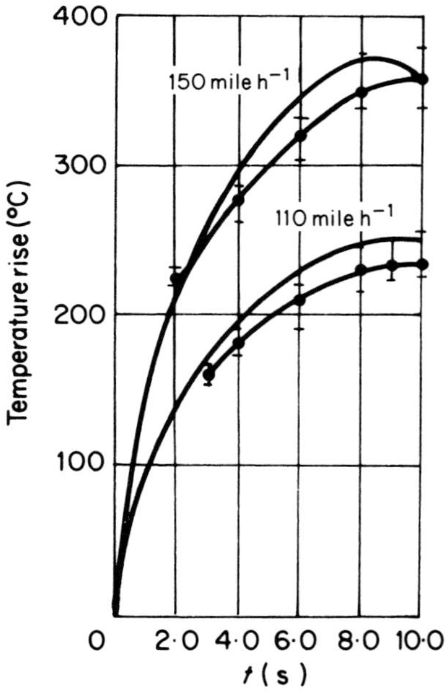

Fig. 13.1 Temperature rise of the surface of a disc during braking from high speed to rest in 10 s. Plain curve represents calculated values. (Newcomb, T. P. (1960). *J. Mech. Eng. Sci.* **2**(3), 167.)

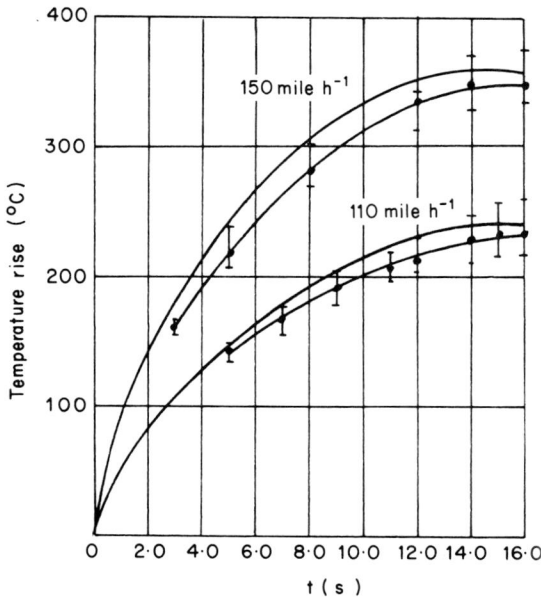

Fig. 13.2   Temperature rise of the surface of a disc during braking from high speed to rest in 16 s. Plain curve represents calculated values. (Newcomb, T. P. (1960). *J. Mech. Eng. Sci.* **2**(3), 167.)

That is,

$$\Delta T = \frac{\alpha y W (1+z) u^2}{4 g J W_1 c_1} \tag{13.13}$$

where c_1 is the specific heat of the drum material. The rise in temperature of the surface of a disc measured experimentally is shown in Figs. 13.1 and 13.2 when the vehicle is brought to a stop in 10 and 16 seconds respectively. As Equations 13.9 or 13.13 show, the higher the driving speed, the greater is the rise in temperature. The overall temperature of the metal counterface is also low when a long time is taken to bring the vehicle to a stop; a case of gentle and possibly repeated braking of short pulses.

    A typical asbestos based brake lining has a low elastic modulus and bulk hardness. The normal brake pressures applied during driving are quite high and consequently the true area of contact at the friction surface is large. The argument is that since the amount of heat generated per unit true area of contact is low, thermal damage will be less serious provided the total heat evolved is not too high. For a similar heat flux, a low load will produce a small true area of contact so that the heat generated per unit area will be large. If this is the case, the true areas of contact will degrade due to excessive temperatures. Obviously, if the braking effort is unusually high, the friction interface will be heated excessively and this will induce high thermal stresses in the areas of contact.

It is possible to calculate the step by step rise in temperature during repeated braking provided the cooling rate of the interface and the energy input per brake application are known. For calculation, the radiation losses are neglected because the surface temperature seldom reaches 500 °C and heat transfer is assumed to be by conduction and convection only. The combined losses by these two modes can be stated as,

$$(\theta - \theta_0) = (\theta'' - \theta_0) \exp(-bt) \tag{13.14}$$

where θ'' is the temperature achieved soon after braking and θ is the temperature at a time t. Cooling rate is denoted by b and θ_0 is the ambient temperature. It can be shown that at the nth stop, the prevailing ambient temperature θ'_n is

$$\theta'_n = \theta_0 + \Delta\theta \left[\frac{\exp(-bt_0) - \exp(-nbt_0)}{1 - \exp(-bt_0)} \right] \tag{13.15}$$

and

$$\theta''_n = \theta_0 + \Delta\theta \left[\frac{1 - \exp(-nbt_0)}{1 - \exp(-bt_0)} \right] \tag{13.16}$$

where θ''_n is the temperature achieved soon after braking and t_0 is the time lapse between the applications of the brake.

As the vehicle travels and is subjected to repeated brakings, the temperature of the drum or disc reaches an equilibrium value θ'' which is given by,

$$\theta'' = \theta_0 + \frac{\Delta\theta}{1 - \exp(-bt_0)}$$

where $\Delta\theta$ is the increase in surface temperature of the counterface immediately after braking.

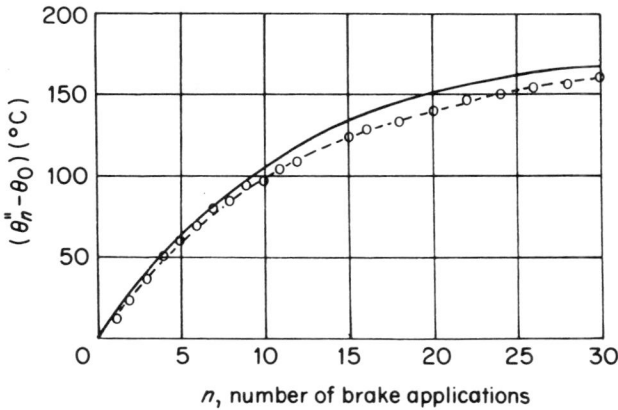

Fig. 13.3 Temperature rise of the braking interface due to n brake applications expressed as $(\theta_n'' - \theta_0)$, calculated from Equation 13.16. —— theoretical curve; – – – – experimental results. (Newcomb, T. P. and Spurr, R. T. (1967). "Braking of Road Vehicles", Chapman and Hall, London.)

In practice bt_0 is very small so that

$$\theta'' \simeq \theta_0 + \frac{\Delta\theta}{bt_0} \qquad (13.17)$$

The rise in temperature due to n brake applications expressed as $\theta_n'' - \theta_0$ as calculated from Equation 13.16 is shown in Fig 13.3. The experimental points are shown as well and agreement with calculated values is good.

13.2.3 Disc and drum brakes

Because of the difference in the value of α between a drum and a disc brake, the temperature rise of the latter is about 4% greater than of the former. Disc brakes are subjected to high contact pressures, resulting in a large true area of contact so that localized hot spots are not abundant. A design advantage is that pads are manufactured flat while linings for brakes in drums are curved. Because of the flat interface, disc brakes can sustain higher temperatures than drum brakes before thermal stresses and distortion become too great. If the volume wear of the two pads is the same, wear should give rise to a greater height loss in a disc brake than in a drum because in the former the pad areas are relatively small. Wear is undesirable since it increases the gap between the pad and the disc. As the pads slide, a surface film builds up with time which alters the coefficient of friction. Wear of the pad prevents the build up of such a film. This is a beneficial aspect of wear although the gap thus produced must be adjusted periodically by mechanical means to ensure uniform pedal pressure.

13.2.4 Braking path

As the brake is applied, the interface temperature is the same for both the disc and the pad. However, the pad material has poor thermal conductivity and, consequently a steeper temperature gradient builds up in the pad than in the disc.

While designing a brake system, it should be appreciated that an optimum braking path must be allowed for good and consistent performance. There are two important criteria which should be considered:

(1) The braking torque must be adequate for the purpose.
(2) The friction component should resist wear.

While the designer makes provision for efficient performance of a brake by choosing the correct friction couple depending upon the application, much depends on the driver regarding the temperature of the interface. An excessive interfacial temperature results in a decrease in the coefficient of friction and hence in the braking torque. Generally, the rate of wear of most materials increases above a characteristic temperature.

13.2.5 Braking torque

A knowledge of the braking torque is important in solving tribological problems concerning friction materials. If F is the braking force exerted on a vehicle of weight W, the force acting on one front wheel during a stop at constant deceleration is,

$$F = YfW/2g \qquad (13.18)$$

where Y is the proportion of the braking force transmitted to the front and g is the acceleration due to gravity. The force P on one side of the disc is given in the following relationship:

$$2PR_m\mu = T = FR_r = YfWR_r/2g \qquad (13.19)$$

where T = braking torque; μ = the coefficient of friction at the pad–

disc interface; R_r = the rolling radius of a tyre; R_m = the effective radius of the pad.

13.3   Experimental methods

Pin–disc types of machines have been used to carry out comparative tests on friction materials. The duration of these tests, however, tend to be low and the normal loads high. A typical example [4] is seen in testing a friction material containing (by volume) 31% copper, 22% mullite, 32% graphite and 15% friction modifiers, viz., iron, lead, tin etc. The running times in seconds were 5, 20 and 60 at loads of 18, 35 and 70 kg respectively at a constant speed of 1750 r min^{-1}. The friction material was in the form of a 19 mm diameter pin rubbing against a hard Cr–Mo–V alloy steel disc with a diameter of 300 mm and a thickness of 8 mm. A variation of the experiment has been the use [5] of a drum as the counterface but the test specimen must be prepared to possess a conforming interface which adds to the expense of an investigation. Not surprisingly, such studies have not correlated well with service performance and hence attempts [6, 7] have been made to produce special equipment which could simulate service conditions. The non-metallic constituents of a friction pad together with high contact pressure and temperature cause the component to behave in a non-Amontonian manner. This has been taken into account when designing special testing equipment and attempts have been made to simulate the following factors as closely as possible:

(1) surface speed;
(2) applied normal pressure;
(3) rate of energy dissipation per unit area of the pad or lining;
(4) interfacial temperature.

In a well-tried equipment, a friction pad rubs on a plate, the dimensions of which are so arranged that the ratio of the weight of the pad to that of the disc is the same as, say, the weight ratio between a commercial friction pad and a disc. Typical operating conditions which must be considered while designing a machine are shown in Table 13.1.

A piece of test equipment (Fig. 13.4) has adapted a drill whose main spindle is provided with a lifting lever operated by a pneumatic cylinder. A thermocouple is provided to measure the temperature of the friction pad which is loaded by variable normal loads. A torque measuring device, not shown in Fig. 13.4, can be mounted below the friction plate assembly and in one piece of equipment it comprises

Table 13.1   Typical operating conditions for friction components [6]

Application	Rubbing speed (m s^{-1})	Normal pressure (kg mm^{-2} × 10^3)	Area of friction pad (mm^2 × 10^3)	Length of friction pad / Length of rubbing path
Car disc brake	9	45	5	0·10
Car drum brake	9	15	15	0·60
Commercial vehicle drum brake	8·5	18	90	0·48

essentially two helical coil springs which resist the frictional torque. The coefficient of friction can be recorded on a chart recorder.

The test programme involves rotating the disc and then applying the friction pad for a very short period of five seconds or so and then releasing it. It is necessary to continue this for a reasonably long period of time, otherwise reproducible results are difficult to obtain. On the other hand, the testing period should not be too long to become uneconomical. At least 0·125 mm loss in the thickness of a pad is aimed at for ease of measurement and experience shows that a test period of five hours per friction pad is adequate. A typical test schedule is shown in Table 13.2.

Table 13.2 shows that for the disc pad, during the first two hours, the load is applied repeatedly for four second periods. During the following

Table 13.2   Test schedule for friction pads [6]

Type	Duration of test (h)	Speed (r min^{-1})	Pressure (kg mm^{-2} × 10^3)	Temperature (°C)	Application time (s)
Car brake lining	2	950	14	125	4
	2	440	11	80	3
	1	1450	17	200	5
Car disc pad	2	950	34	150	4
	2	710	28	100	3
	1	1450	45	250	5

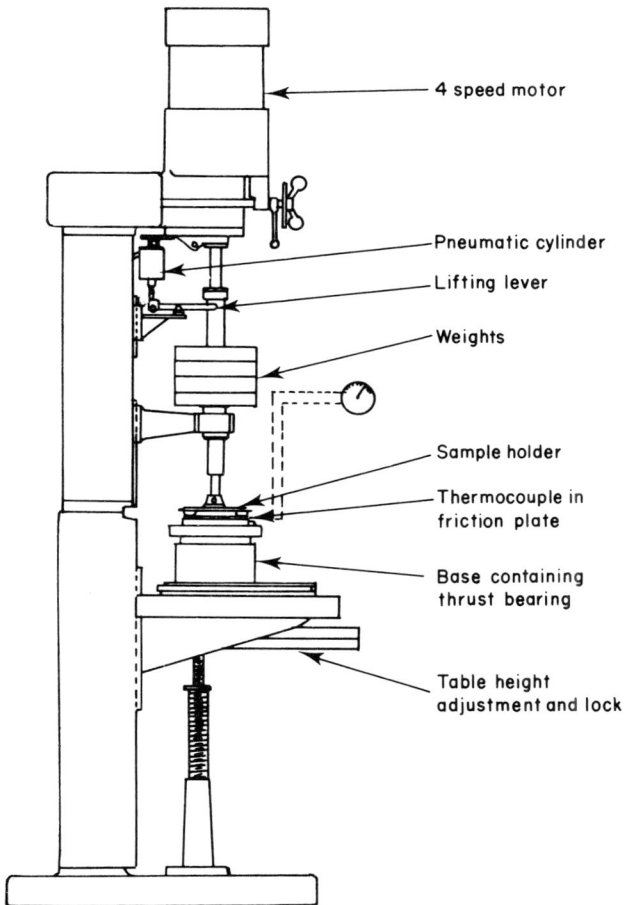

Fig. 13.4   A machine for testing friction materials. (Hatch, D. and Goddard, E. J. (1966). *Proc. Inst. Mech. Eng.* **181**(2A), 61.)

two hours and one hour, the corresponding times are three and five seconds respectively. The time interval between brake applications is adjusted so that the specified bulk temperature is achieved after each application. Correlation between results of such a laboratory machine or other scaled-down test rigs and actual vehicles is very good. In spite of this, it is unwise to state with confidence that a friction material is good or bad based on conclusions from these rigs alone because so much depends on the performance of the driver of a vehicle.

Results from rigs such as that shown in Fig. 13.4 demonstrate that pin–disc machines are unsuitable for evaluation of friction materials for the following reasons:

(1) In a pin–disc machine, the load is constant whereas Table 13.2 shows that it is nearer to what occurs in practice to apply one load for two hours, lower it for the next two hours and increase the load again for the remainder of the test.

(2) A continuous rubbing test at a constant slipping velocity is the opposite to what happens at a disc–pad interface. At any rate, for the same interfacial temperature, an intermittent load application gives a lower coefficient of friction than when the load is constant, as in a drag test such as in a pin–disc machine.

(3) The pin–disc interface ignores the mass effect in vehicle braking systems.

One could say that the only test for any material is in its application. This is a negative approach because the understanding and development of any new design will be slow and expensive and probably dangerous to human life if the only means of evaluating friction materials is from experience in service. It is likely that scaled-down rigs will be of great value for quality control purposes in plants producing friction components but pin–disc types of machine provide us with certain basic information which is necessary for our understanding of the interfacial interaction. Such information has been provided by the Friction Assessment Screening Test [1] abbreviated as the FAST machine.

13.4   Friction and wear

The FAST machine comprises a rotating cast iron disc upon which is clamped a pin prepared from the friction material under investigation. In all experiments both the amount of friction and wear are plotted automatically together with the change in bulk temperature of the disc. A typical plot of a brake lining material (Fig. 13.5) shows the coefficient of friction μ to be high initially. This decreases with time to a steady value of about 0·2. The friction material wears steadily and the temperature of the interface rises, reaching about 300 °C in 90 min. Note that in Fig. 13.5, the coefficient of friction is plotted in a reverse manner, that is, it decreases upwards. An ideal friction material should give consistent values of friction but zero wear. Figure 13.6 shows the characteristics of a brake material which is regarded [1] as unacceptable

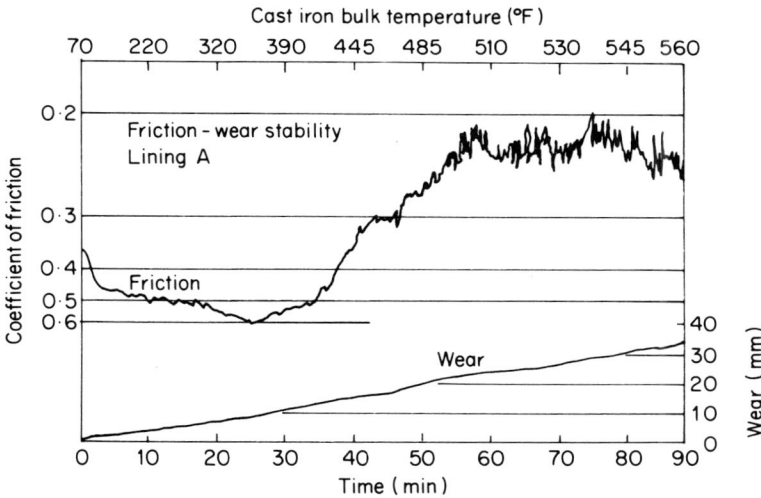

Fig. 13.5 FAST friction and wear traces of lining A, which is acceptable. (Weintraub, *et al.* in Lee, L. H. (Ed.) (1974). "Advances in Polymer Friction and Wear", Plenum, New York and London.)

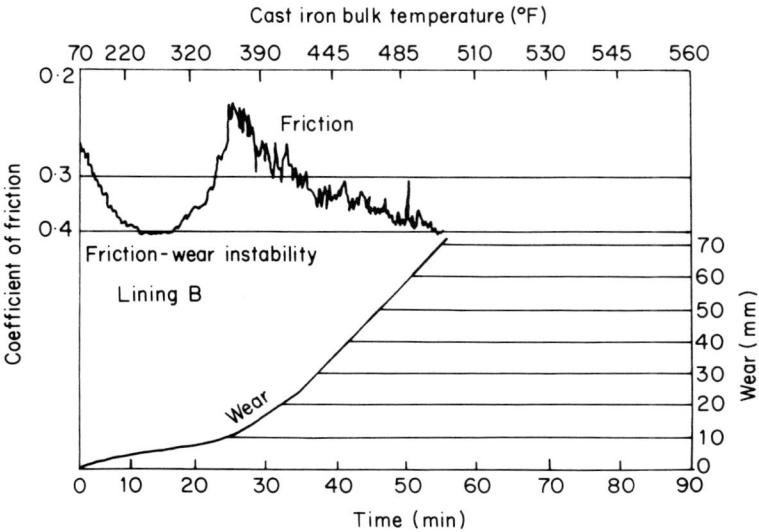

Fig. 13.6 FAST friction and wear traces for lining B, which is unacceptable. (Weintraub, *et al.* in Lee, L. H. (Ed.) (1974). "Advances in Polymer Friction and Wear", Plenum, New York and London.)

for automobiles. The reason given is that the amount of wear is excessive and the friction fluctuates greatly.

The conclusion from such experiments [1] is that the temperature of the hot spots can be as high as 790 °C. This is so even at moderate sliding speeds and low ambient temperatures. A high temperature causes the asbestos fibres to suffer thermal degradation and the surrounding resin bond undergoes pyrolysis. The contact areas orientate in the direction of sliding and the degraded products of asbestos and resin form small patches with a mean equivalent diameter of about 25 μm. These patches accumulate on the friction surface and the particles smear into the shape of platelets as sliding continues. With continued running, these platelets grow to macro-proportions and adhere to the trailing edge of the pin, but re-enter the leading edge later. When the conditions of sliding are gentle, wear occurs by the formation of platelets which are progressively removed from the interface. For a situation analogous to severe braking, a carbonaceous layer appears, and under very violent conditions a glassy phase is seen on the pin interface which progressively breaks up as wear debris. Examination of the cast iron disc shows the rubbing interface to possess a glazed black coating. This has been identified as forsterite, a transformation product of chrysotile asbestos when heated at 810 °C. This suggests at once that the flash temperatures at the contact spots are higher than the bulk temperature of the friction component, as expected.

The phenomenology at the interface of a pad will, of course, be governed by its composition, and a metal phase may have a limiting effect in the sense that its melting point may decide on how severe the braking can be. A material containing mullite, graphite and 31% by volume of copper exhibits the expected increase in wear with sliding distance and load when run continuously [4]. An interesting feature of this material is that the hardness, as measured at room temperature, shows a proportionate amount of fall with sliding distance. The reason for this is attributed to the recrystallization of the copper. At low rubbing times the recrystallization is probably incomplete. The critical temperature when wear is rapid is about 600 °C for this material, and this is further aggravated in the vicinity of the melting point of copper which is 1083 °C. It would appear that the interfacial temperature is decisive in determining friction, and certainly wear, of a friction material. The mean surface temperature T is a function of the applied load W, the sliding time t and the rubbing velocity v [4].

That is,

$$T = \phi(W, t, v)$$

At constant speed,

$$T = \phi'(W, t)$$

The height loss Δh of a friction pad due to wear is a function of the mean temperature T. That is,

$$\Delta h = \phi''(T)$$

Ho and Peterson [4] have carried out a curve-fitting exercise with their results to show that

$$\Delta h = K\{T/(T^* - T)\} \qquad (13.20)$$

where K is a constant and T^*, in this case, is the melting point of copper.

Equation 13.20 has no obvious physical meaning incorporating all the variables. However, it is of value since T can be measured experimentally. The constant K can be obtained for a particular friction material from laboratory experiments and we can speculate on the possibility of obtaining Δh for a particular application if T could be obtained.

An increase in the amount of wear of a purely non-metallic material such as asbestos with organic additives and a resin binder is experienced with increasing load and surface temperature [5]. The wear rate, however, is believed to be governed by the instability of the resin binder with increasing temperature.

13.4.1 *Wear*

Wear of any material is a complex phenomenon and it is unlikely that there will ever be a global formula incorporating all the obvious variables to assist the design engineer in carrying out exact calculations. Nevertheless, as shown in the following chapter, bearing manufacturers have used certain quasi-empirical laws of wear with confidence. There is always the hope that one day approximate formulae will also be devised for friction components and we shall examine one such relationship [8] in what follows.

Consider the interaction between a metallic asperity and a flat component manufactured from a friction couple. If d is the mean diameter of a contact spot, the volume V of friction material lost can be expressed as $V = K_h A_t d$ where K_h is a constant called the specific wear rate and A_t is the true area of contact. Thus

$$K_h = V/A_t d \qquad (13.21)$$

Consider the disc, for a disc brake, to be a rough surface and the friction pad to be a deformable component. As sliding commences the topography of the pad will change. A steady state topography will be established when the wear rate should also be steady. Material from the pad may be removed by a two-body abrasive mode or the mechanism may be that of fatigue wear.

Consider a conical single asperity whose maximum height is h_m. The model asperity in this case is that of the metallic disc and it is assumed to penetrate the friction pad to a depth h from its surface. A term *relative approach ε* can be defined such that $\varepsilon = h/h_m$, which really is a measure of the amount of deformation by the indenting body.

Let $\eta = A_t/A$ where A is the apparent area of contact and η is always less than unity and is usually very small. As a first approximation,

$$\eta = b\varepsilon^{\nu} = A_t/A \qquad (13.22)$$

where b and ν are constants and, depending on the surface finish, b varies between unity and 10 and ν has a value between 2 and 3. The higher the values of b or ν, the finer is the surface finish.

Consider now a light load generating an approach dh and volume dV. Now $dV = A_t dh$ and, by definition,

$$dh = h_m d\varepsilon \qquad (13.23)$$

Thus, the total volume involved is

$$V = \int_0^V dV = \int_0^\varepsilon A_t \, dh$$

substituting for A_t from Equation 13.22 and for dh from Equation 13.23 in the above relationship

$$V = \int_0^\varepsilon Ab\varepsilon^{\nu} h_m \, d\varepsilon \quad \text{or} \quad V = \frac{Abh_m \varepsilon^{\nu+1}}{\nu+1} \qquad (13.24)$$

Since $\varepsilon^{\nu} = A_t/Ab$, Equation 13.24 can be written as

$$V = \frac{A_t h_m \varepsilon}{\nu+1} \qquad (13.25)$$

Substituting the value of V from Equation 13.25 in Equation 13.21,

$$K_{\mathrm{h}} = \frac{\varepsilon h_{\mathrm{m}}}{(\nu + 1)d} \tag{13.26}$$

Define now a linear wear rate $J_{\mathrm{h}} = (V/A)/S$ where A is the apparent area of contact and S is the total sliding distance. That is, the linear wear J_{h} is the height loss of the friction pad per unit sliding distance.

If τ is the shear stress of the junction and W is the applied normal load, the coefficient of friction μ is given as

$$\mu = \tau A_{\mathrm{t}} / W \tag{13.27}$$

If we consider the formation and disruption of one junction only, the sliding distance is now d. Therefore

$$J_{\mathrm{h}} = (V/A)/d = \frac{(V/A)A_{\mathrm{t}}}{dA_{\mathrm{t}}} = K_{\mathrm{h}}\left(\frac{A_{\mathrm{t}}}{A}\right) \quad \text{or} \quad A_{\mathrm{t}} = \frac{J_{\mathrm{h}}A}{K_{\mathrm{h}}} \tag{13.28}$$

Substituting for A_{t} from Equation 13.28 in Equation 13.27,

$$\mu = \frac{\tau J_{\mathrm{h}}A}{K_{\mathrm{h}}W} \quad \text{or} \quad \frac{\mu}{J_{\mathrm{h}}} = \frac{\tau}{K_{\mathrm{h}}\sigma} \tag{13.29}$$

where $\sigma = W/A$ = the applied normal stress.

Substituting in Equation 13.29 for K_{h} from Equation 13.26

$$\frac{\mu}{J_{\mathrm{h}}} = \frac{\tau(\nu + 1)d}{\sigma \varepsilon h_{\mathrm{m}}} = \frac{\tau(\nu + 1)d}{\sigma h} \quad \text{or} \quad J_{\mathrm{h}} = \frac{\mu \sigma h}{\tau(\nu + 1)d} \tag{13.30}$$

In the above relationship, the greater the depth of static indentation h, the greater the amount of wear. A material with a high intrinsic shear strength will give a low amount of wear. Other things being equal, the amount of wear is directly proportional to the coefficient of friction and such a relationship is characteristic of viscoelastic materials. Since a friction pad should have a high coefficient of friction against a steel or cast iron counterface we should note from Equation 13.30 that this is possible if the following factors are considered:
(1) a couple with a high interfacial shear strength;
(2) an originally smooth surface;
(3) a large area of contact but low indentation of the friction pad by the counterface;
(4) a low pedal pressure on the brakes.

A good friction material should produce a high value of μ/J_h. We should note, however, that although Equation 13.30 incorporates physical quantities which are easily identifiable and measured, the parameters are interdependent and temperature sensitive. For example, a low σ should decrease both d and h. A high interfacial temperature will decrease τ and may well reduce ν by depositing material on the counterface.

13.5 The resin bond

A typical friction material which contains asbestos as the main filler material must incorporate some resin to hold the fibres together. Although the asbestos fibres constitute a large part of the friction material, it is known that the greater the thermal stability of the resin binder, the less the wear of the friction pad or of the brake lining. Experiments show that most composites tested under indulgently low stresses and speeds show comparable performance and wear resistance. However, with increased severity of the test variables, some of the binders become inadequate. For example, using a mixture of a drying oil and liquid phenolic resin as the binder results in a friction material which shows a high amount of wear when rubbed at a high speed under heavy pedal pressure. Omission of the oil from the friction material can reduce the amount of wear but the problem is that phenolic resin by itself makes the pad brittle and it can simply fail mechanically under normal and tangential forces. The oil bleeds from the pad as it apparently distils at high temperatures and disrupts the resin bond. The object of adding the oil, however, is to produce a resilient, porous and a non-brittle friction surface.

A porous matrix accommodates thermal expansion but road dust is known to increase the wear rate of phenolic resin by about 250%. It is not known if wear occurs by a three-body mode but sealing of the pad surface has been thought of, although the surface porosity will be reduced. This may be undesirable since a porous pad will also permeate any gas which forms during frictional heating and thus prevent pressure build-up in the subcutaneous zones of the interface. Some rubber is also incorporated and the material probably reduces the permeability of a brake lining or a pad. However, its known role is to impart resilience to the resin binder. Rubber is known to improve the life of a resin-bonded friction component under moderate operating conditions but a rapid increase in the wear rate occurs if the interfacial temperature is high.

A friction material such as asbestos is mixed with additives and a resin and is then pressed in a die to produce the required shape. The green product is then sintered at a suitable temperature to impart mechanical strength by curing the resin bond which holds the filler particles together. Both the time and the temperature of curing are important and it has been shown that there is a decided correlation between the degree of curing and the life of a pad. Tests with the FAST machine and observations on brake linings show that a fully cured component has superior thermal stability but very inferior wear resistance.

13.6 Asbestos fibres

An asbestos which is so mechanically processed that the aggregate comprises a few large fibres, called crudes, is classified as an open fibre material. A fibril of asbestos is about $0 \cdot 03$ μm in diameter and a small asbestos crude contains millions of such fibrils. Impurities such as magnetite are found within the crudes or attached to them. An open asbestos crude wears more than an aggregate containing small fibrils.

Wear of asbestos based friction pads depends on the amount of resin bond and also on the nature and quantity of impurities. Volatiles are generated during severe braking within the friction component and an open asbestos will give very high permeability. This will facilitate the escape of gases and thus minimize localized fracture which is known to occur due to a build-up of subcutaneous pressure.

In-service evaluation of the function of asbestos as a constituent of the friction material becomes elusive because of the presence of organic and inorganic additives in the brake lining, examples being rubber and cashew resin. The effect of adding rubber is to decrease the overall rate of wear and to increase the coefficient of friction, provided the vehicle is brought to a stop by gentle and progressive deceleration. Brake fade, that is a loss of frictional resistance, inevitably occurs if the interface rises to a high temperature. The cashew particles influence the tribology in a similar manner to rubber but a higher concentration of the resin is required to obtain comparable effects. A 15% addition of cashew particles to an asbestos based friction material improves its wear resistance by 50%. In general, the inorganic additives should be softer than the steel or cast iron counterface.

Asbestos bonded with resin forms friction components in aircraft, excavators, oil drilling hoists, racing cars and police vehicles. In fact the precise composition is not known, being proprietary, but the friction component can withstand an interfacial temperature of 1000 °C without degradation. It provides a stable coefficient of friction and the reason has been attributed [8] to a surface layer which forms on the friction component. The surface layer is also wear resistant and is probably formed by some of the additives in the material. Depending on the nature of the additives, the volume wear rate of the counterface can be between 2 and 20% of that of the friction component. A typical abrasive constituent is magnetite but wear of the counterface can be beneficial. This is particularly so in heavy trucks, where by consistent and progressive wear of the cast iron drum, the growth of surface cracks which nucleate due to thermal shock is retarded. The hard particles may already be present in the natural asbestos, some examples being magnetite, quartz and particles of rock. Laboratory studies show that an additive such as litharge is soft and melts at the interface during rubbing so that wear of the counterface can be minimized if so desired.

References

1. Lee, L. H. (1974). "Advances in Polymer Friction and Wear", Plenum Press, London.
2. Geschelin, J. (1961). *Auto Ind.* **124**, 69.
3. Newcomb, T. P. and Spurr, R. T. (1967). "Braking of Road Vehicles", Chapman and Hall, London.
4. Ho, T. L. and Peterson, M. B. (1977). *Wear* **43**, 199.
5. Rhee, S. K. (1974). *Wear* **29**, 391.
6. Hatch, D. and Goddard, E. J. (1966). *Proc. Inst. Mech. Eng.* **181**(2A), 61.
7. Wilson, A. J., Belford, W. G. and Bowsher, G. T. (1968). *Lucas Engineering Review* **4**, 14.
8. Kragelskii, I. V. (1965). "Friction and Wear", Butterworths, London.

14 Dry Bearings

Running a bearing in the wet condition, that is with the aid of a liquid lubricant, is necessary to achieve low friction and wear. There are instances, however, where the very nature of the situation demands bearings which will withstand the service constraints without the assistance of a lubricant. Examples of such situations are the food processing industry and a high temperature or a cryogenic environment. In the former, the use of a lubricant is precluded because of the danger of contamination of the product whereas in the latter the available liquid lubricants may not be stable at high temperatures. A bearing without an externally fed lubricant is a dry bearing. Like a lubricated bearing, the requirements are that it is able to support the duty load at the operating temperature and to give a minimum amount of friction and wear. There are four groups of materials which find applications in dry bearings viz.:

(1) synthetic polymers;
(2) carbons and graphites;
(3) MoS_2;
(4) ceramics or cermets.

14.1 p–v factor

Both the theories of wear and results of experiments or actual experience from applications show that the volume V of material removed from a non-metallic component is proportional to the applied load W and the sliding distance S. Thus if K is a constant of proportionality,

$$V = KWS \tag{14.1}$$

To obtain radial wear for a plain bearing, both sides of Equation 14.1 are divided by the apparent area of contact A. Thus

$$(V/A) = K(W/A)S \tag{14.2}$$

Denoting V/A by h, the radial wear and W/A by p, the apparent contact pressure, from Equation 14.2,

$$h = KpS \tag{14.3}$$

For a sliding time t, $S = vt$ where v is the velocity of the journal. Therefore, Equation 14.3 can be rewritten as

$$h = Kpvt \tag{14.4}$$

Equation 14.4 neglects the effect of any rise in surface temperature. In literature [1] the letters P and V are used for pressure and velocity respectively but we have used the respective lower case letters to remain consistent within this book.

The performance criterion of a bearing is expressed by the p–v factor and takes two forms:

(1) The limiting pv above which wear is excessive either due to heavy loads approaching the elastic limit or due to thermal softening.

(2) The p–v factor for some arbitrary amount of wear for continuous running.

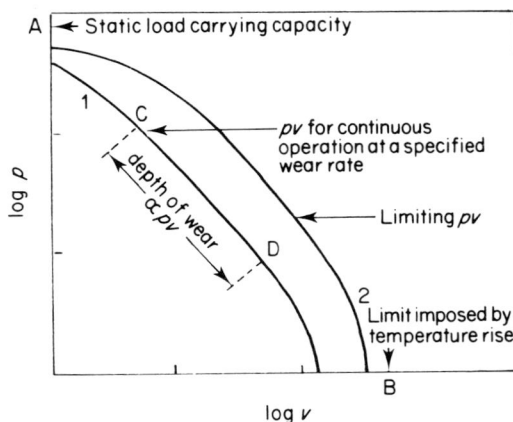

Fig. 14.1   The p–v relationship for dry bearings. (Lancaster, J. K. (1973). *Tribology* **6**, 219.)

Figure 14.1 shows the nature of a p–v diagram where $\log p$ is plotted against $\log v$. The point A on the diagram is the limiting load carrying capacity of the bearing whereas the velocity limit is at point B where the surface temperature becomes excessive. Curve 1 shows that the criterion for choosing a load at a velocity is a specified wear rate under continuous sliding. As expected, only the region CD shows a proportionality of the depth of wear (Equation 14.4) with the product pv because the bearing approaches the limiting load and velocity respectively beyond C and D. Curve 2 shows a similar pattern and is a general case which shows the values of the limiting pressure and velocity either because of thermal softening or of yielding of the material due to high contact stresses.

14.2 Bearing life

The proportionality constant K in equation 14.4 has been called the *specific wear rate* and sometimes the *wear factor*. The specific wear rate is regarded [2] as a material property for the prevailing operating parameters. In fact a specific wear rate K_0 for moderate load and speed is defined and the wear factor K for a particular situation is calculated empirically as follows, using a number of correction factors:

$$K = K_0 C_T C_v C_m C_r \qquad (14.5)$$

where C_T = bearing temperature correction factor, C_v = sliding speed correction factor, C_m = correction factor to take account of the type of motion, and C_r = correction factor for the surface roughness of the journal. The factors C_T, C_v, C_m and C_r are obtained from data provided by experiments on bearings in a similar manner to the data provided by Fig. 14.1. C_r generally increases with surface roughness. For example, for a particular material, the C_r values were 1·6, 2·5 and 4 for R_a values of 0·2, 0·6 and 1·0 μm respectively. The life of a bearing is then expressed as

$$t = \frac{h}{pvK_0 C_T C_v C_m C_r} \qquad (14.6)$$

Sometimes a specific bearing life t_r is required. In that case, while choosing a material, the ultimate choice is made on the basis of cost, with the condition that $t \geqslant t_r$.

For continuous motion, $C_m = 1$ if the load is fixed relative to the bearing and $C_m = 0·5$ if the load rotates relative to the bearing. For oscillatory motion C_m is usually taken as unity.

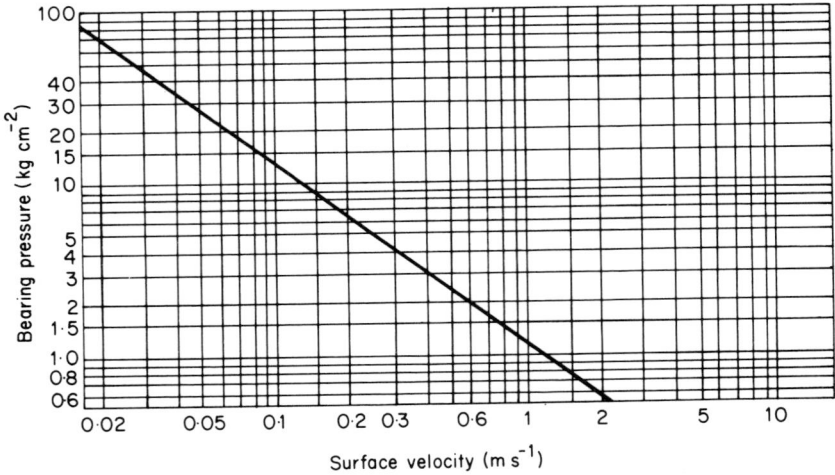

Fig. 14.2   Maximum allowable bearing pressure at a particular surface velocity (log scale). (Courtesy, Polypenco [3].)

14.2.1 *Example*

Equation 14.6 involves evaluation of the correction factors, which is quite straightforward if the appropriate charts are available from the bearing supplier. We shall use an alternative method [3] to first calculate the effective pressure p and then the life of a bearing. We need to consult a $p-v$ diagram (Fig. 14.2) and a chart showing temperature correction factor C_T (Fig. 14.3). The pressure in Fig. 14.2 has been expressed in kg cm^{-2} because of the smallness of p generally. Apart from C_T we need

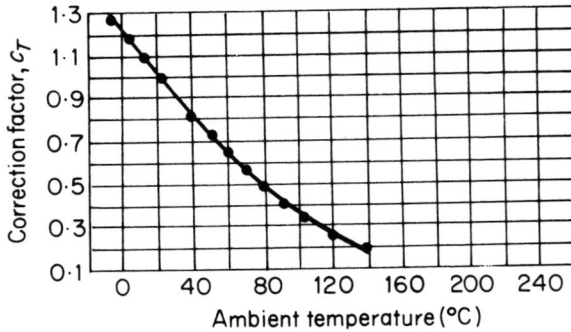

Fig. 14.3   Ambient temperature correction factor, C_T. (Courtesy, Polypenco [3].)

a correction factor C_r for surface finish of the journal, which is also supplied graphically by manufacturers. We can now obtain the probable life of a bearing, given the following information for a plain dry rubbing bearing:

Bearing size:

$$o.d. = 25 \text{ mm}$$

$$i.d. = 19 \text{ mm}$$

$$\text{length, } l = 19 \text{ mm}$$

$$\text{steady load on bearing, } W = 25 \text{ kg}$$

$$\text{journal speed, } N = 110 \text{ r min}^{-1}$$

$$R_a, \text{ of journal} = 0 \cdot 6 \text{ }\mu\text{m}$$

$$\text{ambient temperature} = 23 \text{ °C}$$

$$\text{allowable radial wear} = 0 \cdot 2 \text{ mm}$$

Now

$$p = \frac{\text{Load}}{\text{Projected Area}} = \frac{W}{(i.d.)l} = \frac{W}{19 \times 19} \times 100 \text{ kg cm}^{-2}$$

therefore

$$p = 6 \cdot 925 \text{ kg cm}^{-2} \qquad (14.7)$$

$$v = \frac{\pi(i.d.)N}{60\ 000}$$

assuming the journal diameter equals the i.d. of the bearing. This gives

$$v = 0 \cdot 11 \text{ m s}^{-1}$$

Checking the limiting pressure in Fig. 14.2, $p = 12 \text{ kg cm}^{-2}$ at $v = 0 \cdot 11 \text{ m s}^{-1}$ so that p as calculated in Equation 14.7 is less than this and we can proceed with the correction factors. The ambient temperature is 23 °C which gives a correction factor of $C_T = 1$ (Fig. 14.3). To check the allowable bearing pressure, multiply $p = 12 \text{ kg cm}^{-2}$ by C_T so that the allowable bearing pressure is still 12 kg cm^{-2} and this is greater than the load calculated in Equation 14.7. For this particular bearing C_r is given as 2 for a journal whose R_a is $0 \cdot 6 \text{ }\mu\text{m}$. Multiplying the limiting p of Fig. 14.2 by 2 and comparing this with the pressure in Equation 14.7, we are well within the limiting pv range.

This checks that our design load at the operating temperature does not exceed the *pv* limit, otherwise the bearing may give unstable friction values and very high wear rates. In extreme cases, it may seize or melt. To find the life of a bearing *t*, we may use the following relationship, taking each of C_T, C_v and C_m as unity.

$$t = \frac{h}{pvK_0C_r}; \quad \text{given } K_0 = 2 \times 10^{-15} \text{ m}^2 \text{ kg}^{-1}$$

or

$$t = \frac{0 \cdot 2}{(6 \cdot 925 \times 10^{-2})(0 \cdot 11 \times 10^3)(2 \times 10^{-15} \times 10^6) \times 2}$$

$$= 1827 \text{ h}$$

The life of the bearing, assuming continuous running each day, is about 76 days, which is not a long time for a component.

If the ambient temperature is 90 °C, $C_T = 0 \cdot 42$ from Fig. 14.3. Therefore, at $v = 0 \cdot 11 \text{ m s}^{-1}$, the limiting pressure allowed is $p = 12 \times 0 \cdot 42 = 5 \cdot 04 \text{ kg cm}^{-2}$ which is below the pressure to be sustained, viz., $p = 6 \cdot 925 \text{ kg cm}^{-2}$. This demonstrates how one should be cautious when calculating bearing life.

Another correction factor which is applied is known as the cycle time correction factor. This is to take into account bearings which run intermittently for relatively short periods of time. The correction factor is such that the limiting *pv* value of Fig. 14.2 is effectively increased. The reason is that intermittent running keeps the bearing interface cool.

14.3   Synthetic polymers

The largest group of materials which finds application in the field of dry bearings comprises synthetic polymers incorporating fillers or reinforcements. Typical examples are as follows:

(1) Thermoplastics, e.g., nylon and acetal with or without fillers such as MoS_2, PTFE or glass.

(2) Thermosets such as phenolics and epoxies with such fillers as PTFE and asbestos.

(3) PTFE with glass fillers.

For design purposes, wear data are obtained from laboratory tests and, like the *pv* factors, caution is necessary when considering applications. It has been suggested [1] that, as a rough approximation, the wear rate is proportional to the square or the cube of the R_a value of the journal. During the initial runs at least, the polymer loses material by a two-body abrasion mode and the wear rate for an unchanging counter-face roughness can be expressed as wear rate $\propto (1/\sigma\varepsilon)$, where σ is the breaking strength and ε is the elongation to fracture of the polymer. Although fillers may increase σ by a factor of up to 4, the elongation may fall by two orders of magnitude. It is expected, therefore, that by causing an overall decrease in the product $\sigma\varepsilon$, fillers may cause a deterioration in the wear resistance of the composite. With continued running, a layer of polymer is deposited and sliding then is between like materials. PTFE, acetals and nylons produce a smooth deposit giving rise to a diminished rate of wear in the steady state. The opposite is the case for brittle polymers such as polyesters and polystyrene which produce rough deposits on the counterface. The counterface is seriously roughened if the filler is hard and abrasive. Wear becomes intolerably high as is seen in the case of a 30% addition of glass fibre to PTFE. On the other hand, the presence of graphite and MoS_2 is welcome from the viewpoint of friction and wear. Carbon and graphite filled polymers, however, are inefficient if contaminated by certain fluids. Hence bearings made from such composites become unsuccessful as far as the rate of wear is concerned unless the system is kept dry.

A major limitation of polymers is their poor mechanical strengths, although certain improvements result if fillers are incorporated. Normally, polymeric bearings are designed for light loads or as coatings on metal substrates. The metal component ensures that the whole bearing can withstand the service load while the coating provides the counterface which should give low friction and wear. Wear occurs with time, but it is not a difficult task to withdraw the bearing from service and recoat it for further use. A typical low friction film is a mixture of PTFE and lead, coated on to a layer of porous bronze which is supported on a steel backing by sintering.

The recommended counterface roughness is R_a in the range 0·2–0·4 μm and the p–v relationships of a range of PTFE composites is shown in Fig. 14.4 for a wear depth of 25 μm per 100 h. It is seen that, as expected, the unfilled PTFE is restricted in terms of load and speed when compared with PTFE. The effect of fillers in thermoplastics such as nylons and acetals is shown in Fig. 14.5 for a wear depth of 25 μm per 100 h.

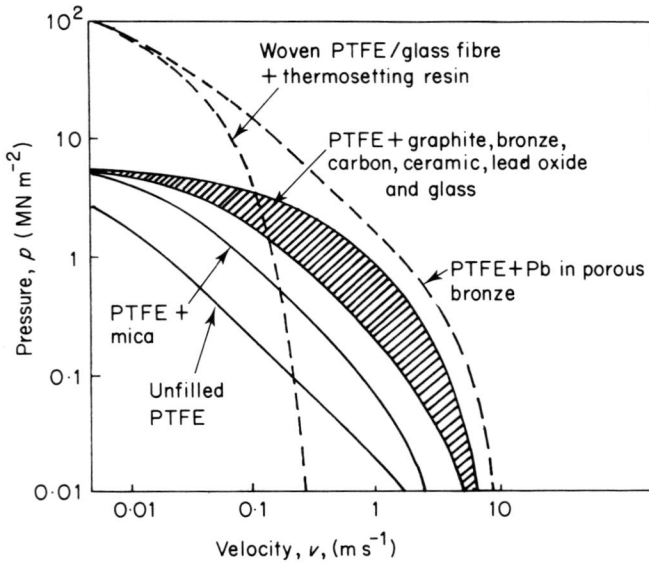

Fig. 14.4   *p–v* relationships for PTFE based bearings. Allowable wear rate 25 μm per 100 h (1 $N = 0 \cdot 1$ kg). (Lancaster, J. K. (1973). *Tribology* **6**, 219.)

One advantage of thermoplastics over thermosets is that the former can be produced cheaply by injection moulding. The raw material costs of some thermosetting resins are less than the thermoplastics, but the cost of producing components in the former is high. A typical technique in producing bearing surfaces using thermosets is to impregnate a fibrous mat or cloth with a liquid resin, followed by pressing and then curing. An alternative method is to produce tubes by a special process, but none of these methods is amenable to mass production. The laminates produced from thermosets show anisotropy in mechanical properties and can have high strength and stiffness. A major area of application of these laminates is in water-cooled systems because heat dissipation is an important factor in bearings. For dry bearings, fillers such as PTFE, graphite or MoS_2 are used. The thermosets are also reinforced with cellulose, asbestos and carbon fibre and find applications at above ambient temperatures, albeit with a limited ceiling because of thermal degradation of the resin.

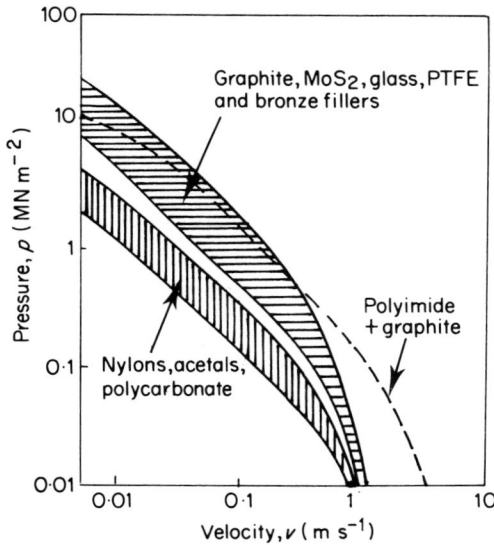

Fig. 14.5 p–v relationships for thermoplastic and filled thermoplastic bearings. Allowable wear rate 25 μm per 100 h (1 $N = 0.1$ kg). (Lancaster, J. K. (1973). *Tribology* **6**, 219.)

14.4 Multifilament woven fabric

Multifilament woven fabric is used as a liner on a metal bearing and its usefulness has been demonstrated over a long period of time [4]. For a plain journal bearing, the shell, manufactured from a suitable alloy, is lined with this multifilament woven fabric forming the liner (Fig. 14.6). The liner comprises essentially two sheets of fabric, the one in contact with the shell being a glass fibre fabric impregnated with phenolic resin. The other layer which is close to the journal is constructed in such a manner that a predominantly multifilament yarn of PTFE forms the bearing surface. The two layers of fabric are bonded together under

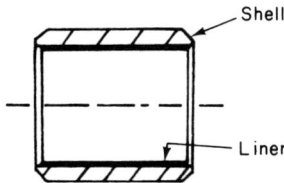

Fig. 14.6 A plain bearing lined with a multifilament woven fabric. (Courtesy, Ampep [4].)

heat and pressure and the advantages of these laminates are quoted [4] to be as follows:

(1) They can operate under dry rubbing situations but offer a high tolerance to ingested solids or liquid contaminants.
(2) They offer a high bearing stress to weight ratio.
(3) They exhibit high resistance to fatigue or impulse load.
(4) They can operate over a temperature range of $-50\,°C$ to $+150\,°C$ with a potential extension to $+250\,°C$.
(5) They provide with a low coefficient of friction and a high wear resistance.

The thickness of the laminates varies from 0.23 mm to 0.53 mm depending on application, and the bearing liner can be bonded to shells manufactured from alloys of steel or aluminium. The former is usually a stainless steel with chromium and nickel contents of the order of 15 and 2% respectively. The light alloy has a nominal composition of 2.5% Cu, 1.5% Mg, 1.0% Fe, 1.2% Ni, 0.1% Ti (by weight) with the remainder being aluminium. Although a plain journal bearing is discussed here for illustration, laminates are used for spherical and other forms of bearings. Apart from use in the aircraft industry there are a large number of other areas, such as fork lift trucks, commercial vehicles, railways and marine engines, where these laminates are used successfully.

14.4.1 Friction
The static coefficient of friction is slightly higher than the corresponding kinetic component up to a temperature of $+50\,°C$ with a negligible difference above that and stick–slip of the interface is not a serious problem. The filamentary structure of the woven fabric is undisturbed when a static load is applied and then removed. As sliding commences, both the journal and the bearing surface show films of PTFE so that rubbing then is possibly between viscoelastic films in shear. The effect of the PTFE film is to cause a decrease in the coefficient of friction with increasing contact pressure or temperature and a rise with increasing speed. Figure 14.7 shows how the coefficient of friction decreases with contact pressure, reaching a limiting value. The effect of temperature is also clear.

14.4.2 Wear
As the lining wears PTFE is ejected from the loaded to the unloaded region of the bearing and the debris appears as small dark brown flakes.

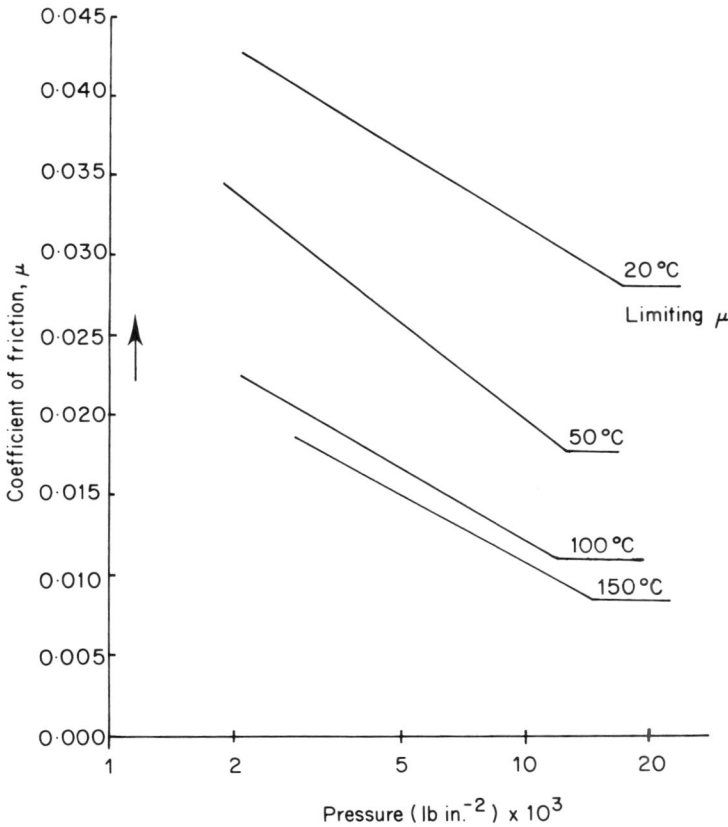

Fig. 14.7   Variation of the coefficient of friction, μ, with the logarithm of applied pressure. (Courtesy, Ampep [4].)

A typical change of radial wear with sliding distance is shown in Fig. 14.8 at a relatively high pressure of 18 kg mm^{-2} and shows four distinct stages:

(1) Initially, there is a rapid bedding-down process when the surface roughness of the fabric is levelled and the shaft is coated by a smooth film of PTFE.

(2) The rate of wear is reduced as the system approaches the steady state regime C.

(3) The steady state wear is slow and is the regime which shows a general loss of bearing material.

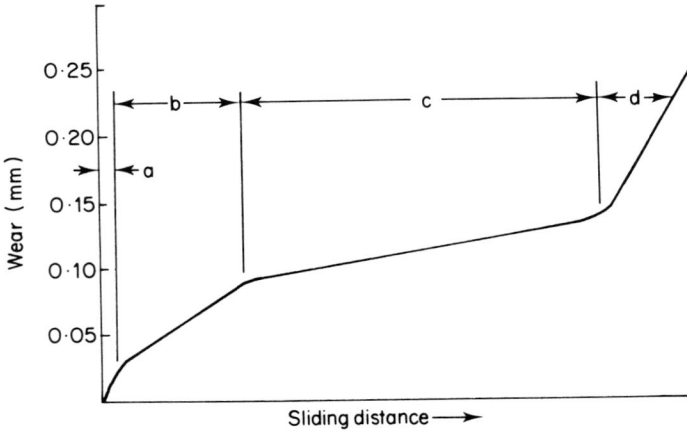

Fig. 14.8   Wear pattern of the lining shown in Fig. 14.6. (a) Bedding in process; typical final wear, 0·025 mm. (b) The running-in stage; typical final wear, 0·075–0·125 mm. (c) A steady-state, at the end of which the lining loss is 0·15–0·20 mm. (d) Accelerated wear, resulting in metal to metal contact. (Courtesy, Ampep [4].)

(4) This is the stage when the bearing material is removed at an accelerated rate and, if not located early enough, will cause destructive rubbing between the journal and the back-up metal of the bearing.

14.4.3 Counterface

The wear rate of PTFE is high at both low and high temperatures but, as discussed in Chapter 10, the physical nature of the counterface is of great importance. A roughly finished journal will plough the laminate and produce thick films of PTFE which will be ejected from the interface with relative ease. A high surface finish is therefore necessary, ideal values being of the order of $R_a = 0·05$ μm. Emphasis is also laid on employing a hard counterface. A corrosion resistant steel, through hardened to 56 Rc has the dual advantage of maintaining a high surface finish and resistance to abrasion due to the ingress of adventitious foreign particles. Abrasion of the journal must be avoided otherwise the counterface will become rough with time, resulting in enhanced wear of the laminate. Electroplated counterfaces provide smooth interfaces but a reduction of fatigue life of the shaft is experienced as with anodized aluminium alloys.

Titanium–PTFE couples produce heavy discoloration and roughening of the metal surface, possibly due to some physico-chemical effect. The damage is particularly severe at high bearing loads but can be obviated if the metal is coated with tungsten carbide and ground subsequently with diamond. Other coatings such as a hard ceramic, glass and enamel provide efficient counterfaces.

14.4.4 *Example*
A spherical bearing (Fig. 14.9) in the actuator of a fixed wing aircraft is required to operate with a bearing incorporating a PTFE based laminate with a thickness $t = 0.4$ mm. The backlash must not exceed 0.1 mm to preclude flutter. Duty cycles of in-flight loads and movements for 900 h endurance are as in Table 14.1, given the distance per

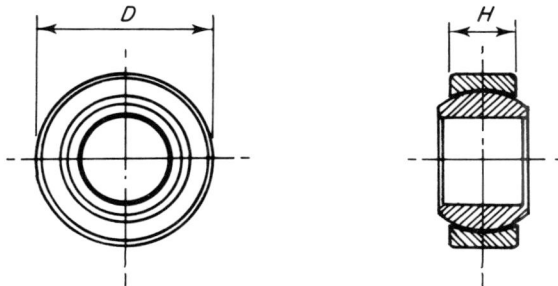

Fig. 14.9   A lined spherical bearing. (Courtesy, Ampep [4].)

degree movement $= 0.3$ mm. The bearing has an outer race width $H = 19.575$ mm and a ball diameter $D = 34.375$ mm. Calculate the total wear.

Table 14.1

Case no.	Total cycles	Frequency (Hz)	Load (kg)	Movements
1	4320	2·0	10591 ± 227	±1·43°
2	4320	0·2	10591 ± 1136	±7·15°
3	6480	2·0	6954 ± 227	±1·43°
4	4320	0·2	6954 ± 2363	±14·3°
5	2160	2·0	−2091 ± 227	±1·43°
6	4320	0·2	−2091 ± 1136	±7·15°
7	1800	0·1	4318 ± 7909	±46·5°

Table 14.2

Case no.	Max load (kg)	$p\,(\mathrm{kg\,mm^{-2}})$
1	10818	17·5
2	11727	18·9
3	7181	11·6
4	9317	15·0
5	−2318	3·7
6	−3227	5·2
7	$\left\{\begin{array}{l}+12227 \\ -3591\end{array}\right.$	19·8 5·8

Solution

We shall follow the method used by a bearing manufacturer [4], but an alternative method would be to make use of Equation 14.3.

The projected area of the bearing surface is

$$A = (H - 4t)D = 617 \cdot 89 \text{ mm}^2$$

The sliding distance for 1° ball rotation is

$$\pi D/360 = 0 \cdot 3 \text{ mm}$$

Step 1. Case 7 needs special consideration because of the magnitude of the reversing load at a ball movement of ±46·5°. For the items 1 to 6, respective maximum loads can be converted to bearing pressure p by dividing the load by A; as given in Table 14.2.

Step 2. Calculate the total sliding distance s in each case, using,

$$s = (\text{degrees swept per cycle}) \times (\text{distance per degree}) \times \text{total cycles}$$

Note that sweep per cycle = 4 × movement in degrees. (See Table 14.3.)

Step 3. The procedure now is to convert all stress and distance values to equivalent sliding distances at a stress of 18·2 kg mm^{-2} by using the following formula:

$$\frac{s_1}{s} = \left(\frac{p}{p_1}\right)^2$$

Table 14.3

Case no.	Cycles	Sweep/cycle (°)	Distance per degree (mm)	Total distance, s in 900 h (mm)
1	4320	5·72	0·3	7413·1
2	4320	28·6	0·3	37065·6
3	6480	5·72	0·3	11119·7
4	4320	57·2	0·3	74131·2
5	2160	5·72	0·3	3706·5
6	4320	28·6	0·3	37065·6
7	1800	186	0·3	100440·0

where s is the sliding distance at a pressure p and s_1 is the sliding distance at a pressure $p_1 = 18\cdot2$ kg mm^{-2} or $s_1 = s(p/p_1)^2$. (See Table 14.4.)

Step 4. Add all distances of the same sign at $18\cdot2$ kg mm^{-2}

+Cases 1, 2, 3, 4, and 7   − Cases 5, 6 and 7
        220 m   13 m

consulting Fig. 14.10, which has been determined for this product at a pressure of $18\cdot2$ kg mm^{-2} and which plots wear in mm against log distance, wear at 220 m = 0·082 mm and wear at 13 m = 0·042 mm.

Therefore the total wear predicted for 900 h endurance at room temperature is $0\cdot082 + 0\cdot042 = 0\cdot122$ mm.

Table 14.4

Case no.	s_1 (mm)	s_1 (m)
1	$7413\cdot1\,(17\cdot5/18\cdot2)^2 = 6853$	= 6.85
2	$37065\cdot6\,(18\cdot9/18\cdot2)^2 = 39971$	= 39·97
3	$11119\cdot7\,(11\cdot6/18\cdot2)^2 = 4517$	= 4·51
4	$74131\cdot2\,(15\cdot0/18\cdot2)^2 = 50354$	= 50·35
5	$3706\cdot5\,(3\cdot7/18\cdot2)^2 = 153$	= 0·15
6	$37065\cdot6\,(5\cdot2/18\cdot2)^2 = 3025$	= 3·02
7	$\begin{cases} 100440\cdot0\,(19\cdot8/18\cdot2)^2 = 118876 \\ 100440\cdot0\,(5\cdot8/18\cdot2)^2 = 10200 \end{cases}$	= 118·87 = 10·2

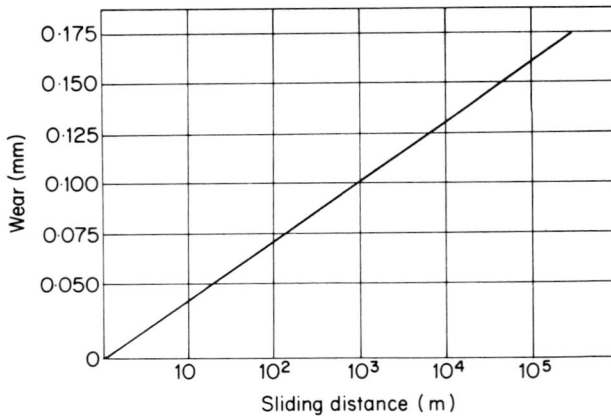

Fig. 14.10   Wear at various log(distance) at a contact pressure of 18·2 kg mm^{-2}. (Courtesy, Ampep [4].)

A temperature factor has also to be used if the operating environment is at or above the ambient temperature. We could also use a relationship of the form given by Equation 14.6 by expressing t in terms of the sliding distance provided the surface speeds could be calculated and the appropriate correction factors and the value of K_0 were known.

14.5 Materials
There is now a plethora of plastics which have been formulated for dry bearing applications but they are broadly classified as thermoplastics and thermosets. Typical plastics which are thermoplastic are polyethylene, acetal, polyamides, PTFE and polyimide. Typical fillers are glass, carbon and textile fibres, the main object being an improvement in mechanical properties. Some thermosets are epoxies, phenolics and polyesters. The usual additives are MoS_2 and graphite to reduce friction and bronze is added to improve thermal conductivity. Selected examples [2] of bearing materials are shown in Table 14.5. Self-lubricating materials are also used in rolling element bearings [5]. The bearing cage is made from a self-lubricating composite which transfers some solid lubricant to the races via the balls. Provided the operating conditions are not hostile this transferred film will prevent metal to metal contact between the balls and the races. Above a threshold load, wear of the steel components occurs. The bearings fail by wear and this distinguishes them from the fluid-lubricated bearings whose failure mode

Table 14.5 Specific wear rates of dry rubbing bearing materials [2]

Material	$*K_0$ (m^2 kg^{-1})	Useful ambient temperature range (°C)
Polyacetal	13·8	−50−+130
Polyamide	43·5	−60−+130
PTFE	593	−210−+250
Polyacetal + PTFE + inorganic fillers	0·79	−45−+130
Polyamide + 15% graphite + 10% PTFE	4·9	−210−+300
Polyamide + 30% asbestos + 20% graphite	8·9	−210−+300
Woven PTFE + glass fibre with a thermoset backing	0·4	−180−+280
Ceramic type filled PTFE	0·4	−210−+270
PTFE + lead impregnated porous bronze on steel backing	0·3	−210−+300

* Specific wear rate under light duty service conditions.

is predominantly that of spalling. The cage materials are composites of PTFE, MoS$_2$ and glass fibres or a composite of polyamide and MoS$_2$.

References
1. Lancaster, J. K. (1973). *Tribology* **6**, 219.
2. ESDU (1976). "A guide on the design and selection of dry rubbing bearings", Engineering Sciences Data Item Number 76029.
3. Polypenco Ltd, U.K. (1978). Private communication.
4. Aerospace Bearing Design Manual (1978). Serial No. 102, Ampep Ltd, U.K.
5. "Self-Lubricating Bearings" (1976). NCT, Risley, U.K.

15   Biotribology

We propose to use the term biotribology to embrace the phenomenon of friction, wear and lubrication of interfaces inside the bodies of animate beings. A typical example is the human hip joint where two articulating surfaces experience motion under external forces during locomotion. Nature is remarkable in its design of such joints in that these articulating surfaces are well cushioned from impact forces and are effectively lubricated, as wear of natural bony surfaces within the body is probably minimal. There is a tendency for wear of bearing surfaces of bones with age when lubrication ceases to be effective. Sliding of unlubricated bony surfaces causes pain, and damage to bone also occurs by disease which results in dysfunction in locomotion. At the moment such inadequacies can only be corrected by surgery to remove the damaged areas and replace them by prostheses, that is by artificial devices. These are made from metals and non-metals and are expected to be tribologically efficient and must be biocompatible. Biotribology is a developing area and there is scope for further work with particular emphasis on geometric design and material selection. We shall discuss all relevant factors in this chapter, but readers unfamiliar with the necessary medical terms which will appear will find the first four references at the end of this chapter useful.

15.1   Biomechanics

Biotribology is only a specialized section in the general field of biomechanics. Anatomy and physiology are two disciplines which are interwoven in both the descriptive and the analytical perimeters of biomechanics. In a limited appraisal, human anatomy deals with segments of a body which can be identified by merely dissecting a cadaver. A study of physiology involves an understanding of the functional patterns of living structure. The living structure may simply be a virus or a macroscopic entity such as a mammalian form, but the basic elements of cell behaviour and systems design are similar in most species. Biomechanics, formerly termed kinesiology, is involved with analyses of motion of living creatures. From a tribologist's point of view, the prime requirement is to obtain quantitative information on the anatomy of motion so that the nature and the movement of the articulating surfaces of joints and the loads sustained by the interfaces during sliding can be evaluated.

Biomechanics attempts to calculate the relevant parameters at the joints in contrast to physiology, which is largely phenomenological, and there is a great need to appreciate the interaction between physiology and mechanics. At least four areas of physiological processes are amenable to analysis by known techniques of applied mechanics, viz.,

(1) the cardiovascular system;
(2) the pulmonary system;
(3) the muscle; and
(4) structures such as bone and cartilage.

Tribologically, the current field of active interest is the biomechanics of injury, or the natural loss of bone by disease and wear. The orthopaedic surgeon is concerned with the examination of skeletal trauma as a result of accidents. Any exploration of the area will be incomplete without quantitative information on the strengths of bone and ligament together with the forces exerted on joints due to the mobile activity of the animals involved. The biomechanics of repair demands that it should be possible to correlate the healing process of a medically treated fracture with the forces acting on the repair. A repair or correction to a joint frequently involves prosthetic treatment such as adding a metallic element to contour a damaged femoral head to interact with a plastic socket inserted into the acetabular cavity. Apart from mechanical strength, the implants must be biocompatible within the environment of the body and must not cause physiological distress in patients. The need is now urgent for investigatory teams to be inter-disciplinary, comprising scientists, engineers and orthopaedic surgeons so that experiments may provide fruitful conclusions. There does not appear to be any other way of alleviating suffering from the ambulatory dysfunction than to remove part of the joint and replace it by artificial materials such as metals, plastics and ceramics. These must be bio-compatible and the broad interrelated areas of study for a tribologist are:

(1) The nature of deformation of biological components such as bones.

(2) The kinematics of human body segments, for example, arms and legs.

(3) Calculation of static and dynamic forces and the corresponding moment relationships in the musculoskeletal system.

(4) The viscoelastic stress–strain relationships in the locomotor apparatus.

Of immediate interest from the viewpoint of quantitative evaluation of wear rates of the articulating surfaces of human joints is the nature and magnitude of forces at the interfaces and this is a pertinent sphere for further development. Primarily, the forces on the components of the bony framework depend on the gravitational pull and the action of the muscles.

15.1.1 *Action of muscles*

The motor system of the human body depends on three interdependent items. The nervous part directs the action of one or more muscles which supply power to move the bony levers in the skeletal enclosure. The muscles are attached across joints at a few points on these levers which move as a result of appropriate muscular contraction. Muscles differ in shape and size according to their function and those designated for speed are long and slender.

The four characteristics of muscles are irritability, contractility, distensibility and elasticity. Irritability describes the response to stimulation. Contractility is the ability to produce tension between the ends of the muscle, which increases monotonically from zero to a maximum value, and unlike its opposite, relaxation, is not instantaneous. An external stretching force such as that exerted by gravity causes an increase in length of the muscle, and this is distensibility. Provided a muscle is not overdistended, the inherent elasticity of it will bring it back to its normal configuration. Muscles rarely act singly and groups co-ordinate to obey the command of the nervous system. Thus a muscle may act as a prime mover or it may be synergic to co-operate in the main movement. The prime mover may become an antagonist if the movement is reversed.

15.1.2 Degrees of freedom

Muscles apply forces to a joint which may have many degrees of freedom. Unlike a plain bearing, a joint may thus slide under load on many planes not necessarily simultaneously. The uniaxial joint has only one degree of freedom so that its movement is limited to a single plane. A hinge joint such as those formed by the elbow, knee or the ankle and the pivot construction as in the radioulner joints in the forearm provide uniaxial movements. The wrist is an example of a biaxial joint with two degrees of freedom so that the mobility of the system is limited to two planes only. The third type of joint axes is termed multiaxial. Movement is in three cardinal planes but the name is multiaxial because the surfaces can articulate in many oblique planes additionally. Examples are the ball and socket joints at the hips and the shoulders. The articular surfaces of the joints between the ribs and the vertebrae can undergo sliding over numerous planes.

Both the type and the range of articulation of a skeletal interface are circumscribed by the number of degrees of freedom and the nature of the joint. Relative motion is also restricted by muscles, ligaments and adjoining tissues so that the intervertebral joints, although triaxial, have a limited range of movement. On the other hand, the forearm can move through 150° from the position of full flexion.

15.1.3 Stressing of bone

Bones are stressed because muscles transmit forces across joints and because the body is under gravitational pull. The musculoskeletal system in a static position must have all the joint forces in a state of equilibrium so that the resultant force will go through the centre of gravity. For a spherical joint, this is identical with the geometric centre. In a uniaxial joint, the resultant intersects the joint axis perpendicularly. For a biaxial joint, the resultant force has its axis perpendicular to both axes. It is thought that, for joints whose shapes are irregular, an instantaneous axis of movement can be constructed. The projections of these axes into the drawing planes are called rotational centres.

Consider Fig. 15.1 where a bone is represented [7] by a vertical rod held at two joints A and B, the articulations being at the two sockets in the two horizontal bars. If an eccentric load *W* is placed, as shown, it can be balanced by a counterweight which, in the human body, is provided by the algebraic sum of the forces produced by the interacting

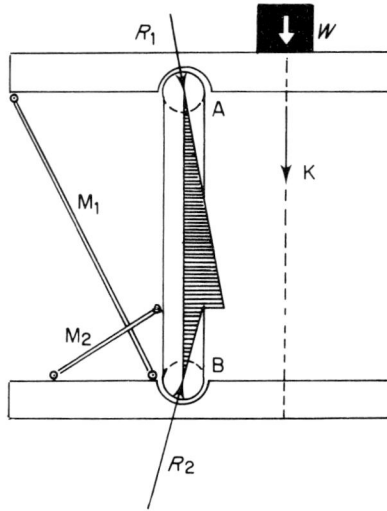

Fig. 15.1   A simplified model of a long bone under stress. (Kummer, B. K. F. (1972). In "Biomechanics", Fung, Y. C., Perrone, N. and Ankler, M. (Eds), p. 253. Reprinted by permission of Prentice-Hall, Inc., Englewood Cliffs, New Jersey.)

muscles. The muscles are represented here by the springs M_1 and M_2 and it is highly unlikely that the direction K of the load and that of the muscle M_1 are parallel to the bone axis. The resultant R_1 for the upper joint will not therefore go through the rotational centre of the lower articulation. Hence the movement of the latter will have to be balanced by the second muscle M_2. The forces R_1 and M_2 will interact, giving rise to a resultant R_2 acting at B. The axis of the bone will be subjected to a bending stress when R_1 and R_2 are divergent, as is usually the case. It is not possible to eliminate bending stresses completely but the magnitude can be reduced if the design could approximate that shown in Fig. 15.2. The bone is bent so that it has two axes which coincide with the forces R_1 and R_2.

15.2   Human motion

The mobile state of human beings involves linear, rotary and curvilinear motion. During the swing phase of walking, the leg moves forward about an axis in the femoral head. It is immobile as it rests

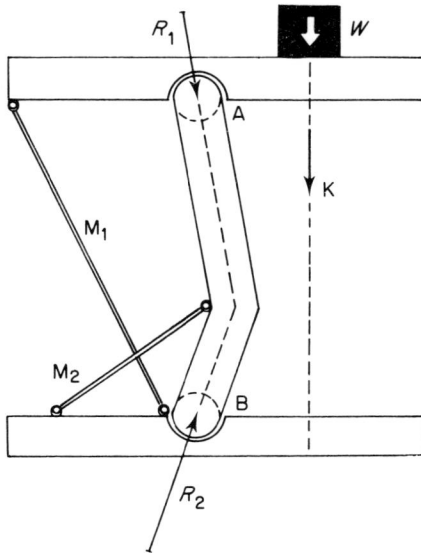

Fig. 15.2   Changing the design of the model of a long bone to minimize bending stresses. (Paul, J. P., Hughes, X. X. and Kenedi, R. M. (1972). In "Biomechanics", Fung, Y. C., Perrone, N. and Ankler, M. (Eds), p. 535. Reprinted by permission of Prentice-Hall, Inc., Englewood Cliffs, New Jersey.)

during the stance phase and the foot bears the body weight. As the thigh and the leg carry the rest of the body over, movement occurs at a number of joints involving the toes and the ankle. While the lower limb undergoes rotary motion, the trunk moves curvilinearly and this latter mode becomes accentuated as the walking phase transforms to running. Locomotion thus causes sliding of articulating surfaces under load giving rise to a potential situation of wear.

For an analysis of movement, it is necessary at the start to know the posture for that moment. That is, the subject may stand on his toes or be in a state of equilibrium by being simply balanced on his feet. The next step is to identify the joint of support and calculate the load acting on it due to gravity and muscle action. The calculations are tedious when carried out manually and a complete movement such as in walking a distance and standing again involves obtaining as many basic poses as possible. Each pose will produce a different load on a joint depending on the body configuration. A complete movement is the sum of many sequential poses which can be isolated by examining stills of motion pictures [6].

An engineering plain bearing supports a load which is reasonably constant over a known area and the motion is consistently uni-directional. A human joint gives rise to a complex situation as both the magnitude of the load and the style and the degree of movement may vary from one pose to another during a single cycle of movement. A body has also got to be considered as a system of links forming a kinematic chain so that movement of a proximal segment in it will give rise to movements in adjacent links. Thus consider a person starting the forward swing while kicking a ball. We can immediately identify three different actions:

(1) flexion of the hip;
(2) extension of the knee;
(3) planter flexion of the ankle.

It will be the skeletal levers which will transmit movements to the joint interface. To isolate the levers and to calculate forces, the body segments involved in the movement must be identified. Thus flexion of the hip joint moves the femur on the pelvis. A common mistake is to think that the only segment moved by this action is the thigh. In fact, shortening of the hip flexors involves an open kinematic chain composed of three links viz., thigh, leg and foot. This is so because the contracting muscles due to flexion of the hip must move all the segments attached to the distal end of the femur as well.

15.3 Forces on joints

Forces acting on a human body may be internal or external. Internal forces in the musculoskeletal system are provided by tensions of the muscles while the external force is largely the gravitational pull. The force of gravity always acts downwards and other external forces arise due to kicking, throwing, etc.

It is clear that human joints undergo oscillatory and tangential sliding in the presence of a lubricant which is kept at a consistent temperature and the force on a joint will achieve a maximum value during a walking cycle. Tests [9] on subjects walking on level surfaces show a variation of joint forces with the leg position during a walking cycle (Fig. 15.3). Expressing the ordinates by J/W where J is the load on the joint and W is the body weight, the dimensionless ratio rises to a peak as the heel strikes both at the beginning and towards the end of the stance phase.

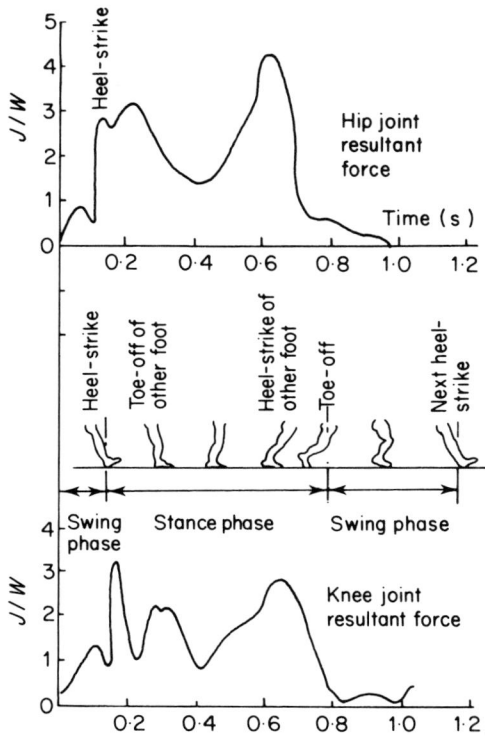

Fig. 15.3   Typical patterns for hip and knee joints of variation of the ratio
(joint force/body weight) during walking. J = load on the joint;
W = body weight. (From Paul, J. P. (1966). *Proc. Inst. Mech. Eng.*
181(3J), 8.)

The pattern is similar for both the hip and the knee joint. With one leg
stance, the load on the femoral head is about 2·5 times the body weight
of a person [10]. During gait at walking speed, forces are of the same
order but loads equalling four times the body weight have been
recorded during running. Regarding speed, the peak values of hip joint
velocities in young adults vary between 2·5 and 7·5 cm s^{-1}, the motion
being a combination of rolling and sliding. As an average, a human
being may take 6000 steps a day so that there will be about one million
walking cycles in a year. It is possible to have an estimate of the load,
speed and the sliding distance of articulating surfaces.

Forces at the hip and the knee joints are shown [12] to vary linearly
with the ratio WL/H (Fig. 15.4) where W, L and H are the weight,

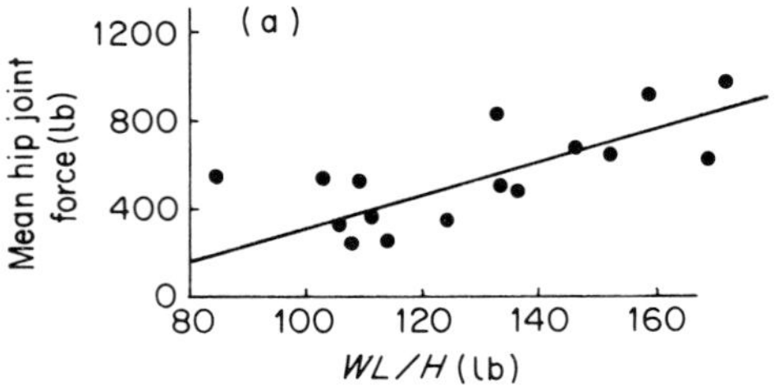

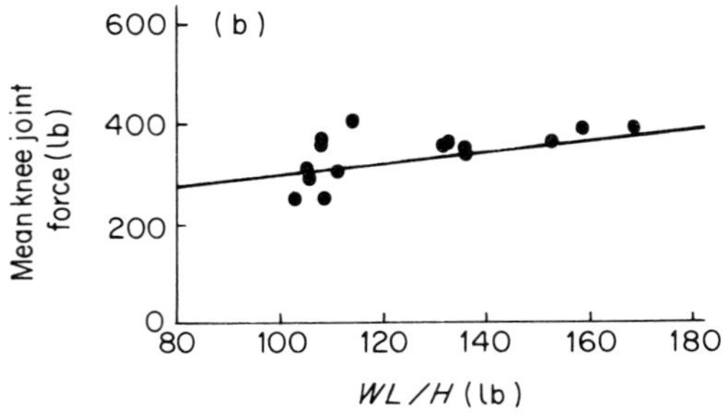

Fig. 15.4   Variation of mean joint forces with body weight, *W*, stride length, *L* and height, *H*. (a) Hip joint; (b) knee joint. (Paul, J. P. (1970). *Proc. Roy. Soc. Med. A* **63**, 200.)

stride length and height of the person respectively. That is apart from the weight of the body, a long stride increases the joint forces. For a particular situation the knee joint forces are generally lower (Fig. 15.4(b)) than the hip joint forces. The inverse relationship between the joint forces and the height of a subject needs explanation.

15.4   Prosthesis

A prosthesis is becoming increasingly common when the natural joint malfunctions as a result of disease or injury. It would appear that the unfortunate areas needing replacements most commonly are the hips and the knees which become painful as a result of arthritis. One form of the ailment is *rheumatoid arthritis* which is a systematic disease when the synovial membrane is first attacked. The result is a simultaneous change in the cartilage and the underlying bone, the latter losing its mineral contents. The most common form is *degenerative arthritis* which does not produce inflammation and pain and is inevitable with advancing years. The binding material that holds the bundles of cartilage fibres is destroyed and the fibres stand erect like the pile of a carpet. As these wear due to movement, fissures and pits appear in the articulating joints. As the bony surfaces interact without a protective layer of cartilage, the subject experiences stiffness and pain during gait. Appearance of knobs called Heberden's nodes at the end joints of the first and second fingers is the principal sign of degenerative arthritis. Some other joints involved are knees, sacroiliac, hips and shoulders. Arthritic disability can also occur due to infection by micro-organisms and to trauma.

A prosthetic device is a replacement for a missing part of the body. An endo-prosthesis replaces parts within a body and is referred to the removal of a damaged part and its replacement by an implant. The joints which are of particular interest to tribologists are only partially removed and then reshaped with the object of restoring the prior geometric form. As shown schematically in Fig. 15.5(a), a damaged part of a bone is identified and removed, say, by sawing along the line AB. A carefully prepared artificial component is then interdigitated by cementing along the plane AB (Fig. 15.5(b)).

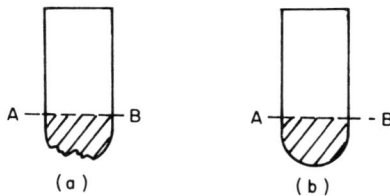

(a)   (b)

Fig. 15.5   (a) A damaged bone is sawn off along AB. (b) A prosthesis cemented along AB (schematic).

In the total hip joint replacement an artificial implant in the form of a sphere forms the femoral head and this is coaxed into a cup which is fitted into the acetabular cavity. The femoral component is a curved intermedullary stem fitted within the bone and is secured in place by cement [14] after the joint has been excised. Consider a ball and socket device [15] as in Fig. 15.6 representing a total hip replacement, although not a very representative configuration. A ball of radius r is the femoral head which is free to swivel in a socket. The latter is held with a cement layer, forming a radius R in the incised bony matrix. The ball itself will be able to rotate freely around a vertical axis or slide due to flexion and

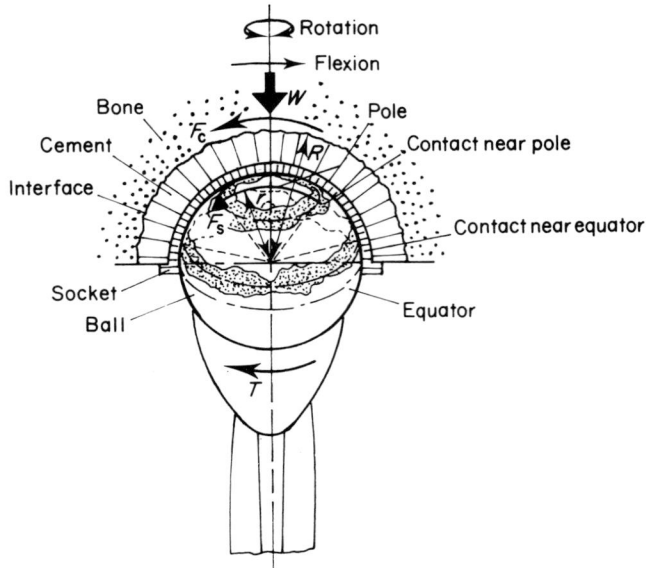

Fig. 15.6   A ball and socket device representing an idealized version of a total hip replacement. (Walker, P. S. and Bullough, P. G. (1973). Symposium on Interposition and Implant Arthroplasty, Orthopaedic Clinics of North America, **4**(2), 275.)

extension. This latter movement produces a maximum amount of torque T so that if the frictional resistance of the ball–socket interface is F_s,

$$T = F_s r \qquad (15.1)$$

The interface described by the excised bone which is also the interface with the cement has a radius R. If F_c is the frictional resistance at the cement–bone interface,

$$T = F_c R \qquad (15.2)$$

comparing Equations 15.1 and 15.2,

$$F_c = r(F_s/R) \qquad (15.3)$$

For a given value of F_s and R, the frictional resistance at the cement–bone interface is seen to be directly proportional to r, the radius of the ball. A high frictional force will be experienced when the ball fits tightly into the socket so that contact is near the equator. The condition for a low frictional torque is that contact should be aimed to occur within a region of a solid angle of 45° from the pole (Fig. 15.6).

At present there are two types of friction couples which have been tried for some time in total hip joint arthroplasty viz., metal–plastic and metal–metal combinations. In the Charnley prosthesis, the femoral part is manufactured from stainless steel and the socket for the acetabulum from ultra-high molecular weight polyethylene (UHMWP). The respective components in the Charnley–Mueller design are Co–Cr–Mo alloy and UHMWP. For metal–metal combinations, both parts are produced from a Co–Cr alloy.

There are two main designs in the replacement parts of knee joints, viz., the one embodying a hinged construction and the other an unconstrained configuration [16]. The latter, termed condylar replacements, uses a convex metal component in stainless steel or in a Co–Cr alloy for the lower femur while the upper tibia incorporates a concave counterface made from UHMWP. Polymethyl methacrylate cement polymerized *in situ* interdigitates each component to the bone.

A total replacement arthroplasty of the knee has been designed [8] in which a stellite shell fits on the femoral condyles and a high density polyethylene (HDPE) is cemented to the tibial plateaux. The plastic component is produced by heat forming in a die. Whereas by and large the friction couples comprise metal–plastic combinations, the hip prostheses usually have conforming bearing surfaces. On the other hand, knee joints have been designed with both ball-on-flat and totally conforming surfaces.

Like metal combinations are notorious for galling and seizure under sliding. Dissimilar metal couples may exhibit low friction and wear but there is a danger of corrosion due to the electrochemical action of the body fluid. Combinations employing similar alloys have been used but there is little or no published information about their suitability. One is tempted to suggest that these like metal combinations survived because some internal body fluid provided lubrication.

In the natural state, the load bearing capacity of the lower limbs of human beings is astonishingly large and galling during movement does not occur probably because of an efficient lubricant at the interface. The articular cartilage is the bearing material in human joints and is a permeable elastic matter, being known to have the capacity of holding the lubricating fluid in large quantities. The present state of knowledge regarding joint lubrication is inadequate but it has been suggested [13] that, as with intermittently run engineering components, joints operate under both fluid film and boundary lubrication depending on the load, speed and the viscosity of the synovial fluid. It appears that the mode of the fluid film lubrication is elastohydrodynamic in nature. Unfortunately, a full fluid film is unlikely [13] to develop with artificial implants and the lubrication mode is boundary and mixed-film in nature. For a metal–plastic combination, a film of the polymer transfers on to the metal with a resultant low coefficient of friction. Fluids can inhibit transfer films and if secretion of the synovial fluid is not generous, as in old age, a plastic–metal couple should be expected to function more effectively than when the subjects are young.

Recently the possibility of using a high alumina high density ceramic (HAHDC) in combination with plastics is being explored seriously. Ceramics, however, are brittle. Although complex, biological materials broadly comprise two or three phases so that one element reinforces the other. It is not surprising that an analogous material such as a reinforced plastic is being considered for endo-prosthetic devices.

Although an internal transplant such as a hip joint must have favourable tribological properties, four other important areas must be taken into account while choosing a prosthetic material. It is imperative that the biochemistry is favourable so that the material does not lose its mechanical properties or fail completely due to corrosion or general degradation. An adverse physiological reaction such as edema is not tolerable and the factor of histology demands that the shape and the surface condition of the implant is favourable to the natural growth of tissue adjacent to it. Pharmacologically, the waste product of the prosthetic material as a result of wear must not be toxic.

15.5 Friction and wear

Human locomotion would be seriously impaired if the skeletal joints did not perform adequately. The hip joint undergoes flexion–extension and abduction–adduction together with internal and external rotation during walking and running. Thus, from the locomotion point of view, an understanding of the synovial joints includes a study of the articular cartilage and the synovial fluid, the bearing component and the inter-facial lubricant respectively. Movements cause the surfaces to slide or roll under load, giving rise to friction and wear. Both these parameters are influenced by surface topography and this aspect has been examined

[17], providing much needed information in the area of tribological phenomena at the articulating surfaces. Apart from gross wear, clinical failure of artificial hip joints occurs due to a physiological rejection of wear debris escaping into the surrounding tissue and due to loosening of the prosthesis from the bone because of high frictional forces [18]. Note that Equation 15.3 shows that the larger the radius of the femoral head, the greater is the frictional force at the cement–bone interface. The friction can be reduced by designing in such a manner that the ball makes an essentially polar contact with the acetabular cup. The cement of course must be such that the socket (Fig. 15.6) is held firmly in the bony matrix. It is also necessary to consider the femoral head, one end of which is fixed by cement as well.

There are three ways of evaluating wear of prostheses:

(1) *in vivo* observation;
(2) use of simulators such as a full-size model undergoing the articulating actions of joints;
(3) basic study of materials, e.g., a pin rubbing on a disc.

Pins of diameter 9·5 mm machined from high density polyethylene were run on stainless steel or chromium–cobalt alloy discs at an ambient temperature of 42 °C and keeping the R_a values consistent at 1 and 0·05 μm respectively [8]. The coefficient of friction was of the order of 0·1 when a bovine synovial fluid was used, being about 50 per cent higher when the couples were run dry. Scanning electron microscopy showed that the metal discs became rough if particles from the pin transferred onto them. The volume, V, lost for a sliding distance S and a normal load W is expressed as

$$V = KWS \qquad (15.4)$$

where K is the specific wear rate in $m^2\,kg^{-1}$. Note that the equation is the same as that used for dry bearings in Chapter 14. The wear coefficient for HDPE for these experiments was $0·06 \times 10^{-15}\,m^2\,kg^{-1}$ which can be compared with various other plastics in Table 14.5.

Modification of Equation 15.4 as we have seen produces a relationship for the wear coefficient in terms of height loss h due to wear and the contact stress σ as,

$$h = K\sigma S \qquad (15.5)$$

Values of h have been obtained [19] for a number of non-metals in an apparatus producing oscillatory motion at a speed of $75\,mm\,s^{-1}$ and a contact stress of about $0.4 \, \text{kg mm}^{-2}$, using blood plasma as the lubricant. The following materials were compared using a stainless steel counterface:

(1) Polyethylenes of three different densities with respective molecular weights of 10^6, 1.25×10^6 and 3×10^6 were examined; some measurements were also taken on an intermediate density polymer after irradiating pins with Co_{60} γ-rays in an argon atmosphere to doses of 20, 200 and 1000 Mrad.

(2) Pins were also made from resin impregnated graphite and carbon mixed with 8 and 15% silicon respectively to form silicon carbide; silicon carbide is itself wear resistant when slid against metal but the latter may undergo severe two-body abrasion.

It was observed that the wear rates of the polymers decreased as the pins were made from materials with increasing molecular weights. The friction remained sensibly constant between materials, which is a pity because viscoelastic materials tend to show a correlation between friction and wear, unlike metals. The carbon samples showed a decrease in wear rates with increasing amount of silicon.

Experiments [5] under unidirectional sliding also show a lack of correlation between friction and wear for a range of materials using recirculating water at $37 \, °\text{C}$ as the lubricant. The conclusion is that ceramics show a high rate of wear. This conclusion should be treated with some caution because a ceramic with a high alumina content is now being explored which shows encouraging trends. It seems unlikely that a ceramic component will be successful against a similar counterface for endo-prosthetic devices. The most probable combination is ceramic–plastic, for example, a high alumina ceramic femoral head to fit a cup made from UHMWP. About 25% graphite powder, provided it is biocompatible, can be incorporated in the latter with a great improvement in friction and wear. A self-lubricating total arthroplasty will improve the quality of life for patients of advancing years whose joints suffer from fluid starvation.

A low coefficient of friction is noted in blood plasma for metal on metal, Co–Cr, couples using a hip simulator [20]. The intriguing suggestion is that there is a strong possibility of the formation of metal–protein complexes on the articulating surfaces, providing an effective film of lubricant with low frictional resistance. Equipment has been developed to simulate flexion–extension of a leg, ensuring that the femoral head interrogated the acetabular cup at the compound angles

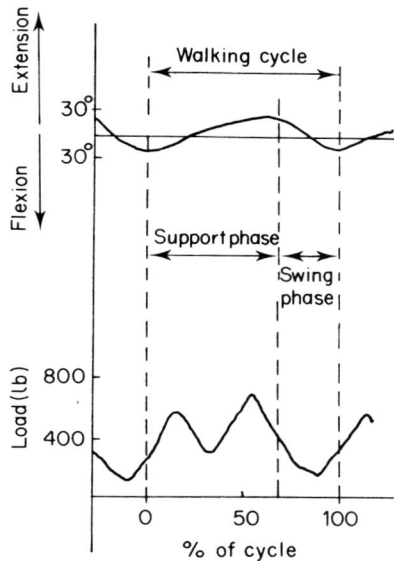

Fig. 15.7 Load and motion cycles for a hip simulator. (Weightman, B., Simons, S., Paul, I., Rose, R. and Radin, E. (1972). *J. Lubrication Technol.* **94**, 131.)

found in a human body during its motor activity [18]. The loads and the cycles of movement are shown in Fig. 15.7, the former being the mean values obtained from measurements during walking of a human subject. The load varies during the support and the swing phase as expected and the flexion–extension is near sinusoidal with the walking cycle. The results are useful as they confirm that the coefficient of friction between metal–metal couples is not very high at 0·12 and is fairly consistent for the lubricants studied, viz., bovine and human serums, albumin and bovine synovial fluid. Lubricated studies with a plastic–metal couple show that the coefficient of friction is halved.

Laboratory studies simulating known characteristics of *in vivo* situations of human joints would seem to be the only rapid method of assessing new designs or materials. Observations on implants in human beings have also been made providing much welcome information. There is a report [14] on the long term progress of total hip replacements on a large number of subjects. The arthroplasty employed a stainless steel head, 22 mm in diameter, which mated with a socket of high density polyethylene, HDPE, both the components being fixed to the bone with self-curing acrylic cement. A marker was placed in the plastic socket and using this as the datum line, wear was followed from radiographs taken annually. There was no logical pattern regarding the amount of wear and the author [14] suggests that this was due to a variation in the quality of the plastic because little wear has been observed after a certain time on vigorous patients and vice versa. A number of prostheses examined at post-mortem showed that the plastic, originally white, assumed a cream colour although there was no other change in its physical condition. Measurements on seven sockets gave an average wear of 0·13 mm per year.

Current theories of friction and wear place emphasis on surface topographic parameters such as asperity heights, the linear density of the peaks and their inclinations. It is encouraging to see, therefore, an interest in the nature of surfaces of the pertinent components in the human body. Natural joints of human beings at post-mortem have been examined *in vitro*, the parts including femoral-acetabular couples between 30 and 80 years of longevity and femoral condyles from cadavers of persons up to ten years of age. The samples were studied by optical and scanning electron microscopy and with the aid of a Talysurf. The areas of both wet and dehydrated adult articular surfaces showed a regular distribution of depressions, 20–40 μm in diameter usually in pairs, the frequency of occurrence being 190–900 mm^{-2}. These characteristics of the articular cartilage were similar in shape and frequency to the underlying cell lacunae, suggesting the structure of the cartilage to be formed from these underlying cells. The peak to valley heights of the surfaces fell in the range 2·2–18·2 μm, and the depths of the depressions observed microscopically were about 5 μm. In a similar manner to that with soft metal components, the stylus cut a groove into the cartilage and the author [17] suggests that the best method for obtaining surface topographic parameters is by using scanning electron microscopy where the interface under observation does not undergo mechanical damage. Hopefully, published work will continue to appear on the surface topography of bone joints which may be used with advantage to explain the nature of friction and wear of *in vivo* articulation.

15.6 Design of prostheses

A total hip joint prosthesis is a permanent implant with a hopeful life of some 30–40 years. The replacement may fail by infection or by permanent deformation. Fracture may occur due to overloading or as a result of the cumulative effect of cyclic stressing. Failure may also occur by corrosion, by loosening of the cement and by excessive wear of the interface.

It should not be difficult for engineers to formulate materials which will resist corrosion in the body environment. However, one problem with structural implants is the interface between the prosthesis and the original biological material. Representing a femoral head prosthesis with a straight stem (Fig. 15.8), a normal load will compress the

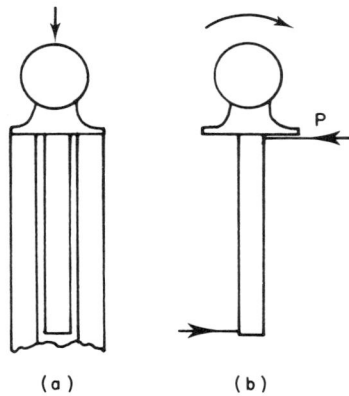

(a)   (b)

Fig. 15.8   An idealized assembly of total hip prosthesis shown in (a). (b) Shows that contact stresses will open up the interface at P. (Fung, Y. C., Perrone, N. and Ankler, M. (Eds) (1972). "Biomechanics, Its Foundations and Objectives", Prentice-Hall, Englewood Cliffs.)

interface between the prosthesis and the original bone. A bending moment will be transmitted to the bone end if loading is oblique and relative movement at the interface may occur. If the flexural rigidity given by the product of the Young's modulus and the second moment of area of the prosthesis does not match that of the bone, contact stresses will tend to open up [7] the interface at a position P in Fig. 15.7(b). Cementing the interface should prevent relative movement but the stiffness of the implant should match as far as possible that of the host otherwise high stress gradients will be set up across the interface. Routine activities of human beings such as running, walking, carrying loads, etc., usually involve high joint loads. The stresses are impulsive in nature and are intermittent. If a compliant cement is introduced at the prosthesis–bone interface, the peak dynamic forces may be significantly attenuated. The polymethylmethacrylate cement is potentially labile and it also acts as a matrix for distribution of stresses between the stiff implant and the relatively elastic bone. Metal–bone systems will cause steeper stress gradients than plastic–bone interfaces.

Actual wear of the replacement components results in a malfunction of the arthroplasty. One reason is simply the mechanical disadvantage as a result of a large bearing clearance and the other is the tissue response to the wear debris. In an all-metal prosthesis, metal particles have been found in the joint fluid and the adjacent tissue while debris of polyethylene deposits in the periarticular cartilage [19]. From the design point of view, wear rates by and large depend on the contact stresses (Equation 15.5), so that the size of the femoral head should be considered while planning hip joint replacements. High contact stresses give rise to high rates of adhesive or two-body abrasive wear but the rate accelerates if wear debris is unable to escape due to geometric peculiarity as in an elbow joint. Wear debris are trapped by closely conforming couples and this aspect has been studied [16] in detail by simulated tests and by observations on clinical *ex vivo* implants. Five types of knee prostheses were implanted on eleven patients and these were examined after surgical operations at times between seven and thirty months later. The various designs of the implants, each comprising a metal alloy mating with UHMWP, were as follows, the first three being relatively non-conforming prostheses:

(1) duo-condylar;
(2) uni-condylar;
(3) marmor;
(4) Freeman–Swanson, a prosthesis which conforms when the knee is straight and gives partial conformation on flexion;
(5) geomedic, a prosthesis conforming closely at all positions of the knee.

The failure of the prosthetic devices was largely due to wear which had a three-body abrasive mode by the cement which became loose during articulation. The cement debris were in the form of balls, 30 μm in diameter, and produced large dents and craters together with numerous shreds of plastic on the surface. Pits, 0·02–0·1 mm deep, and cracks transverse to the direction of sliding, were observed. According to the author [16], the nature of surface destruction of the knee implants depends both on the contact stresses and on the design. As the conformity improves, smearing and stretching of the surface of the polymeric component with cracks in the highly stressed regions become predominant. For joints with a low degree of conformity, failure was characterized mainly by cracking and pitting.

It is impossible to pin down such variables as load in service from *ex vivo* examinations, and all that can be done is to attempt a statistical conclusion based on the number of subjects chosen. To seek more controlled information, it is quite common to experiment with monkeys, sacrifice them, and then carry out *ex vivo* examinations. The pain inflicted on the animals is presumed justifiable on the grounds that the experience gained will ease the life of *homo sapiens*. Variables, however, can be controlled and quite fruitful design information can be collected from laboratory studies using simulators.

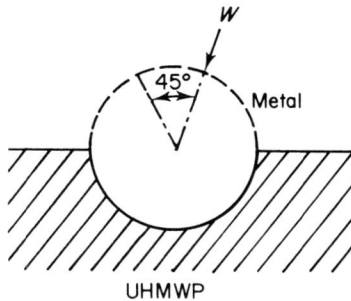

Fig. 15.9   A metal cylinder oscillating in a UHMWP block where the conformity between the bodies can be varied. (Trent, P. S. and Walker, P. S. (1976). *Wear* **36**, 175.)

The stance phase has been simulated [16] in which Co–Cr cylindrical specimens, 9·5 mm in diameter, have been subjected to oscillatory motion against concave blocks of UHMWP for one million cycles at 0·5 Hz. The bearing length was 9·5 mm and the diametral clearance was varied in such a manner that the conformity became more inferior as the clearance changed from 0 to 9·53 mm in steps of 1·6 mm. The cylinder was oscillated through an angle of 45° (Fig. 15.9) under a load of 45 kg for one half of the cycle and under zero load for the other half. The experiments were carried out in distilled water at 37 °C in a sealed chamber. An analysis of the results from this simulator showed that the failure mode was analogous to fatigue in non-conforming interfaces but highly conforming pairs may entrap wear debris or loose cement in a prosthesis in a human body, resulting in three-body abrasion.

High conformity devices provide much needed ambulatory stability by transmitting the load directly across the interface. Unfortunately, the shear forces transmitted across the cement are large and hence a high propensity to loosening of the implant. In low conformity designs, if the joint cement is not strong, the force may act on the edge with eventual loosening. It follows that the conformity must be such that there will be an absence of three-body abrasion and the contact stresses must be below the endurance limit of the plastic chosen. A highly conforming couple will impede the ingress of body fluid at the articulating interface starving it of lubrication. This will increase wear and both metal and plastic debris have been detected in tissues surrounding implants in human beings [15]. The metal debris show greyish black discoloration and are round and rod-like particles with an average size range of 1–2 μm.

Tissue response to polyethylene debris gives local inflammatory reaction in animal experiments. The possibility of carcinogenic effect of polymers exists since many types of these materials tested on rats have produced tumors [15]. No malignant change has so far been reported on human subjects. However, implants are being used increasingly in young patients and danger to health as a result of physiological antagonism by tissues to endo-prosthetic devices must be studied.

References
1. "Gray's Anatomy", Davies, D. V. and Davies, F. (Eds) (1962). 33rd edn, Longmans, Green & Co. Ltd., London.
2. Rasch, P. J. and Burke R. K. (1974). "Kinesiology and Applied Anatomy", Lea and Febiger, Philadelphia.
3. Smout, C. F. V. (1967). "Basic Anatomy and Physiology", Edward Arnold, London.
4. Perrott, J. W. (1959). "Anatomy", Edward Arnold, London.
5. Galante, J. O. and Rostoker, W. (1973). *Acta Orthopaedica Scand.* Supplementum no. 145, Munksgaard, Copenhagen.
6. O'Connell, A. L. and Gardner, E. B. (1972). "Understanding the scientific bases of human movement", Williams and Wilkins Co., Baltimore, MD.
7. Fung, Y. C., Perrone, N. and Anliker, M. (Eds) (1972). "Biomechanics: Its Foundations and Objectives", Prentice-Hall, New York.
8. Seedhom, B. B., Dowson, D. and Wright, V. (1973). *Wear* **24**, 35.
9. Paul, J. P. (1966). *Proc. Inst. Mech. Eng.* **181**(3J), 8.

10. Rydell, N. W. (1966). *Acta Orthopaedica Scand.* Supplementum no. 88, Munksgaard, Copenhagen.
11. Grieve, D. W. (1966). *Discuss. Proc. Inst. Mech. Eng.* **181**(3J), 150.
12. Paul, J. P. (1970). *Proc. Roy. Soc. Med. A* **63**, 200.
13. Kenedi, R. M. (Ed) (1973). "Perspectives in Biomedical Engineering", Macmillan, London.
14. Charnley, J. (1972). *J. Bone and Joint Surgery* **54B**, 61.
15. Walker, P. S. and Bullough, P. G. (1973). Symposium on interposition and implant arthoplasty, Orthopaedic Clinics of North America **4**(2), 275.
16. Trent, P. S. and Walker, P. S. (1976). *Wear* **36**, 175.
17. Clarke, I. C. (1972). *Ann. Biomed. Eng.* **1**, 31.
18. Weightman, B., Simon, S., Paul, I., Rose, R. and Radin E. (1972). *J. Lubrication Technol.* **94**, 131.
19. Dumbleton, J. H., Shen, C. and Miller, E. H. (1974). *Wear* **29**, 163.
20. Duff-Barcley, I. and Spillman, D. T. (1966). *Proc. Inst. Mech. Eng.* **181**(3J), 104.

16 Dental Tribology

Cosmetic and restorative dentistry are two important branches of the discipline of attending to human teeth. In the merely literal sense, cosmetic dentistry should deal with the aesthetic quality of teeth while restoration implies repair so that a tooth is again functional. As society gets more affluent, the aesthetic domains of human beings receive increasing attention and cosmetic treatment in dentistry becomes widespread by demand of the populace. In that situation, it is not easy to draw a sharp dividing line between cosmetic and restorative dentistry except to say that the primary objective for cosmetic treatment is to enhance a person's appearance when the teeth are exposed. On the other hand, a restoration such as fillings used to cover cavities in teeth must be functionally efficient in that the damaged tooth will act as if undamaged. However, a filling or any other form of repair to a tooth must have an aesthetically appealing appearance. An aesthetic treatment is not now regarded as a luxury only as this often forestalls failure and thus avoids sudden and very elaborate repair. An example is aesthetic contouring which is in fact a reshaping process of natural teeth. There is a certain aesthetic benefit in this but the contouring often facilitates self-cleaning of the teeth which by itself is good dental care. Dental tribology, however, must study the interaction of tooth surfaces under load with a counterface. The oral environment is not always a compatible surrounding for human teeth considering the variations in diet and drinks. The forces on teeth are largely as a result of biting during eating and the motion undergone is that of sliding when the upper teeth chew food against the lower teeth. Wear of teeth, and in particular, that of artificial restorative materials is fully discussed in what follows but, prior to this, the reader may wish to consult reference [1] at the end of this chapter or a similar book to familiarize himself with dental histology and anatomy.

16.1   Historical

Cosmetic dentistry was known by early man and there is evidence [2] that, about 4000 years ago, the Japanese used to stain their teeth brown or black. The Aryan society in India was established *c.* 2500 BC and black tooth stain is prevalent even today among some pre-Aryan ethnic groups there. Among the Indian peoples of the Americas, the contemporaries of the Mayan civilization decorated their teeth with jadeite inlays. Etruscans [3] used either human or animal tooth to shape as replacements for their own missing teeth. Prosthetic dentistry was practised probably in Egypt about 2500 BC as indicated by the find in a cemetery near the great pyramids of two molar teeth encircled with gold wire. Around the fifth century BC, one of the Roman laws stated, "If a man's teeth are joined with gold, it shall be no violation of the law to bury or burn the body along with that gold". In those days, the attachment of a replacement was achieved by fastening a gold wire around the adjacent teeth. The prosthetic component was shaped from a human or a bovine tooth, and even as early as the first century AD, the Roman dentist Cascellius probably developed some form of slip-on dentures. This would appear to be so from the statement of the Roman satirist Martial: "She, at night, lays down her teeth as does she her silken robes".

After these early beginnings, there appeared to be a long hiatus until around the sixteenth century. There is now a zealous interest in the whole gamut of dental technology, and whereas a good appearance is very desirable, efficient mastication of food is the all important factor. Hoffman's description [4] of the teeth system in the oral cavity as a microcosmic mill is quite appropriate.

16.2   Loss of tooth material

Loss of tooth material can occur for a variety of reasons, but it is of fundamental importance to observe the masticating process which involves impact, albeit very gentle, followed by grinding of the involved dental areas. Many individuals have a habit of grinding their teeth even while not eating and particularly when asleep. When food is ground in the mouth, most of the maxillary teeth have a sliding action against the surfaces of those which are embedded in the mandible. The motion is largely unidirectional with a sequence of movements, viz., sliding forward a short distance, retracting without the opposing surfaces being in contact and followed again by gentle impact, sliding, and so on. Depending on the habit of the individuals there will be variations to this pattern, but essentially, the tribological interaction is that of a very low energy impact and sliding at small velocities, althouth the impact action is regarded as negligible by many workers.

The position of a tooth in relation to its whole family in the oral configuration may give rise to some nearly exclusive functions, the tribological implications of which are by no means insignificant. Thus the incisors are used for cutting vegetables and meat while the molars undergo three-body abrasion, the severity of the abrading medium depending on the nature of the food. The teeth are perpetually in a fluid environment and, apart from the saliva, all kinds of liquids of varying degrees of acidity join the abrasive food particles during mastication. There are a number of ways therefore by which loss of tooth material may occur, but the modes can be broadly classified under erosion, attrition and abrasion, all of which are distinct tribological situations.

16.2.1 *Erosion*
There are two forms of erosion [1], viz., idiopathic and chemico-mechanical. The factors responsible for loss of material by the former mode are not fully understood but idiopathic erosion has the effect of giving localized areas of teeth a polished appearnce. As the name implies, chemico-mechanical erosion occurs due to mechanical inter-action such as impact or sliding or both in a hostile chemical environment. This has often been called chemical wear when experienced in engineering mechanisms and we may reserve the terminology chemico-mechanical erosion for the loss of tooth material in the mouth. The phenomenon is similar to dental caries [5] which occurs as a result of decalcification of a tooth. However, erosion is distinguishable [2] from caries in that the former is a result of chemical wear without a bacterial action as in carious lesions.

In vivo observation [6] with rats has shown that their dental tissues are attacked if the animals are forced to drink citrate fluids. Experiments show [7] that the upper teeth of human beings are more seriously eroded than the lower teeth and a suggested reason is that it is the parotid saliva which is the more damaging since the parotid ducts open in the vicinity of the upper first molars. The chemical contribution to the erosion phenomenon is the acidity of the environment. The most destructive acid is citric followed by acetic, hydrochloric, nitric and sulphuric acid in decreasing order of erosive attack. Apart from the type of acid, the magnitude of erosion clearly increases with a decrease in the pH value of the solution. The degree of erosion of dental enamel must therefore be related to the hydrogen ion concentration of the tooth environment. Experiments show an inverse relationship of enamel erosion with the pH value for all acids. The pH of the oral environment should be a fair guide to the erosion propensity of teeth provided it is appreciated that the degree of decalcification varies between acids. Sulphuric acid produces insoluble calcium sulphate so that loss of enamel from a tooth is low, while it is high for acids such as nitric and hydrochloric which form soluble compounds with the appropriate elements of the tooth.

The most disastrous chemical is known to be citric acid which causes rapid dissolution of the enamel. One of the sources of citric acid is lemon, so that sucking of the fruit should be discouraged. The citric acid content of the saliva, unfortunately, increases with age and must have a deleterious effect, as it is known that the teeth of older people are quite vulnerable to erosion. It is unlikely, however, that adults will suffer from very serious erosion of teeth, provided their acid intake is low. In children, clinical observations [9] have recorded serious erosion damage because of the habit of giving children dummies to suck which have been previously dipped in various fruit syrups. Similarly, decalcification by the chemico-mechanical effect is common in the teeth of those babies whose feeding bottles have been filled with fruit concentrates.

The fact that teeth are highly sensitive to the acidity of the oral environment from the viewpoint of enamel loss is highlighted by interesting published information on clinical observations [10]. A 46 year-old patient often vomited three times a day due to a duodenal ulcer. The vomit had a low pH value, indicating its acidic nature with consequent erosion of the patient's teeth. The incisors, especially those of the maxilla, were the most affected being in the path of the vomit, as evidenced by a general loss of enamel and dentine. It is noteworthy that an acidic fluid which is free flowing does not necessarily attack all the teeth in the mouth and the reason has been attributed to the presence of saliva. It is assumed that the teeth which escape attack are protected by the trapped saliva between their buccal surfaces and the corresponding cheek. The diluting effect of the saliva is beneficial if it is alkaline even for such a harsh medium as citric acid [11] and erosion probably depends [12] on the acidity of the fluid directly in contact with the tooth. That is, the general fluid stream which passes through the mouth is not as harmful as the acid medium which remains in contact with the enamel or dentine for prolonged periods of time as in the case of babies' dummies sucked after being dipped in acidic juices. Decalcification due to acidic fruits has been reported as early as 1892 [12] from studies on the teeth of monkeys and dogs, but it seems that abrasion, as in chewing of food [13], is necessary for the effect to be appreciable. In other words, chemico-mechanical wear is aggravated as a result of abrasion and it is known that sliding with metal friction couples giving rise to adhesive wear can deteriorate to various degrees if abrasives intrude at the interface.

Most animals are expected to trap abrasive particles from the surrounding environment [14] in their mouths. An intake of dust from the atmosphere will be harmful because of the abrasive effect, but in addition, one such environment has been shown [15] to contain 1·0 mg of tartaric acid per cubic metre of the atmosphere giving a pH of 3·8. This low pH solution in the mouth combined with abrasion by the dust particles will result in erosion. Erosion lesions by chelation brought about in this manner occur also because of the presence of natural agents in the saliva apart from animal and plant tissues taken from external sources at intervals. Thus, a factor in the aetiology of dental erosion is the accumulation of pyrophosphate in the human salivary debris [16]. Unfortunately, the biological or chemical factors which facilitate accumulation of pyrophosphate in oral cavities do not appear to be known.

16.2.1.1   Shapes of erosion lesions
Erosion lesions are found on the labial, proximal and lingual surfaces. If erosion is progressive, it leads to the formation of a crater in the enamel, thus exposing the underlying dentine, but the surface of such a lesion appears smooth and polished making it difficult to identify from the surrounding unattacked portion of the tooth.

Three types of lesions have been identified, viz.,
(1) dish shaped;
(2) wedge shaped;
(3) irregularly shaped.

Very often, the labial surfaces of the incisors show shallow saucer-shaped depressions, and are usually arrested if conditions change to make their growth unfavourable. If, however, growth occurs, the shallow depressions become U-shaped and the cause is believed to be associated with uncongenial dietary fluids or oral saliva washing over the affected area. It is understood that the washing action is probably providing a sliding motion, carrying with it any particles which will cause abrasion. However, for the fluid to be effective, a layer must remain adjacent to the lesion to contribute the chemical factor, as in the discussion earlier regarding the patient vomiting due to a duodenal ulcer. That is, the acid content which attacks the tooth is not diluted in the particular spot where a lesion is formed and is growing.

The wedge or V-shaped lesions appear commonly on the buccal surfaces of posterior teeth as thin straight lines, which may spread rapidly and become very deep. The irregularly shaped lesions are common on proximal and lingual surfaces and are believed to be associated with environmental disorders such as an atmosphere laden with chemical fumes and dust. The air from such surroundings will have acidic constituents and when inhaled will cause decalcification which, combined with the mechanical movement in the mouth, will accelerate wear of the enamel and dentine. Of course, such dissolutions occur at a critical concentration of hydrogen ions, and foods and other externally ingressed matter with pH values between 5·5 and 9·0 should be innocuous. Possibly, because of the variations in the occlusal forces and the production of pyrophosphates in the duct, the degree of damage seems to vary between individuals as would appear from comments by some clinicians that some patients are erosion-prone. Erosion rates can be reduced by 50% if the lesions are restored quickly by filling with a polymer composite.

16.2.2 Attrition
Attrition is that process which requires mastication and is defined in this text as wear without the presence of a third body at the interface. It must therefore be likened to the adhesive wear process in metals where the interface is clean, and the rate of wear can be related to the normal load divided by the flow stress of the metal. In a similar manner to the adhesive mode, as long as the couple is loaded and one surface moves relative to the other, wear by attrition is progressive and manifests itself in decreasing cusp height and in the formation of hard, smooth and

shiny surfaces, often showing brown discoloration. Wear can be seen on the proximal and occlusal surfaces of posterior teeth, and on the lingual and labial areas of maxillary and mandibular incisors, respectively.

It is believed by many dental surgeons that attrition is not a problem of the young, although there is no documented evidence to conclude that the magnitude of wear increases with age. There is indeed a body of opinion that attrition arises not so much because of physiological ageing but as a result of disease which weakens tooth material. The weakness as stated is not quantified but one would think intuitively that the material loses toughness and becomes brittle with age. Another important consideration is that certain behaviour patterns in the oral cavity contribute greatly to attrition. The factors which have been included in these patterns are bruxism, that is clenching, and malocclusion, which is a mismatch of the mandibular and maxillary teeth when closed.

Psychic disturbances such as emotional imbalance or neurotic conditions are presumed by some to be the reasons for bruxism. In a clinical study of about 222 school children in the 11–13 year age group, Lindqvist [17] defines bruxism as atypical dental facets produced by grinding which extend to the dentine, being more than 3·5 mm from the intercuspidation position. Bruxism, of course, occurs during the conscious moments either while chewing food or due to involuntary cusping of teeth. Like the latter reaction, bruxism occurs during sleep involving chiefly masseter activities. Studies [17] show that, whereas motor disorders in children have no effect, bruxism may be accentuated in frequency and intercuspidation pressure by nervous diseases and mental stress arising out of discord with their social environment.

In what is known as the physiological stage of attrition, only the enamel layer is removed. The cusps, which are round or conical protuberances on or near the masticating surfaces of teeth, wear with time as a result of sliding under load. Usually in persons over the age of 40 the dentine is exposed, a symptom being the not altogether pleasant sensation felt by the victim when having hot drinks. In older patients, the senile or the final stage of attrition develops when the dental pulp is exposed. This causes pain and the affected teeth, if not treated, will be lost forever.

16.2.3 Abrasion

In the dental literature, the line of demarcation between the definitions of attrition and abrasion is diffuse, but in this text abrasion will be defined as being similar to the three-body mode, that is when sliding occurs in the act of chewing foreign particles between the maxillary and the mandibular teeth. Apart from in chewing food, humans lose tooth material by biting such articles as cotton thread, pins or if they smoke, the stem of their pipe. Crunching and masticating betelnuts, as is customary among the people of the Indian subcontinent, can give rise to serious wear by abrasion. A foreign body chewed wilfully, or inadvertently as when inhaling dusty air, can cause severe wear of localized areas of a tooth, eventually exposing and damaging the pulp itself.

The deleterious effect of acid in the oral cavity on the abrasion of teeth has been known a long time [18] and this aspect has been fully discussed in Section 16.2.1. Natural abrasion of human teeth occurs unavoidably during chewing of food or when lips are rubbed [19] over labial surfaces. Most teeth develop envelopes of films or plaques which are either bacterial in nature or are denatured protein. Although dental hygienists recommend removal of the bulk of the deposit for fear of formation of subgingival calculus, a thin layer of plaque on each tooth would appear to be beneficial from the tribological point of view. It is stated that these surface films are acid resistant and should thus stifle attack by the erosion mode but abrasion would destroy their coherency, exposing the enamel of a tooth.

Abrasion is a pathological process and is known to be initiated on an exposed cementum, progressing into the dentine of the root and spreading to the enamel with time. The exposed enamel of course is directly vulnerable to abrasion and in general terms the process can occur in three stages, viz., physiologic, transitory and senile. The last stage implies that either the tooth material loses some intrinsic abrasion resistance or that the teeth have been damaged with time and the parotid duct is secreting excess citric acid. Whatever the reason, if the abrasive action is interrupted by, say, a change in the subject's habits, secondary dentine forms subjacent to the affected part. The physiological and transitory stages of abrasion do not usually give rise to pain, but defects due to complete loss of enamel are irreversible and they are normally treated by one of the aesthetic dental restorative materials.

16.3 Requirements of restoratives

Partial damage to teeth due to wear can be repaired by using metallic and non-metallic materials. Restoration in dental terminology is the replacement of a part of the tooth structure which has been denuded, from a tribologist's point of view, by erosion, attrition and abrasion. Historically, metallic restoratives have been greatly used, but polymeric materials, especially composites, are currently receiving much attention. While the aesthetic appeal is quite important, an ideal dental restorative material should closely resemble the properties of the teeth and generally satisfy the following criteria. The restorative material should

(1) be easily manipulated during the repair of a tooth;
(2) resist wear;
(3) be mechanically strong;
(4) resist stain;
(5) have a low thermal conductivity; and
(6) be resistant to chemical attack by the salival environment.

A jacket crown can be described as a restoration process to replace all the lost enamel of a tooth. The material used must satisfy many conditions [20]. The prosthetic material must be appealing to the patient from the anatomical, local physiological, general physiological and aesthetic points of view. To fulfil these requirements, the restorative material must have comparable mechanical and physical properties to the natural tooth and should match its colour very closely so that the repaired patch is not detectable.

Even if these conditions are successfully maintained, the restorative material will be useless if the local physiological requirement is not satisfied. That is, the reaction of the dentine, the pulp and the periodontal membrane to the reparation must be favourable. Among the criteria used to evaluate pulpal response are the degree of inflammation and the amount of edema [21]. Adverse physiological response is of course a possibility because of the dentist's method of repair. The affected area is drilled to produce a cavity in which a near putty-like material is gently pushed, which is then finished flush with the surrounding tooth surface. It is evident that the restorative material is in contact with fresh dentine and can well react adversely, giving rise to pain and inflammation. However, rather than being in direct contact, the dentine can be soaked [22] in a silver nitrate solution. This gives a dense black coating that contains metallic silver, which ensures that the restorative putty adheres to the cavity. The restorative materials commonly employed, including porcelain, adhere well to dentine thus treated but not to enamel. However, it would appear [23] that the safest procedure is to prepare the cavity and insert a 25 μm thick polystyrene liner before filling with the restorative material to ensure complete absence of pulpal irritation. In this case the treatment with silver nitrate solution is unnecessary as the polymeric liner attaches firmly to the exposed dentine, as does the restorative material to the liner. Such details of procedure by the dentist should be known to tribologists in the same way as the method of manufacture of engineering components should be known to assess the suitability of application or to examine the causes of failure in service.

16.4 Metallic restoratives
Of the various metals used in dental repair are silver, tin, copper and zinc. Whereas these are prepared as amalgams of mercury, the noble metals such as gold and platinum are used directly.

16.4.1 Amalgams
An amalgam is a mixture of mercury plus an appropriate metal, and about 75% of dental restoratives used are amalgams of silver. Another prominent constituent of an amalgam is tin, but copper and zinc are also added. A typical analysis of the metal part of an amalgam is 65% silver, 27% tin, 6% copper and 2% zinc. To prepare an amalgam with mercury, these metals are pre-alloyed by melting and casting into ingots. The cold ingots are machined to produce filings, and these powder-like alloys are pressed into small pellets which are mixed with mercury in an appropriate manner to form the amalgam. This is then used to fill the cavities in teeth. To understand the friction and wear properties of these amalgams, it must be realized that not only are the alloys at least quarterneries, but they exist in a powder-like mass and form compacts of low density when *in situ*. A laboratory study of wear of the amalgams will therefore need detailed planning of the material alone from the point of view of composition and manufacture.

The silver–copper alloys are aesthetically appealing to most, the role of copper being to resist corrosion in the oral environment. One would expect then that such an alloy will be a suitable restorative material to combat chemico-mechanical wear, viz., erosion. As well as copper, there is experimental evidence to show that zinc enhances the erosion resistance of these dental alloys markedly by improving the life of the restorative by a factor of about two [24]. It is generally believed that these amalgams resist wear by attrition or abrasion, but there is scope for experimenting with the alloy content and method of manufacture to improve erosion resistance. An attendant disadvantage is that the products of erosion can cause staining of teeth.

16.4.2 Gold and its alloys

Gold is a very efficient restorative material provided the patient does not reject it from an aesthetic point of view. It is well known that some people find it very satisfying to show a mouthful of teeth with gold coverings, while to others anything that does not match the colour of their original teeth is unacceptable. Thin sheets of 24 carat gold or its alloys are cold formed into pellets, as in amalgams, or into ropes for use by the dentist to carry out a restoration. Pure gold has the advantage of being malleable; hence its popularity as a filling material for tooth cavities. Although lacking in the ease of being worked into shape, cast gold alloys are used for restoration of such areas as cusps because of their toughness. The gold alloys are of two colours, viz., yellow and white, and are nominally mixtures of silver and copper with additions of palladium, platinum and zinc. The white alloy was developed in 1935 with the idea that it has better aesthetic appeal and possesses comparable mechanical properties to the yellow alloys. The yellow colour of gold can be progressively changed to white by the addition of platinum, palladium and silver. The metal nickel has also been incorporated with the same object, but it makes the alloy brittle and reduces its resistance to staining when exposed to the oral environment.

16.5   Non-metallic restoratives

Of the non-metallic materials, porcelain is still popular for crown restoration and was probably first used by Wood [2] in 1862. Apart from porcelain, other non-metals used for dental work are silicate cement, polymers and their composites. However, these materials are weak and are not as durable as gold but are more attractive to human beings for dental use because their colour can match that of natural teeth.

16.5.1 Porcelain

The material described by the generic term porcelain consists of china clay, silica and feldspar, the latter being a mixture of silicates of sodium, potassium and aluminium. There are two types of porcelain which have been used for dental work, viz., high-fusing and low-fusing [25]. The former containing about 4% clay, 15% silica and 80% feldspar fuses at a temperature of 1300–1400 °C. The most commonly used dental material is the low-fusing porcelain which contains about 25% silica, 60% feldspar and 50% of a low-fusing flux such as borosilicate glass. In the restorative material as received by the dental technician the constituents of the low-fusing porcelain are mixed and fused together at a temperature in the range 850–1100 °C, formed into a frit and then ground to a fine powder. To this are added about 50% of alumina crystals with an average diameter of 10–25 μm to improve the rupture strength of the restorative material. Alumina addition has the beneficial effect of improving the thermal shock resistance of the porcelain but may impair the eye appeal if proper precautions are not taken during construction [2]. Porcelain is well tolerated by dental tissues, being used extensively for forming dental crowns, and is supplied to the dental technician in the form of a proprietary powder which probably incorporates a binder such as sugar or starch to give the necessary amount of green bond to the shape prepared from porcelain.

To fit an artificial crown, the damaged tooth is finished to a suitable shape by grinding and a crown is prepared which is like a thimble to slip on to this damaged tooth. The method of producing the crown, or jacket as it may be appropriately called, is to add water to the mixture of porcelain, incorporating alumina and a binder. The crown is then prepared by hand with the object of producing a shape which will slip on the prepared tooth. The green product once formed is vibrated lightly or patted with a spatula to remove as much water from the component as possible. It is then dried and fired again at temperature to form the final crown.

A crown should be a perfect fit to the prepared tooth. Since it always has to be finished by hand and because of the uncertainty of the amount of shrinkage during drying and firing, a good fit is not possible. Hence a core or a thimble which slips over the finished tooth is made from a suitable metal. The front portion of the thimble, that is the area which shows when the teeth are exposed, is covered with porcelain to complete the crown and once fully constructed is fired at a suitable temperature in the same manner as that described for an all-porcelain crown. The composition, moisture content and the firing conditions are so adjusted that the porcelain overlay gives a slightly smaller coefficient of expansion than the metal substrate so that the former is in a state of slight compression after firing. The materials used for the thimble are gold based palladium alloys and those of nickel and chromium.

Table 16.1 Properties of natural teeth [25]

Properties	Enamel	Dentine
Hardness (KHN)	280–360	60–70
Compressive strength (kg mm^{-2})	14–28	28–32
Modulus of elasticity under compression (kg mm^{-2})	$(2 \cdot 7) \times 10^3$	$(1 \cdot 5 – 1 \cdot 8) \times 10^3$
Ultimate tensile strength (kg mm^{-2})	$1 \cdot 0$	$(3 \cdot 5 – 5 \cdot 0)$
Transverse breaking strength (kg mm^{-2})	$8 \cdot 0$	$\cdot 27 \cdot 3$
Thermal conductivity (cal s^{-1})	$2 \cdot 23 \times 10^{-3}$	$1 \cdot 36 \times 10^{-3}$
Coefficient of thermal expansion (per °C)	$(8–11) \times 10^{-6}$	$(8–11) \times 10^{-6}$

16.6 Filling materials

Some of the main requirements of a dental filling material are that it should have a similar expansion characteristic to the tooth substance and that it should adhere adequately to the portion of the tooth under repair. The filling material must be bacteriostatic and should prevent carious attack of the tooth matrix in which it is embedded. An ideal situation is that where the filling material has similar mechanical and physical properties to natural teeth. Unfortunately, a tooth is a composite construction in which there is an outer layer of enamel whose properties vary from the supporting body of dentine (Table 16.1). While the enamel is much harder than the dentine, the former is also brittle so that the filling deposited in the tooth cavity may be surrounded by two different materials. However, although there is a difference in thermal conductivity, the coefficient of thermal expansion is similar for enamel and dentine, which is a great advantage as the oral temperature of human beings is subject to change.

Dental filling materials are traditionally amalgams of mercury and silver. Whereas the mercury is a health risk for operatives producing the filler, the amalgam is inert [25] to oral fluids and other extraneous food and drink, once prepared and installed in the cavity of a tooth.

However, the amalgam shows low adhesion to both enamel and dentine and exhibits poor resistance to *in vivo* erosion in human teeth. Both the coefficient of expansion and the thermal conductivity of an amalgam are higher than those of natural teeth. The restoration is poor aesthetically which is one of the main reasons why the mercury–silver preparations are unacceptable to patients in affluent societies. The amalgams cause leakage in the restorative margins due to their mismatching thermal properties *vis-à-vis* a tooth and the silicates, being soluble in oral fluids, lead to secondary caries as well. A major limitation of gold is its high thermal conductivity. Nevertheless it has many favourable properties as a restorative material but it is expensive and may not be acceptable to some patients because of its colour.

Like porcelain, the use of certain polymers is an attractive proposition. As a dental restorative material, methylmethacrylate is a logical choice because of its matching colour, good adhesion to tooth substrate and its low solubility in oral fluids. Unfortunately, both its polymerization shrinkage and coefficient of thermal expansion are high, the two factors contributing to leakage in cavity margins.

Incorporation of additives, that is fillers, in the form of beads or fibres in methylmethacrylate produces composites with low polymerization shrinkage and coefficient of thermal expansion. Some of the fillers used are glass, quartz, alumina and beta-eucryptite, a synthetic mineral. Up to 85% of filler is incorporated in the polymer, and one such proprietary restorative material is being used widely [26]. The composite has very good aesthetic appeal, is insoluble in oral fluids and possesses a high wear and stain resistance as *in vitro* studies show. Detailed clinical trials have yet to be made.

16.7 Forces on teeth

A tribologist should be equipped with a knowledge of dental anatomy and of the physical, mechanical and chemical properties of both natural teeth and synthetic materials. Without such background knowledge, a study of sliding merely provides data, albeit useful on an *ad hoc* basis, but an insight into the mechanisms becomes almost impossible. With a knowledge of these materials including their composition, the next question to ask is what is the nature of the mechanical forces in the mouth?

Although much is made of aesthetic appeal, the reason for the existence of teeth in all animals is to tear and chew food. It has already been stated that when the cusps approach for crushing food, the teeth undergo very gentle shock due to impact, which is followed by sliding as chewing commences. The eating process involves the mechanical action of sliding but details such as the actual amount of force vary depending on the nature of the food being chewed and the personality of the individual.

Although the mechanics of interaction between the maxillar and the mandibular teeth would appear to be complex, chewing forces exerted by human teeth have been measured [27]. As the food is first introduced in the mouth and chewing commences, the biting force pattern shows peaks and troughs indicating that the action is that of gripping followed by relaxing. Initially, the gripping forces are high, as shown by the heights of the peaks, but they diminsh progressively with time as the food becomes crushed and softened with saliva. Typical mean normal loads exerted while chewing biscuits, apples and chewing gum were 2·35, 1·34 and 1·21 kg respectively. It seems that, being hard and brittle, the biscuits require greater effort than the softer apples and chewing gum. In a separate measurement [28], the normal loads for peanuts, coconut and raisins were 3·67, 4·06 and 4·85 kg respectively.

A report on the nature of the biting force exerted in the mouth has been produced from measurements made over a period of four years on a group of patients with immediate denture treatment. The biting force was measured using strain gauges fitted with hard rubber pads. The pad was held between the teeth and prior calibration could give the exact load exerted by the patient. The biting forces as observed on female and male subjects are shown in Figs. 16.1 and 16.2 respectively. The interesting feature of these results is that in the first week or so after having a denture fixed, the patient is hesitant and the biting force is small. Figures 16.1 and 16.2 show a trend where both female and male subjects use a larger biting force as the months go by. It is thought that the patient becomes confident with time and a higher steady force prevails. The male biting force exceeds that of the female in the ratio 3 : 2 which is similar to the results obtained from dentate persons. This shows that even edentulous patients develop biting forces comparable to those possessing their natural teeth provided other conditions such as the nature of the food chewed remain the same.

Apart from the eating process, forces are exerted on tooth surfaces during cleaning with brushes and dentifrices. Fraleigh *et al.* [30] have obtained useful data regarding forces exerted by tooth brushes from

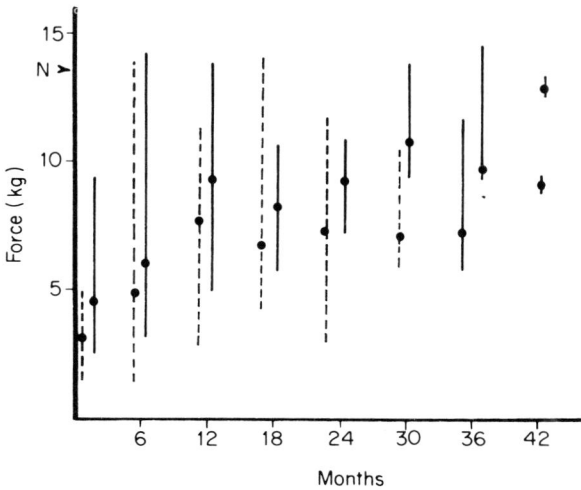

Fig. 16.1 Average and range of maximum biting force. Two groups of females: – – – complete upper denture opposing natural mandibular teeth; ——— complete upper denture opposing natural mandibular anterior teeth and a posterior partial denture. *N* is the average biting force. (Atkinson, H. F. and Ralph, W. J. (1973). *J. Dent. Res.* **52**, 225.)

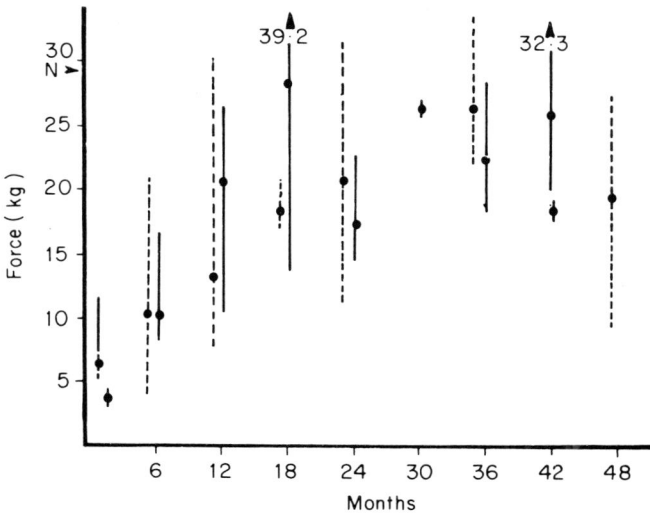

Fig. 16.2 Average and range of maximum biting force. Two groups of males. Other details as in Fig. 16.1. (Atkinson, H. F. and Ralph, W. J. (1973). *J. Dent. Res.* **52**, 225.)

studies on 208 subjects in the age range 4–65 years. Tooth-brushing is a
two-body abrasion process and the loss of enamel will probably depend
on the hardness and stiffness of the bristles. This variable was studied
by testing various types of bristles incorporating three modes of motion,
viz., manual and two electrically operated brushes giving rotary and
reciprocating movements respectively. Strain gauges mounted between
the brushing end and the handle of a brush measured the normal force
plus the torque developed at the teeth during brushing and the two
forces were termed by the authors as scrub and sweep respectively. It is
not clear as to what the authors mean by a torque force but it is
recognized that the stiffness of the tooth-brush may be a variable which
will affect the wear of tooth enamel. It is difficult to devise sophisticated
tests to measure stiffness but the authors simply held a brush against a
scale pan and noted the load at which the bristles buckled so that the
tooth brush could be assigned a property, stiffness, measured in kg.

The measured normal load during brushing of teeth varied from a
value of 0·045 to 1·31 kg, depending on the nature of the bristles and the
type of motion of the brush. For example, a rotating polyamide brush
gave a normal force of 0·26 kg whereas the corresponding force for a
manually operated multitufted soft nylon brush was 1·3 kg. The forces
in descending order for other tests were as follows, the normal force in
kg being given in the parenthesis and the number outside the parenthesis being the buckling force in kg:
  (1) reciprocating cordless nylon brush (0·973), 1·270;
  (2) rotating action nylon brush (0·868), 1·110;
  (3) rotary, battery operated nylon brush (0·564), 0·650;
  (4) rotating polyamide brush (0·260), 0·610.
The buckling force for the multitufted soft nylon manual brush
which gave the highest normal load was 1·94 kg and the values of the
normal force decreased with the buckling force, as seen above.
Unfortunately, these tests show a wide variation in results between
individuals regardless of the nature of the tooth-brush. The variations
are in the degrees of force exerted by individuals, but the authors do
show that the nature of the tooth-brush does determine to a large extent
the pattern of the tooth-brushing force a human being will apply during
cleaning of teeth. The maximum pressure exerted with a soft brush was
that required to collapse the bristles but that for a hard brush was what
was comfortable to the individual. Whatever the object is for cleaning of
teeth, one should try to achieve this with a minimum amount of enamel

loss. Much work needs to be done, but at the moment one would conclude that the normal load during tooth brushing varies between individuals, however with some general patterns. If the normal load determines the amount of wear, the bristles should be as pliable as possible and a rotating polyamide brush will be ideal for cleaning teeth. A study of tooth-brushing force on human subjects remains difficult, as Fig. 16.3 shows that the vigour of tooth brushing increases up to the age of about 30 and falls again to the original value at 10 years of age when a patient is over 50.

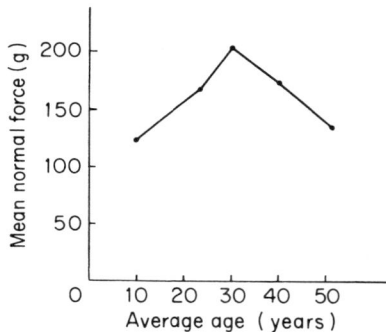

Fig. 16.3   Brushing force exerted by persons from various age groups. (Fraleigh, C. M., McElhany, J. H. and Heiser, R. A. (1967). *J. Dent. Res.* **46**, 209.)

Since wear of teeth should increase with age, an attempt has been made to establish a parameter which will give an estimate of the total sliding distance each tooth undergoes. The parameter is the number of chews during the eating activities of the subject under study. This varies again according to the personality of the subject, the type of food and the size of mouthful, but a mean value of eighty chews per minute has been quoted.

16.8   Wear studies

Dental restorative materials have been assessed by simulated studies in the laboratory using mainly three types of tests:

(1) rubbing under load in abrasive slurry [31, 32];
(2) use of a tooth-brush simulator [33];
(3) simulated mastication using impact loads [26].

16.8.1 Dentifrice

The role of the tooth-brush in terms of normal load on enamel has been discussed in Section 16.7. Cleaning of teeth, of course, involves a third medium which is a dentifrice or a toothpaste. An important property of a toothpaste or powder is its abrasiveness but two aspects must be considered. Firstly, a dentifrice must be able to eradicate coatings and stains to make the teeth look clean. Secondly, the abrasive component in the toothpaste must not cause catastrophic wear of tooth enamel. It is recognized that the force applied to teeth during cleaning will certainly be a variable in deciding on the rate of enamel loss, and any wear study must examine the effect of normal load, among others. Ideally, freshly extracted human teeth should be the best components to study using a standardized friction and wear machine. This has been done to compare the abrasiveness in terms of the wear magnitude of a few dentifrices. Unfortunately, the experimental data show wide scatter of results and the reason has been attributed [34] to the variation in hardness between teeth. The hardness of the enamel is known to vary even between different parts of an individual tooth.

There is always the possibility of using a counterface of metal of standardized chemical composition and mechanical properties, in particular hardness, to make a comparative study of the relative abrasiveness of dentifrices. Rather than standardize on one metal, pure copper, silver and antimony surfaces have been used to evaluate the abrasiveness of a large number of proprietary dentifrices using an abrasion machine. Abrasiveness of a dentifrice was assessed by noting the weight losses of these plates after running the machine for a set number of strokes. The dentifrice was held between a tooth-brush and the plate under study. Weight losses of tooth enamel were also obtained using the same machine. Table 16.2 shows the loss of weight of the metal plates on each day for 6 days' running of the machine, so that for each product, six readings for each plate were obtained.

Examination of Table 16.2 shows that the values for weight losses were consistent between days, each day constituting 20 000 strokes of the machine. The loss of metal was the greatest for antimony and the least for copper. Antimony and copper have comparable hardnesses and the reason for such a variation in weight loss is not clear unless copper, a metal with an f.c.c. structure, work hardens during abrasion. The authors showed by hardness measurements on abraded surfaces that this is not so. Using a Rockwell hardness tester, the hardness of silver

Table 16.2   Weight loss of copper, silver and antimony plates due to abrasion by a denti-
frice over a period of 6 days

Day	Average weight loss of plates due to abrasion (mg)		
	Copper	Antimony	Silver
1	8·0	19·2	16·8
2	7·1	20·0	14·8
3	6·7	20·4	14·3
4	7·8	19·2	13·9
5	6·6	20·1	12·7
6	7·0	21·8	12·4

is about one half of that of copper and it shows a greater amount of wear.

The next stage was to abrade human teeth under similar conditions and attempt to correlate the loss of enamel and loss of metal with a particular dentifrice. If that was successful, it would be relatively easy to evaluate the abrasiveness of a toothpaste by running against a metal plate rather than a human tooth.

The wear rates of metals, however, did not correlate with enamel loss under comparable conditions. In two tests, typical weight losses (in mg) for metal and enamel were 0·7 and 17 or 5·5 and 158 respectively. In another test, the corresponding values were 63·5 and 69 mg. Undoubtedly, if the properties within an enamel layer of one tooth vary, it is not surprising that reproducible results are difficult to obtain. The bulk of wear studies of dental materials would appear to have been carried out by running a test for a while and then measuring the loss of mass. As we have discussed before, the possibility of obtaining reproducible results becomes remote if a wear run is not long enough to give a steady state wear. That is, the rate of wear should be taken as the steady state wear for comparison purposes. The steady state wear is usually reproducible between replicate measurements but the experimental technique is tedious as a complete graph needs plotting.

Some of these difficulties, it would appear, prompted many investigators to undertake well-designed planned statistical experiments. In this way the effect of a variable in a system on some sought-after parameter such as wear can be isolated but it is not always easy to draw a physical model from results of statistical tests. For example [35], a factorial design experiment was planned using five variables, viz., hand brushing, mechanical brushing, two concentrations of a dentifrice and three levels of applied force using human teeth as samples in a tooth-brushing machine. Analyses of results showed that the effect of increasing the applied load was to increase the amount of wear, but this is true of most materials. The interesting conclusions were about the interaction between the concentrations of abrasives in the toothpaste. At a normal load of 0·1 kg, there was no significant wear between the two concentrations of the dentifrice. As the normal load was increased by 100%, enamel loss increased by a factor of 2 as the concentration of dentifrice was increased from 20 to 50%. A further increase in the wear rate occurred at a high dentifrice concentration and a normal load on the tooth of 0·5 kg. This is useful information in that an inadvertent increase in the amount of abrasives may not be disastrous provided the pressure exerted on the teeth is not excessive. Note (Section 16.7) that a polyamide rotary brush gives a reasonably low load and it may be a fruitful area of investigation to see if a suitable design of bristles could be made which will exert low normal pressures on teeth without impairing the cleaning action.

A factorial design experiment allows a large number of variables to be altered simultaneously. For dental research, this can be carried out on human volunteers. It is difficult of course to obtain quantitative wear rates *in vivo*, but a factorial design experiment will allow a random sample of extracted teeth to be tested in a tooth-brushing machine. The problem is that a large number of replicate measurements are necessary and the number of readings for a project becomes large, being 19 584 in the above design [35]. This is very lengthy and it is doubtful if investigators will always consider statistical methods unless there is no alternative.

16.8.2 *Basic studies*
As with polymers, one is never certain if the weight loss technique is an ideal way of evaluating wear of teeth. Reproducible results, however, have been obtained by taking Talysurf traces at intervals on the same wear track. The experiments were conducted in a simulated tooth-brushing machine on which were mounted human incisors, the enamel of which was brushed in a toothpaste slurry. The normal load was kept constant at 0·0315 kg, as was the reciprocation of the tooth-brush at

90 strokes per minute. The slurry temperature was maintained at 37 °C.

Another approach to obtain quantitative information is to use an irradiated tooth which undergoes wear in an abrasive slurry. In a particular experiment [33], the tooth was mounted and brushed in a machine under a normal load of 0·15 kg, while immersed in a slurry containing 8·5 μm diameter $CaHPO_4 \cdot 2H_2O$, 11·4 μm diameter anhydrous $CaHPO_4$, 7·4 μm diameter $NaPO_3$ and 1·8 μm diameter $CaCO_3$. After a suitable period of time, a 2 ml sample of the slurry was withdrawn and dried at 105 °C. The radioactivity of the dried sample was determined and the magnitude of enamel loss on a weight basis was calculated per 100 brush strokes. After completion of this run, the same tooth was again brushed under the same conditions using a standard abrasive of $CaCO_3$ and an abrasive ratio AR was defined such that,

$$AR = \frac{\text{Abrasiveness of the unknown sample}}{\text{Abrasiveness of the standard}} \qquad (16.1)$$

The abrasiveness of course was the rate of wear of a tooth for a predetermined period of abrasion and expressed as mg per 100 brush strokes.

This is a quantitative way of expressing the abrasion resistance of teeth or the abrasiveness of a dentifrice. The method could be combined with profilometric studies, and the technique, together with surface examination under a microscope, should provide the means for fruitful research on dental wear.

A great deal of information of a basic nature has been provided [37] by employing a radioactive tracer technique in which human teeth are irradiated to convert the phosphorus content of the tooth to the radionuclide, P^{32}. This is a high energy β emitter and the method is capable of measuring worn dental tissue weighing as little as 10^{-7} g using a Geiger–Müller counter. The apparatus for studying wear is essentially a tooth-brush simulator (Fig. 16.4) where a brush scrubs a tooth under a normal load of 0·2 kg at a stroke of 38 mm, reciprocating at a frequency of 2·5 Hz. The tooth is mounted on an epoxy resin holder, and in some experiments, the labial surface of an incisor is held against the brush to estimate the wear rate of the enamel. Teeth from other positions in the mouth have also been studied. The mounting mechanism for the tooth under study can be varied and a perspex trough holds the dentifrice containing the abrading medium. In order

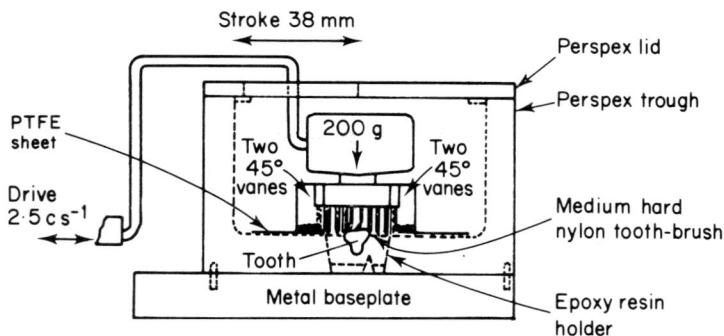

Fig. 16.4   A tooth-brush simulator. (Wright, K. H. R. and Stevenson, J. I. (1967). *J. Soc. Cosm. Chem.* **18**, 387.)

to maintain a consistent quantity of the abrasive over the tooth, the slurry is stirred, and the apparatus has four perforated vanes which allow the slurry to pass through the brush fibres which assist in preventing sedimentation of the solution. After a typical brushing test lasting a pre-determined period of time, 1 ml of the solution is withdrawn from the trough and dried before measuring the β activity.

The wear rate for both enamel and dentine in μg per 10 strokes and μg per stroke respectively are shown in Fig. 16.5. In these experiments,

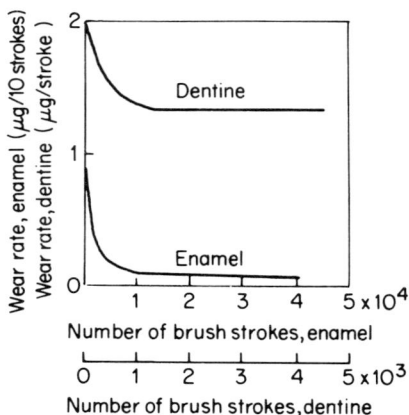

Fig. 16.5   Variation of the wear rates of enamel and dentine with the number of brush strokes. Note two different scales. (Wright, K. H. R. and Stevenson, J. I. (1967). *J. Soc. Cosm. Chem.* **18**, 387.)

a sample was run for a predetermined number of strokes and the loss of material was expressed as μg per stroke or μg per 10 strokes for enamel because of its low wear. The wear rates were then plotted against the number of strokes. For both tissues, the wear rate decreases with time, as happens in the running-in stage with metals and many other materials, eventually reaching a steady state. In this particular example, it was necessary to remove a minimum of 50 μm of dentine and an 8 μm layer of enamel before steady state wear was achieved. The running-in period depends on the prior condition of the tooth surface and the nature of the dentifrice, but 1000 and 10 000 brush strokes are usually extreme values for dentine and enamel, respectively. Unlike the tests where weight loss has been the criterion, a tracer technique such as this is useful because it provides a steady state, and, if reproducible, can be used with confidence for comparative studies. Unfortunately, it is not possible for every investigator to work with irradiation techniques.

If it is assumed [37] that the fibres of a tooth-brush do not contribute to the material loss of the counterface, the rate of wear will be governed by the concentration of the abrasives trapped at the fibre–enamel or fibre–dentine interface. One assumes for an ideal case that the whole interface is packed with jostling particulate matter. This is unlikely, and the particle concentration at the interface will be governed by the probability of trapping a number of solids at the tip of each fibre. The general term approximating the situation is derived from the Poisson approximation to the binomial theorem and is

$$\exp{(-nV)}n^rV^r/r!$$

where n = population density of the particles; V = volume of the region which may trap particles; and r = number of particles trapped by a fibre.

The wear rate expressed on a weight basis can be related to S, the sliding distance, or in this case the number of strokes, by

$$\frac{\mathrm{d}w}{\mathrm{d}S} = wN[1 - \exp{(-nV)}] \tag{16.2}$$

where w = the intrinsic wear rate of a fibre armed with abrasives; N = total number of fibres in the brush; and $\exp{(-nV)}$ = probability of trapping no particles when $r = 0$.

The population density of the abrasive particles at the rubbing interface will be proportional to the dentifrice concentration, C.

Therefore the wear rate can be expressed as,

$$\frac{dw}{dS} = \alpha\,[1 - \exp(-\lambda C)] \qquad (16.3)$$

where α = maximum possible rate of wear when all the fibres are loaded with abrasives; and λ = a measure of the ratio of the trapped volume of slurry at the fibre tip and the volume of an abrasive particle.

If R is the radius of the fibre cross-section and d the diameter of an abrasive particle, the trapped volume of the slurry is approximately equal to $10/9\pi Rd^2$ when $R \gg d$ and $\lambda \simeq 7R/d$. Typical particle diameters for $CaHPO_4 \cdot 2H_2O$ are about 12 μm with $R = 100$ μm.

Assuming the above particle size, for a monodisperse system where all the particles are assumed to have the same diameter and abrasiveness, a maximum of 5 particles will be trapped by each fibre for the usual dentifrices. Thus a brush with 1500 fibres is thought [37] to remove soft tissues to a thickness of 1–3 μm in one stroke.

16.8.3 Composites

Undoubtedly filled polymers, that is composites, are favoured as dental restorative materials because of their physical and mechanical properties which are superior to unfilled resins. The advantages are greater compressive and tensile strengths, a high modulus of elasticity and greater hardness. The composites show less shrinkage upon polymerization and a low coefficient of thermal expansion.

Like most dental materials, it is difficult to obtain correlations between results obtained in clinical trials and those from *in vitro* studies. A clinical examination has been conducted [38] to compare a proprietary composite with an amalgam by placing 124 paired restorations in 73 patients. At the end of two years, a moderate change in the anatomical form was noticed in the composites while the amalgams were unaffected, the reason being attributed to occlusal wear of the composites.

Contrary to the observations of a few workers, the results of a laboratory study [39] show that the effect of fillers is to cause an excessive amount of wear in a polymer. The wear experiments involved two-body abrasion when materials were run against an abrasive grinder. The wheel incorporated vitrified bonded silicon carbide particles with an average particle size of 85 μm and the experiments were conducted at a surface speed of 300 m s^{-1} and a normal load of 0.5 kg. Tests were carried out in both dry and wet conditions, the latter being achieved by immersing the polymer in water for a period of time long enough to give a constant weight. The polymer specimens were then dried by blotting the adsorbed moisture before rubbing against the wheel, each sample being slid for 100 cycles. Figure 16.6 shows a typical volume loss–sliding distance curve with a linear relationship, whereas filled polymers generally wear more while sliding against abrasives (Fig. 16.7). If the polymers are saturated with water, the rate of wear can be lowered, but not always to any great extent. One may be tempted to infer that fillers induce brittleness which should increase wear but the authors [39] suggest a different mechanism. This is that the filler materials, once detached, become trapped at the interface, giving rise to three-body abrasion. By implication, this means that a three-body mode is more severe than a two-body wear process, but it is unlikely that there is any experimental evidence for this. A three-body abrasive process is the most common situation producing wear in the mouth, but it is well to note that an experiment using a grinding wheel is severe and does not simulate the conditions in the oral cavity during mastication. It is not

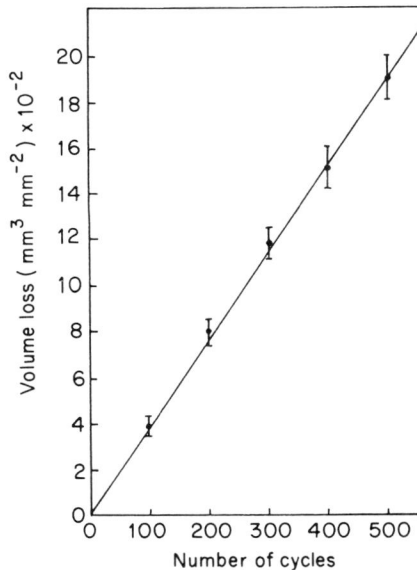

Fig. 16.6   Variation of volume loss with the number of abrading cycles for filled resin samples. (Lugassy, A. A. and Greener, E. H. (1972). *J. Dent. Res.* **51**, 967.)

Volume loss $(mm^3 \, mm^{-2}) \times 10^{-2}$

27
24
21
18
15
12
9
6
3
0

0   100   200   300   400   500

Number of cycles

Filled
Dry

Wet
Unfilled
Dry

Wet

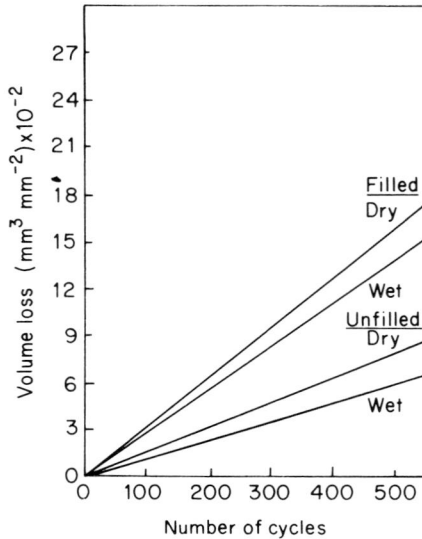

Fig. 16.7   Variation of volume loss of filled and unfilled polymers with the number of abrading cycles. (Lugassy, A. A. and Greener, E. H. (1972). *J. Dent. Res.* **51**, 967.)

surprising therefore that a correlation with clinical trials has not been obtained from this study.

For the filled polymers, wear should be influenced not only by the shape, size and the hardness of the additives but also by the density of the resin matrix supporting the fillers. The other parameters to consider are the physical quantity of the filler material in the composite and the degree of adhesion between the particles and the polymer matrix. A large number of variables has thus to be rationalized in an experimental programme and often there is a justifiable reluctance among practising dentists to accept the conclusions of *in vitro* observations unless the experiments are planned realistically.

Simulated studies must be carried out, and use has been made of an impact abrading machine together with a tooth-brushing simulator to study filled polymers [26]. In the former, the polymer underwent both impact and sliding, while in the latter a reciprocating tooth-brush cleaned experimental samples 25 mm in diameter. The results indicated that a silver–mercury amalgam gives inferior results from the point of view of wear resistance under tooth-brushing conditions compared with composites. It must be noted of course that the amalgams last a long time in *in vivo* environments but become clinically unacceptable only because of leakage in the margins. Furthermore, clinical studies show a greater degree of occlusal wear with a proprietary composite than with an amalgam.

This supports the speculation that optimization of composition, physical nature of the particles, manufacturing technique and the density of the resin matrix must be sought. In the time-scale available, this can only be done before clinical trials by simulated accelerated tests *in vitro*. There is evidence to suggest that the volume fraction of the fillers employed in laboratory studies, viz., silica, quartz and wollastonite, is very important [26]. The amount of wear decreases with increase in the filler content at small concentrations but increases again when the filler volume is raised above a critical value. Particle size, shape, hardness and the integrity of the bond would appear to be serious variables regarding wear resistance of dental restorative materials.

16.8.4 Fluid environments

Mastication of food and tooth-brushing both take place in fluid environments, the sources of fluid being saliva and extraneous liquids. The fluid intake can be acidic, as can the saliva. We have discussed already that an acid medium coupled with the mechanical movement of one tooth over another will give rise to chemico-mechanical wear known as erosion in the dental parlance. Acids increase the wear rates of teeth and a study of the hydrogen ion concentrations in the mouth regarding wear of teeth is necessary.

In such an experiment [19], the acid solution was made up from acetic acid plus sodium acetate buffer, the pH varying in the range 1·5–4·0. The enamel of freshly extracted human incisors was rubbed against strips of acrylic dental resin in a specially built machine and three types of experiments were carried out.

In one set, a group of five teeth was placed in a tube containing the acid, maintained at a constant temperature of 37 °C while being stirred using air bubbles. The teeth were left in this state for 24 h, after which samples were withdrawn from the bath so that the calcium content of the solution could be estimated.

In the second set, two different types of experiments were conducted. In a series of tests, the enamel was brushed with a toothpaste for 10 s while still exposed to the acid. The object was to remove the plaque and expose a fresh enamel surface to the acid. The second series in this second set of experiments had the enamel surfaces merely washed in running water before being exposed to the chemical solution. In the third set, the enamel was rubbed against the resin while immersed in water for 24 h. As on previous occasions, the calcium content of the solution was estimated to establish the degree of decalcification of a tooth under study.

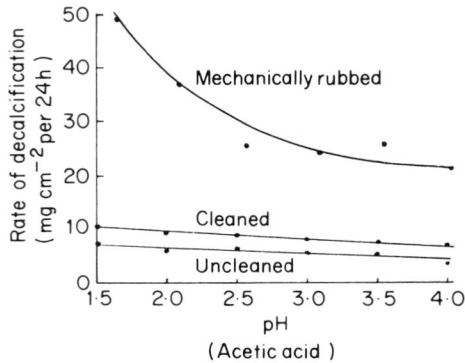

Fig. 16.8   Effect of acidity on the decalcification of tooth enamel. (Steel, J. and Browne, R. C. (1953). *Brit. Dent. J.* **94**, 285.)

The rate of decalcification in mg cm^{-2} per 24 h is plotted against the pH values of solutions in Fig. 16.8. Decalcification is a measure in this case of erosion and the pattern of all three curves in Fig. 16.8 is familiar in that the rate of wear of a tooth increases with increased acidity of the environment. The high amount of wear of cleaned teeth shows that the presence of surface films on teeth is a natural design with the object of prolonging the life of enamel. It is clear that the acid used in these experiments destroys the integrity of the enamel and an abrasive action compounds the wear process. It is known [19] that once the outer surface layer of a tooth enamel is destroyed, the more soluble inner enamel is attacked readily and wear accelerates. Incidentally, the normal load in these experiments is not quoted but rubbing against the resin pad in a neutral solution produced little wear and this demonstrates the importance of the acidity of the environment in the chemico-mechanical wear processes in these materials.

The effect of saliva has been studied [41] by rubbing hemispherical riders of various materials against a counterface. A surface grinder was adapted to form the apparatus for the experiments and the speed could be varied in the range 1–750 mm s^{-1}. A test could be arranged so that the rider covered a fresh track at each traverse, and both dry and saliva-lubricated studies were undertaken using various combinations of like and unlike couples, the normal load varying between 0·1 and 1 kg.

Table 16.3 Coefficient of friction of bovine enamel rubbed against various counterfaces [41]

Counterface material	Coefficient of friction	
	Dry	Lubricated with saliva
Gold	0·12–0·20	—
Amalgam	0·18–0·22	—
Acrylic denture base	0·19–0·65	—
Cr–Ni alloy	0·10–0·12	—
Porcelain	0·10–0·12	0·50–0·90
Bovine dentine	0·35–0·40	0·45–0·55
Bovine enamel	0·22–0·60	0·50–0·60

As Table 16.3 shows, there were variations in the coefficient of friction when the counterface was altered or when comparisons were made between dry and wet conditions. Most of the counterface materials can be regarded as suitable dental restoratives except the bovine dentine and enamel. The values of the coefficient of kinetic friction, μ, in saliva are of interest for all the materials listed in the table except perhaps for the Cr–Ni alloy. The wet coefficients of friction for gold and amalgam were so low that they could not be measured accurately. Except for a very high maximum value for porcelain, the differences in μ between bovine dentine and enamel and porcelain are not great.

A low value of friction does not always mean a correspondingly small rate of wear with metals but friction does play a part in this respect when the material is viscoelastic. On the basis of Table 16.3, however, it will appear that a gold filling or an acrylic denture will require less masticating effort than, say, a porcelain crown.

It is a pity that there is no apparent reason as to why the coefficient of friction is generally higher with saliva than for dry runs, but the wear

patterns in these experiments are encouraging. The coefficient of
friction increased with the contact pressure for bovine enamel rubbing
against itself which, however, is the opposite trend to polymeric
friction. This was also true for the enamel when the counterface was an
acrylic resin and in both cases there were large amounts of wear debris.
Another observation was that the coefficient of friction was independent of the sliding velocity which was in the range 0.2–37 mm s^{-1} in
these experiments. An increase in friction was observed if sliding
continued on the same track. The reason for this may be that like
materials, by an analogous situation to the solid solubility effect in
metals, will adhere more. It is also likely that a deposit will contribute to
the ploughing effect, which should increase the frictional resistance. In
separate experiments where materials were rubbed against an abrasive
wheel for the purpose of comparison, it was observed that the abrasive
wear resistance decreased as the material became softer (Table 16.4). It
is to be expected that a soft material will wear readily against an abrasive
counterface. However, a grinding wheel is a severe test and laboratory
studies should find a more suitable apparatus such as a tooth-brush
simulator.

Table 16.4   Abrasive wear resistance of materials rubbing against an abrasive wheel

Material	Wear (mm^3 per 1000 revolutions) $\times 10^{-3}$
Acrylic resin	25·40
Zinc phosphate cement	16·00
Amalgam	11·00
Cr–Ni alloy	0·67
Porcelain	13·00

16.9   Impact and sliding wear model

A wear model has been proposed for the case of mechanical printing
when a piece of type is pressed against paper supported on a hard
platen, and the essential action is a combination of impact and slip. That
is during the process of typing, the piece of type is undergoing impact
combined with a small amount of tangential sliding. Mastication is
probably a similar process in that a maxillary tooth engages with the

opposing mandibular part to bite a mouthful of food and then slide to break the mass into small pieces. Sliding of the interface during eating is probably semi-rotary with most animals but the wear model [42] described here is for impact combined with simple unidirectional tangential sliding.

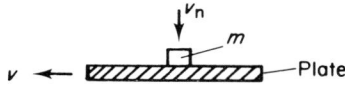

Fig. 16.9 A mass *m* striking a plate of infinite volume.

Consider a small component, say with a rectangular cross-section, of mass, *m*, striking a plate of infinite volume at a speed whose normal component is v_n (Fig. 16.9). Let the plate travel a very small distance to the left at a speed *v*. Suppose that, upon hitting the plate, the component is accelerated sideways, assuming a velocity *u* to the left due to the frictional force at the interface. There is a dynamic normal force, $W(t)$, because of impact and a coefficient of friction μ. It follows that,

$$m\left(\frac{du}{dt}\right) = \mu W(t) \tag{16.4}$$

Equation 16.4 gives a measure of the interfacial frictional resistance and note that the normal force varies with time *t*. The pattern is as shown in Fig. 16.10 where the force reaches a peak value W_0 before falling to zero and, to a first approximation, the impact force pulse can be regarded as sinusoidal. That is,

$$W(t) \simeq W_0 \sin\left(\pi t/t^*\right) \tag{16.5}$$

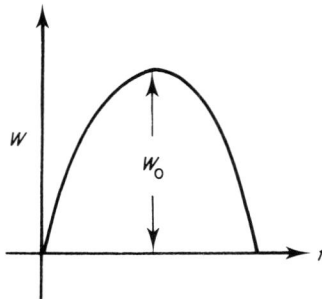

Fig. 16.10 Variation of normal force with time when a small mass strikes a plate.

where t^* is the period of time for which two occluding teeth are in contact during each of the biting and the chewing process and $0 \leqslant t \leqslant t^*$. The instantaneous tangential speed of the projectile can be obtained by integrating Equation 16.4. Thus,

$$m \int_0^u du = \mu \int_0^t W(t)\,dt \quad \text{or} \quad u = \frac{\mu}{m} \int_0^t W(t)\,dt \qquad (16.6)$$

It is quite likely that at some time $\bar{t}$, the tangential slip of the component, say the upper tooth, equals the lateral movement of the plate, say the lower matching tooth, when $u = v$. The component and the plate will cling together and slipping will stop in which case Equation 16.6 will not apply. Substituting for $W(t)$ from Equation 16.5 in Equation 16.6,

$$u = \frac{\mu}{m} \int_0^t W_0 \sin(\pi t/t^*)\,dt \qquad (t < \bar{t} < t^*)$$

Integrating the above,

$$u = \frac{\mu W_0 t^*}{\pi m} \left(1 - \cos\frac{\pi t}{t^*}\right) \qquad (16.7)$$

For the condition of no slip, $u = v$ and $t = \bar{t}$. Substituting these in Equation 16.7,

$$v = \frac{\mu W_0 t^*}{\pi m} \left(1 - \cos\frac{\pi t}{t^*}\right) \qquad (16.8)$$

or

$$\bar{t} = \frac{t^*}{\pi} \cos^{-1}\left\{1 - \frac{\pi m v}{\mu W_0 t^*}\right\} \qquad (16.9)$$

The authors [42] define a slip factor f such that,

$$f = \frac{\pi m v}{\mu W_0 t^*} \qquad (16.10)$$

The criteria are as follows:
   (1) $0 < f < 2$: slipping stops and the component and the plate cling together;
   (2) $f > 2$: slipping continues throughout the contact.

Substituting for f from Equation 16.10 in Equation 16.7,

$$u = \frac{v}{f}\left(1 - \cos\frac{\pi t}{t^*}\right) \qquad [0 < t < \bar{t}] \qquad (16.11)$$

For a sliding distance dS, the differential equation incorporating u and v is,

$$\frac{dS}{dt} = v - u$$

Substituting for u from Equation 16.11,

$$\frac{dS}{dt} = v - \frac{v}{f} + \frac{v}{f}\cos\frac{\pi t}{t^*} \quad \text{or} \quad \frac{dS}{dt} = v\left[1 - \frac{1 - \cos(\pi t/t^*)}{f}\right] \qquad (16.12)$$

The equation for the abrasive wear law in this case is,

$$\Delta V = \int_0^{t^*} \frac{\beta}{H} W(t)\, dS(t) \qquad (16.13)$$

where ΔV = volume wear of the component per cycle of impact; H = its hardness; and β = probability of producing a wear fragment.

For the case when slip takes place, that is putting $f = 2$, for simplicity, Equation 16.12 can be written as,

$$\frac{dS}{dt} = \frac{v}{2}[1 + \cos(\pi t/t^*)] \qquad (16.14)$$

Substituting for dS in Equation 16.13 from Equation 16.14,

$$\Delta V = \frac{\beta W_0 v}{2H}\int_0^{t^*} [\sin(\pi t/t^*) + \tfrac{1}{2}\sin(2\pi t/t^*)]\, dt$$

Integrating the above,

$$\Delta V = \frac{\beta W_0 v t^*}{\pi H} \qquad (16.15)$$

The above equation gives the volume loss per cycle of mastication. That is, the teeth engage by impact and slip tangentially. Note, however, that the slip in the oral cavity is more complicated than that presented in this model. If the number of chews, that is the number of impacts, is n, the volume of tooth material removed, V, is

$$V = \beta \frac{n W_0 v t^*}{\pi H} \qquad (16.16)$$

The volume loss depends on the maximum normal force and the time of contact. As expected, the wear of teeth will depend directly on the lateral velocity, which is very small and allows the computation of the total sliding distance if n is known. Like metals, the harder a tooth is the less it will wear and there is some evidence of this in Table 16.4. The effect of impact may well be failure by a fatigue effect, but note that for Equation 16.16 to have any credibility $v \neq 0$. It does not follow that at every impact followed by slip, there will be a wear particle and hence the probability of producing a wear debris, β, is included. β must be found from experiments but as yet no such values are available simply because the model has not been devised for wear of tooth materials.

Substituting for $W_0 t^*/\pi$ from Equation 16.10, in Equation 16.16,

$$V = \beta \frac{n m v^2}{H \mu f} \qquad (16.17)$$

Expressed in terms of μ it is seen that the higher the frictional resistance, the lower the amount of wear, which is a characteristic of viscoelastic materials.

Both the orthodontist and the tribologist are interested in knowing the nature and magnitude of translational and rotary movements of teeth during mastication. The load–deformation characteristics of teeth have been elucidated as early as 1933 by Synge and co-workers [43]. Dental scientists hold the work as elegant and informative but the deduction is abstruse. The relationship shown in Equation 16.17 from the work of Engel *et al.* [42] is relatively lucid. The model, however, is based on the action in a typewriter and we are not really sure if this is comparable with the phenomenon of impact in an oral cavity. It is possible that impact forces, if any, are very small. Nevertheless, a very useful exercise will be to examine the validity of the impact and sliding wear model by planned experimentations on wear of dental materials.

Published information [44–46] shows that the speed of approach of upper and lower teeth while chewing is about 5 mm s^{-1}. Frequent tooth to tooth contact occurs during mastication and the sliding distance varies from 1 to 6 mm. Generally, in a typical chewing stroke, starting with the open position, the mandible moves in an upward direction with relatively high velocity. The speed of approach decreases as the teeth contact the bolus. There is a pause in mastication of about 0·3 seconds and the mandible then drops rapidly. As a bolus is introduced into the mouth, mastication begins automatically and is almost involuntary. It has been stressed that if the subject is conscious that his chewing activity is being followed, the process is then voluntary. It is likely therefore that a chewing simulator if designed may not correlate with *in vivo* situations because of the controlled mastication, as is inevitable in a mechanical system. The need to devise a simple test which can be used, at least for ranking of dental restorative materials from the point of view of wear, is pressing. The task is daunting if we consider also that a number of interrelated factors such as the size of the jaws, posture of the lips, size of the tongue, habits and even the language spoken by the subject, establish early in life the neuromuscular pattern of jaw movement during chewing, which is both automatic and involuntary.

References

1. "Dentalman" (1969). (USA) Bureau of Naval Personnel, No. 0500-193-0750.
2. Goldstein, R. E. (1976). "Esthetics in Dentistry", Lippincott, Philadelphia.
3. Anderson, J. N. (1965). *Brit. Dent. J.* **98**, 119.
4. Hoffman, F. (1756). "A Treatise on the Teeth", Davis and Reymers, London.
5. Stephan, R. M. (1966). *J. Dent. Res.* **45**, 1551.
6. McClure, F. J. and Ruzicka, S. J. (1946). *J. Dent. Res.* **25**, 1.
7. Zipkin, I. and McClure, F. J. (1949). *J. Dent. Res.* **28**, 613.
8. Elsbury, W. B. (1952). *Brit. Dent. J.* **93**, 177.
9. James, P. M. C. and Parfitt, G. (1953). *Brit. Med. J.* **2**, 1252.
10. Allan, D. N. (1969). *Brit. Dent. J.* **126**, 311.
11. Ericsson, Y. (1953). *J. Dent. Res.* **32**, 850.
12. Darby, E. T. (1892). *Dent. Cosm.* **34**, 629.
13. Holloway, P. J., Mellanby, M. and Stewart, R. J. C. (1958). *Brit. Dent. J.* **104**, 305.
14. Rost, T. and Brodie, A. G. (1961). *J. Dent. Res.* **40**, 385.
15. Elsbury, W. B., Browne, R. C. and Boyes, J. (1951). *Brit. J. Ind. Med.* **8**, 179.
16. Rapp, G. W., Prapuolenis, A. and Madonia, J. (1960). *J. Dent. Res.* **39**, 372.
17. Lindqvist, B. (1973). *Quintessence Int.* **4**, 67.
18. Bloom, G. F. H. (1919). *Brit. Dent. J.* **40**, 305.
19. Steel, J. and Browne, R. C. (1953). *Brit. Dent. J.* **94**, 285.
20. Lehman, M. L. (1967). *Brit. Dent. J.* **123**, 419.
21. Adams, R. J. and Lord, G. H. (1971). *J. Dent. Res.* **50**, 474.

22. Cadwell, D. E. and Johannessen, B. (1971). *J. Dent. Res.* **50**, 1517.
23. Brännstrom, M. and Nyborg, H. (1972). *J. Prosth. Dent.* **27**, 181.
24. Jorgensen, K. D. and Esbensen, A. L. (1973). *Tandlaegebladet* **77**, 162.
25. Anderson, J. M. (1972). "Applied Dental Materials", Blackwell Scientific, Oxford.
26. Lee, L. H. (Ed.) (1974). "Advances in Polymer Friction and Wear", Plenum, London.
27. Bearn, E. M., (1973). *Brit. Dent. J.* **134**, 7.
28. Howell, A. H. and Brudevold, F. (1950). *J. Amer. Dent. Ass.* **20**, 133.
29. Atkinson, H. F. and Ralph, W. J. (1973). *J. Dent. Res.* **52**, 225.
30. Fraleigh, C. M., McElhany, J. H. and Heiser, R. A. (1967). *J. Dent. Res.* **46**, 209.
31. Hollenback, G. M., Villanyi, A. A. and Shell, J. S. (1966). *J. California Dent. Ass.* **34**, 250.
32. Shell, J. S., Hollenback, G. M. and Villanyi, A. A. (1966). *J. California Dent. Ass.* **42**, 521.
33. Grabenstetter, R. J., Broge, R. W., Jackson, F. L. and Radike, A. W. (1958). *J. Dent. Res.* **37**, 1060.
34. Tainter, M. L. and Epstein, S. (1943). *J. Dent. Res.* **22**, 381.
35. Manley, R. S. and Foster, D. H. (1967). *J. Dent. Res.* **46**, 442.
36. Ashmore, H. (1966). *Brit. Dent. J.* **120**, 309.
37. Wright, K. H. R. and Stevenson, J. J. (1967). *J. Soc. Cosm. Chem.* **18**, 387.
38. Phillips, R. W., Avery, D. R., Mehra, R., Swartz, M. L. and McCune, R. J. (1972). *J. Prosth. Dent.* **28**, 164.
39. Lugassy, A. A. and Greener, E. H. (1972). *J. Dent. Res.* **51**, 967.
40. von Fraunhofer, J. A. (1971). *Brit. Dent. J.* **130**, 243.
41. Tillitson, E. W., Craig, R. G. and Peyton, F. A. (1971). *J. Dent. Res.* **50**, 149.
42. Engel, P. A., Lyons, T. H. and Sirico, J. L. (1973). *Wear* **23**, 185.
43. von Fraunhofer, J. A. (Ed.) (1975). "Scientific Aspects of Dental Materials", Butterworths, London.
44. Harrison, A. and Lewis, T. T. (1975). *J. Biomed. Mater. Res.* **9**, 341.
45. Bates, J. F., Stafford, G. D. and Harrison, A. (1975). *J. Oral Rehab.* **2**, 349.
46. Bates, J. F., Stafford, G. D. and Harrison, A. (1975). *J. Oral Rehab.* **2**, 281.

Index

A

Abradants, 68
Abrasive factor, 85
Abrasive wear, 2, 64, 198, 246, 386
Abrasive wear law, 66
Adhesion, 12, 33, 154, 245, 268
Adhesive wear, 1, 33, 197
Adhesive wear law, 37
Alumina, 239
Aluminium, 34, 125
Aluminium–silicon alloys, 204
Amalgams, 389
Amontons' laws, 7
Applications of polymers, 264
Arthritis, 366
Asbestos, 88
Asbestos fibres, 339
Asperity height, 240
Asperity radius, 240
Atmosphere, 152, 198
Atom plucking, 35
Attack angle, 67, 105
Attrition, 317, 385
Autoradiography, 53

B

Ball and socket, 367
Ball bearings, 61
Ball mill, 82, 84
Balsam wood, 295
Banded structure, 241

Bearings,
    cast iron, 201
    of bone, 6
    of wood, 7, 293
Bearing life, 175, 343
Bed sheets, 301
Beilby layer, 48, 199
Biomechanics, 358
Biotribology, 358
Blasting, 81
Brake fade, 321
Brake pad, 320
Brake temperature, 321
Braking path, 328
Braking torque, 328
Brass, 53, 194
Brass bearing, 242
Bruxism, 386

C

Cadmium, 60
Cage slip, 184
Cannon, 231
Carbides, 89, 201, 312
Carbon black, 88
Carbon fibres, 239
Carbon filled resin, 243
Cassiterite, 81
Cast iron, 81, 200
Cast polymer, 254

417